List of Symbols

π (pi)	Population proportion
P_0	Probability of no occurrences
P_1	Probability of one occurrence
P_n	Probability of n occurrences
$P(A)$	Probability of event A
$P(A\|B)$	Conditional probability of event A, given that B has occurred
$_nP_r$	Number of permutations of n items taken r at a time
r	Sample correlation coefficient Random variable in the binomial distribution
r^2	Coefficient of determination in bivariate regression
R^2	Coefficient of determination in multiple regression
s	Sample standard deviation
s_p	Sample standard error of the proportion
S_{YX}	Standard error of estimate in regression
S_b	Standard error of regression coefficient
s_d	Sample standard error of difference between means
σ (sigma)	Population standard deviation
σ_d	Population standard error of difference between means
σ_p	Population standard error of a proportion
$\sigma_{\bar{x}}$	Population standard error of the mean
s^2	Estimate of variance taken from a sample
$s_{\bar{x}}$	Estimated standard error of the mean
σ^2	Population variance
Σ (sigma)	Summation
t	Value used for Student's t distribution in hypothesis testing
X	Independent variable in regression analysis Variable in the normal or Poisson distribution Value of item in a sample
$\bar{X}$	Arithmetic mean of items in a sample
Y	Dependent variable in regression analysis
Y_c	Computed value for Y in regression analysis
z	Number of standard deviation units under a normal curve

Business Statistics

Business

Paul R. Winn
Millikin University

Ross H. Johnson
James Madison University

Statistics

Macmillan Publishing Co., Inc.
New York
Collier Macmillan Publishers
London

**To Scott, Erin, and their mother
P. R. W.**

**To Marjorie
R. H. J.**

Copyright © 1978, Macmillan Publishing Co., Inc.

Printed in the United States of America

All rights reserved. No part of this book may be reproduced or transmitted in any form or by any means, electronic or mechanical, including photocopying, recording, or any information storage and retrieval system, without permission in writing from the Publisher.

Macmillan Publishing Co., Inc.
866 Third Avenue, New York, New York 10022

Collier Macmillan Canada, Ltd.

Library of Congress Cataloging in Publication Data

Winn, Paul R
  Business statistics.

  Bibliography: p.
  Includes index.
  1. Statistics. I. Johnson, Ross H., joint author.
II. Title.
HA29.W56   519.5   77-7891
ISBN 0-02-428700-8

Printing: 1 2 3 4 5 6 7 8   Year: 8 9 0 1 2 3 4

Preface

Owing to the growing complexity of the business world and to a more extensive use of computers, statistics has become an increasingly important field in recent years. The instructor in statistics is faced with multiple difficulties, including the selection of an appropriate text, the development of realistic examples, and the important problem of motivating students to learn the material.

Many students express difficulty with statistics, and often this difficulty is both real and justifiable. It is not a rare phenomenon that just as beginning students are becoming convinced of the merits of statistics, they are suddenly faced with numerous situations and examples in their text that just do not seem relevant. For example, many students state that they cannot relate to examples in which cards, dice, coins, and/or urns are used.

To help motivate students to learn the important subject matter of statistics, we have done two things not commonly found in other first-level texts. First, wherever possible, we have attempted to use realistic or even actual business examples. For instance, much of the material in Chapters 2, 3, 8, 9, and 10 is taken directly from real business situations. In other places examples are developed from actual business experiences. Both authors have extensive business or consulting experience. While the text is not devoid of some of the more commonly used examples, most are real and are designed to make problems and their solutions more understandable.

Second, we have approached our writing in a conversational way in order to make the material more readable. This is not done at the expense of rigor of the subject matter but to enhance the presentation and help maintain student interest. We do this with the knowledge that most students who take an introductory statistics course are not going to become statisticians. Rather, they need the material as a basis for other courses and as a background when employed later in business. For this reason we feel that understanding the *concepts* of statistics is as important as being able to utilize the *mechanics*.

In order to be consistent with the philosophy of the text, the exercises are designed to be realistic and help the student gain insights into business situations. A student study guide is available to provide extensive practice problems and some help in the interpretation of the text material. The study guide corresponds with the text chapter by chapter. The authors of the study guide, Donald W. Kroeber and R. Lawrence LaForge, who also gave valuable technical assistance with the text, have presented alternative explanations of the material for many chapters. The study guide also contains additional problems, with their solutions. The text can be used either with or without the study guide.

The material in the text and study guide has been developed and tested in our teaching. In addition, undergraduate students at the sophomore level have read the chapters and worked the examples and exercises. Many modifications, based on their suggestions, were made.

The text is oriented toward students taking their first statistics course beyond a basic math course. No knowledge of calculus is assumed; nor is calculus introduced in the text. Because business applications and examples are stressed, only some background in basic algebra is necessary.

The chapters are arranged in a logical sequence, but we have left leeway to the instructor because of variations in the length of time over which the course is taught, as well as variations in the background of the class. For example, in a college or university quarter, the basic topics could be covered by the core chapters 1, 2, 3, 4, 5, 6, 7, and 9. If the time available and the preparation of the students allow, the instructor could add any or all of the following chapters, either in the sequence in the text or at points limited by the following:

Chapter Number and Topic	Prerequisite
8. Tabulations, Chi-Square Analysis, and Analysis of Variance	Chapter 7
10. Multiple Regression Analysis	Chapter 9
11. Nonparametric Statistics	Chapter 7
12. Decision Making Under Conditions of Uncertainty	Chapter 7
13. Index Numbers and Time Series	Chapter 9

Most chapters have been divided into sections for ease in making daily assignments. Exercises follow each major topic so as to correlate the exercises to the assignment.

The writing of any text involves many more people than the author or authors. Sometimes it seems that these people are even vital to the work. One author (P. R. W.) wishes to thank W. A. Neiswanger, who was a graduate-level professor in the Economics Department of the University of Illinois and also a pioneer in classical statistics who patiently taught the subject matter.

We are greatly indebted to the following whose scrutiny of the manuscript in whole or in part, and resulting suggestions are appreciated: Paul D. Berger, Boston University; Thomas W. Jones, University of Arkansas; Edward L. Ranck, University of North Alabama; John B. Snyder, Mohawk Valley Community College. Excellent assistance in checking problems and solutions was provided by Raymond Stone and Carmond Decatur.

Once again, as was the case on another text, we were able to work with Mrs. Mary Lou Travers; she not only acted as a typist for most of the material but also as a technical assistant and coordinator. Mrs. Mary Lou Glick and Ms. Bonnie Givens also handled part of the manuscript typing. They did their work with a sense of humor, exceptional accuracy, and exemplary professional competence.

The other people behind the scenes were our understanding wives, Susan and Marjorie. It is only through their patience that we could complete this text.

Paul R. Winn

Ross H. Johnson

Contents

1 The Meaning of Statistics 3

Learning Objectives 3

An Example 3
   Definition of Terms 3
   Data Collection Methods 4
   Other Definitions 4
   Types of Samples 6
   Sampling and Measurement Error 8

Uses of Sample Data 9
   Description Versus Inference 9
   Inferential Uses 9
   Other Uses of Sample Data 10

Other Business Examples 11

2 The Scientific Decision-Making Process 15

Learning Objectives 15

Definition of the Problem 15

Selecting Alternative Courses of Action 16

Testing Alternative Solutions 17
   Use of Secondary Information 17
   Observation 18
   Experimentation 18
   Models 19
   Surveys 19

Selecting the Best Alternative(s) 21

Implementation 21

A Business Example 21
   Bank-Problem Definition 22

Selecting Alternative Bank Advertising Strategies   22
Testing the Alternative Bank Advertising Strategies   22
Selection and Implementation of the Advertising Alternative   28

Need for Computer Analysis   28

Additional Definitions   29

3
Frequency Distributions and Summary Measures   33

Learning Objectives   33

Need for Simplification in Data Handling and Display   33

Constructing a Frequency Distribution   35
   Examination of the Data Encountered   36
   The Array   38
   Class Limits   39
   Visual Presentation   43

Other Frequency Distributions   46
   Cumulative Distribution   46
   Relative Frequency Distribution   48

Measures of Central Tendency   54
   Arithmetic Mean   55
   Mean for Grouped Data   56
   Median   58
   Mode   59
   Characteristics of Measures of Central Tendency   61
   Representativeness   62
   Influence of Extreme Values   63
   Examples Where Measures Are Misleading   64

Measures of Dispersion   69
   Distance Measures   70
   Average Deviation   71

Variance and Standard Deviation   72
Variance and Standard Deviation for Grouped Data   74
Coefficient of Variation   78
Constant-Cause System   79
Interpretation of Results: Possible Pitfalls   80

Probability   87

Learning Objectives   87

Importance of Probability to Management   87

Probability As a Relative Frequency   88

Important Concepts Used in Probability   89
   Sets and Events   89
   Sample and Population   91
   Replacement   92
   Venn Diagrams   92
   Independence   93

Basis for Determining Probabilities   94
   Objective Probability   95
   Subjective Estimates   95

Types of Probability   97
   Marginal Probability   97
   Joint Probability   98
   Independent and Dependent Events   99
   Conditional Probability   100
   Example Illustrating Marginal, Joint, and Conditional Probabilities   101

The Laws of Probability   106
   Addition Laws   106
   Multiplication Laws   109
   Probability Trees   114
   Bayesian Concepts   117

Business Applications of Probability   122

5

Probability Distributions 127

    Learning Objectives   127

    Expected Value   129

    Normal Distribution   130
        Standard Deviation Units   132

    Bernoulli Processes   139

    Binomial Distribution   140
        Review of Counting Methods   141
        Calculating Probabilities with the Binomial   143
        Binomial Expansion   145
        Summary of Binomial Distribution Assumptions   148
        Tables of the Binomial Distribution   148
        Binomial Distribution: Mean and Variance   150
        Normal Approximation to the Binomial   151

    Poisson Distribution   156
        Applications of the Poisson Distribution   156
        Poisson Calculations   157
        Poisson Distribution As an Approximation to the Binomial   158

6

Sampling and Sampling Distributions 163

    Learning Objectives   163

    Sampling   163

    Advantages of Sampling   164
        Other Factors Related to Sampling   165

    The Population   166

    Random Versus Nonrandom Sampling   167

Simple Random Sample   167
   Replacement Versus Nonreplacement   168
   Table of Random Numbers   169

Other Random Sampling   171
   Systematic Sampling   171
   Stratified Sampling   172
   Cluster Sampling   173

Nonrandom Sampling   174
   Single, Double, and Sequential Sampling   175

Sampling Distribution of the Mean   177
   Normal Population   179
   Distinguishing Between s and σ_x and Other Measures   182
   Population Nonnormal   183

Small Populations: Sampling Distributions   189

Sampling Distributions of Proportions   191

7 Estimation and Hypothesis Testing   197

Learning Objectives   197

Statistical Inference   197

Estimation   198

Point Estimates   199
   Point Estimates of the Mean   200
   Point Estimates for the Proportion   200

Interval Estimates   200

Frequently Used Procedures for Interval Estimation   202
   Large Sample: Population σ Known   203
   Estimating μ 205
   Determining n 206
   Large Sample: Population σ Unknown   207
   Finite Population   208
   Interval Estimate: Proportion   208
   Interval Estimates of μ from Small Samples   210

Properties of Good Estimators 212
    Unbiased Estimator 212
    Efficiency 213
    Consistency 213
    Selecting the Correct Test Equation 213

Hypothesis Testing 217

Basic Need for Hypothesis Testing 218

Steps in Testing a Hypothesis 219
    Rationale for the Decision 223

One-Sample Hypothesis Tests 224
    Population Normal with Standard Deviation Known 224
    Two-Tailed Tests 227
    Population Normal with Standard Deviation Unknown 229
    Population Distribution and Standard Deviation Both Unknown 230
    One-Sample Tests for Proportions 230

Two-Sample Hypothesis Tests 234
    Difference Between Two Means (Sample Size 30 or Larger) 234
    Difference Between Two Means Based on Small Samples ($n < 30$) 237
    Difference Between Two Proportions 238
    Selecting the Correct Hypothesis-Test Equation 241

Risks and Errors in Hypothesis Tests 243
    Risks (or Errors) 243
    Operating Characteristic Curve 245

Design of the Hypothesis Test 249
    Significance Level 249
    Balancing α and β 250

Setting the Critical Value: An Example 250

Summary of Sample-Testing Procedure 253

Variation in Boundary Value 253

Sample-Size Variation 254

Limitations in the Use of Hypothesis Tests 256

8
Tabulations, Chi-Square Analysis, and Analysis of Variance   259

Learning Objectives   259

Introduction   259

Tabulation   259
   Simple Tabulation   259
   Cross Tabulations   264

Chi-Square Analysis   271
   Purposes   271
   Procedure   271
   Steps in the Hypothesis Test   275
   Other Numerical Examples   276
   Cautions in Using Chi-Square Analysis   278

Analysis of Variance   280
   One-Way Analysis of Variance   281
   Two-Way Analysis of Variance   284
   Assumptions   287

Appendix   289

9
Regression and Correlation Analysis   291

Learning Objectives   291

Introduction   291
   Variables   292
   Other Terms   293

Data Collection and Use   294

Bivariate Regression   294

Interpretation of Regression Estimates   297
   Constant Term a 298
   Regression Coefficient or Slope b 298
   Other Interpretations   299
   Standard Errors of a and b 300

A Numerical Example   300
    Formation of Specific Y_c and e Values   303
    Standard Error of the Estimate   303
    Sampling Error of b 304
    Confidence Interval for b 305
    Confidence Interval Around Y_c 306

Correlation Analysis   307

Assumptions of Regression Analysis   310
    Assumption 1: Linearity   310
    Assumption 2: Random Error Terms   310
    Assumption 3: Uniform Variance   311
    Assumption 4: Normality   313

Cautions   313

Appendix—Derivation of a and b 318

10
Multiple Regression Analysis   321

Learning Objectives   321

Introduction   321

Interpretation   322

Assumptions   323

Evaluation   326
    Net Regression Coefficients   327
    Significance of the Independent Variables   328
    Other Interpretations   328
    Collinearity   329
    Further Analysis of the Results   330
    Interpretation of Beta Weights and Y_c Values   332

General Cautions in Multiple Regression   333

A Case Example   334
    Problem   334
    Output   334

Steps in Interpretation   335
   Remedy   337
Categorical Data   338

Validation   340
   Forecasting with Regression   342
   Summary   343

Appendix   355

11 Nonparametric Statistics   357

Learning Objectives   357

Introduction   357

Meaning and Advantages of Nonparametric Statistics and When to Use Them   358

Test for Significance of Difference Between Two Proportions   358

Sign Test   361

Mann-Whitney U Test   363

Ordinary Length-of-Runs Test   369
   Cautions and Comments   371

Tests for Goodness of Fit   371

Rank-Order Correlation   375
   Spearman's Rho Coefficient   375
   Tests of Significance of Spearman's Rho   377

12 Decision Making Under Conditions of Uncertainty   385

Learning Objectives   385

Structure of Decision Making   385

Decisions and Probability 387
  Weighted Averages 388

Subjective Probabilities and Decisions Under Uncertainty 389

Payoff Tables for Decision Making 391

Expected Opportunity Loss 394

Expected Value of Perfect Information 399

Bayesian Analysis 404
  Types of Bayesian Analysis 405

Example Using Bayesian Analysis 406
  Prior Analysis 407
  Posterior Analysis 408
  Bayes' Theorem: A Review 408
  Preposterior Analysis 411
  Benefits of Bayesian Analysis 415
  Limitations of Bayesian Analysis 416

13
Index Numbers and Time Series 421

Learning Objectives 421

Index Numbers 421
  Types of Index Numbers 421
  Price Indexes 422

Considerations in Constructing an Index 423
  Selection of Items 423
  Weighting 424
  Base Period 424

Construction of Price Indexes 425
  Simple Aggregative Index 425
  Weighted Aggregative Index 426
  Simple Average of Price Relatives 427
  Weighted Average of Price Relatives 428

Other Indexes 429
   Quantity Indexes 429
   Value Indexes 430

Shifting the Base and Splicing 431

Deflation of a Series 432

Review of Index Uses 433

Time Series 436

Trends 437

Components of Time Series 439
   Seasonal Variations 439
   Cyclical Variations 440
   Irregular Variations 440

Techniques in Time-Series Analysis 441

Appendixes 445

A Normal Probability Distribution Areas 446
B t Distribution 447
C Binomial Distribution: Individual Probabilities 448
D Binomial Distribution: Cumulative Probabilities 458
E Poisson Distribution: Individual Probabilities 472
F Poisson Distribution: Cumulative Probabilities 476
G Chi-Square Distribution 482
H F Distribution 483
I Squares and Square Roots 485
J Random Digits 486
K Common Logarithms 488
L The Greek Alphabet 490

Index 491

Business Statistics

Learning Objectives

An Example
   Definition of Terms
   Data Collection Methods
   Other Definitions
   Types of Samples
   Sampling and Measurement Error

Uses of Sample Data
   Description Versus Inference
   Inferential Uses
   Other Uses of Sample Data

Other Business Examples

Learning Objectives

In this chapter, which introduces the topic of statistics, the primary learning objectives are:

1. To be able to identify a sample and a population.
2. To become familiar with some of the basic terminology of statistics.
3. To understand some of the business uses of statistics.
4. To recognize on an elementary level some of the errors in statistics and the ways in which statistics can be misused.

An Example

The Meaning of Statistics

The topic of statistics often causes some students to be apprehensive about their abilities. In many cases this apprehensiveness can be traced to a feeling of "I just don't understand what's going on," or "I don't see how statistics can help me to become a manager." The objectives of this chapter are to point out some of the basic concepts and definitions used in statistics. We also include several practical examples to demonstrate the application of statistics to business and management. In this process some of the misunderstandings and potential misuses of statistics will also be discussed.

The best way to get a preliminary feeling for statistics is through an example. Assume that the governing board of your college or university has hired your class to collect some data that it hopes will be useful in securing additional funding in order to avoid a tuition increase. One basic question to be asked is: What is the "average" income for the student body? We shall concentrate our attention on this one question, although others must also be asked. Our question appears to be very straightforward; so think to yourself for a few minutes about how you would collect the information to answer it. . . .

Definition of Terms

Some of the following points may have come to mind as you thought through this problem. First, how do we define "student body"? Second, is this definition consistent with that used by the governing board? For example, do we include graduate students, correspondence students, extension students, and/or part-time students? There is no single, correct answer to these questions, but we do need to know what we mean by the term "student body" before proceeding with our data collection. For now, let us assume that we define "student" as anyone enrolled in the university, regardless of status. Note that this definition will have to coincide with that of the governing board if we want our answer to match its needs. Further, any comparisons that the board may want to make across other colleges or universities will have to be made with data collected using the same definition.

We must also be very careful to define the term "income." For example, do we mean household income, before or after tax income, yearly income, or what? For our current purposes, let us define "income" as pretax, yearly, household income. For married students and those who are self-supporting, this will be their own household income. For those students who are not married and either wholly or partially supported by some outside party (parents, guardian, relative), it will be defined as the outside party's household, yearly, pretax income.

Data Collection Methods

Second, we must decide *how* we are to collect the data. At this point, we shall not debate the merits of the various methods of data collection, such as personal interview, telephone interview, or mail questionnaire. These are all common means of collecting *primary data*. Primary data are data not currently available in books, records, or files of a company, the university, or any other source. Data that *are* available from these sources are called *secondary data*. For now, let us assume that we have to collect primary data. After investigation of our needs and resources as well as of the comparative advantages of each data collection method, the telephone interview is selected.

The next question involves how many people to call and how they will be selected. These questions are critical to statistical analysis. We all know that colleges and universities range in enrollment from 300 to well over 50,000 students. Needless to say, as the institution becomes larger, it becomes more difficult, or in some cases almost impossible, to talk to each and every student. For our example, let us say that the institution has 10,000 students. If we wished to, or could, talk to all 10,000 students about their or their parents' income, the result would be a *census* of the students. If we talked to a subset of these 10,000 students, say 500, the result would be a *sample* of the students. In many, if not most, business situations a sample, rather than a census, is taken, owing to financial constraints, time limitations, the lack of need for a census, and even the flat impossibility of taking a census. For example, it may be impossible to reach *all* students because of their varying class schedules, work schedules, and time away from campus.

Prior to continuing with our example, let us establish a few definitions that are common to statistics and that will be used throughout this text. It is important to understand the basic terminology early, since it will serve as the foundation for mastery of the topic.

Other Definitions

Population. In our example the term *population* refers to the entire 10,000 students. More generally, it is the entire collection of objects, people, or other things in which we are interested. The term *universe* is sometimes used interchangeably with population. We should be careful not to limit our definition of population to people as it may pertain to ball bearings coming off a production line, incomes, elephants in a circus, or a colony of ants. The letter N is often used to represent population size. In our example, $N = 10,000$.

Sample. A *sample* is a subset of a population. In our case it is the 500 students. Note that there are many, many different samples of 500 students that could be taken from the 10,000. If a selection of a different group of students is made, the reported incomes of one or more respondents will be different, giving us new results. Imagine all the different combinations of 500 students. Throughout most of statistics the usual samples taken are *random samples*. Random samples are of two types: (1) an *unrestricted* random sample, in which every member of the population has an equal chance of being selected in the sample, and (2) a *restricted* random sample, in which each person has a known but not necessarily equal chance of selection. Unrestricted random samples are most frequently discussed in statistics. The letter n is used to represent the size of the sample, which in this case is 500.

Parameter. A *parameter* is a measure pertaining to a population. For example, it could be the average, or *mean,* income for which we are searching. It could be found by adding all the incomes in the population and dividing by N:

$$\frac{\text{Sum of all incomes pertaining to the 10,000 students}}{10,000}$$

There are other parameters that will be discussed later, but the most commonly used one is the one mentioned here, the mean of the population, which is called μ (the Greek letter "mu"). In this text, generally, whenever we see a Greek letter, we are talking about a parameter.

Statistic. A *statistic* is a measure that pertains to a sample. Any measure calculated from our sample of 500 students would be a statistic. One commonly used measure would be the average of all items in the sample:

$$\frac{\text{sum of all incomes pertaining to the 500 students}}{500}$$

This measure is called $\bar{X}$ (X bar). We shall spend considerable time dealing with $\bar{X}$ in later chapters. In general, whenever we see a Roman letter it refers to a sample. Other parameters and statistics will be discussed later.

There is a difference that we should note between *statistic* and *statistics*. A statistic is a measure calculated from a sample, while statistics refers to the area of knowledge pertaining to *all* the types of measures that can apply to a sample or a population.

Average. *Average* is one of those general words which is commonly misused, as will be shown later in this chapter. It pertains to some measure of centrality calculated from a population or sample. The most commonly used measure is $\bar{X}$, which more correctly is called the *arithmetic mean* of a sample. At this point we should note that there is more than one method of calculating an average; these methods will be discussed more fully later. For the purposes of our example, $\bar{X}$ will be called an average. We shall see that we need $\bar{X}$ to help us describe our outcome from the sample of 500 incomes.

Types of Samples

Now, to return to our example, we must decide whether to take a census or a sample. As mentioned previously, we shall most often take a sample, for the reasons noted. The question becomes: How? Of the two types of random samples, an *unrestricted* random sample is a useful approach for our purposes. For example, if we had a master list of all 10,000 students and selected a random starting point and drew each 20th student to appear in this sample, this would be an unrestricted random sample. This is also called an nth-item sample, as we have selected every 20th student. However, in some cases we might wish to select more heavily from one group than another. This would depend on the similarity of the answers coming from groups. For example, if we felt that the seniors were fairly homogeneous while the freshmen were rather heterogeneous in terms of income, we might wish to sample more heavily from the freshmen and less heavily from the seniors.

This type of sample procedure differs from nonrandom procedures that are used in some survey and sampling work. For example, much "street corner" and "straw poll" sampling is not really random. When we in a community are asked to call a television station regarding preferences for a political candidate, typically only those very pro or con call, while those who feel less strongly do not call. The result is not a valid representation of community feelings.

In order to take effective random samples, among other factors respondents have to be selected correctly, times of day when contacts are made have to be varied to allow for people who are typically not at home at that time, and there must be a sufficient number of attempts to get in touch with the respondent.

Before proceeding to discuss sampling error and measurement error, we should pause to make sure that we know why samples are taken. In our case, it would be difficult, if not impossible, to talk to all 10,000 students. In addition, it would be very expensive in terms of data collection and processing. Thus, if we could collect data from a sample that represented the 10,000 students fairly well, this would be a preferable procedure. On a more general basis, data collection is very expensive and time-consuming. To collect data through personal interviews can often cost as much as $10 per interview. This does not include sampling, questionnaire design, or data-processing costs. Telephone interviews can often cost $2 to $3 each, also making this type of data collection expensive. Further, samples, if selected and handled properly, can most often yield results that are fairly representative of the population.

For now let us make one final assumption that will help in our comprehension of the terms presented here. Unknown to us or the governing board, the registrar of the University has collected information from each of the 10,000 students at the time of enrollment regarding the pertinent income figures. He knows that the average incomes for the 10,000 students (μ) is $11,000. We do not know this and proceed with our study. From the list of 10,000 students, we select 500 using the nth-item procedure. We then proceed to collect information from the 500 by telephone interviews.

We shall discuss the determination of sample size in a later chapter, so for now let us accept 500 as being an adequate size. Our interviewers are careful to call each of the 500 students in the sample. For those who cannot be reached on a call, other calls are made on different days at different times. Interviewers are instructed to call up to 5 times, as necessary. In the end, 475 students are contacted, 20 cannot be reached and 5 refuse to participate. The response rate is 475/500 or .95, with nonresponse rate .04 and the refusal rate .01. These rates represent good work on the part of our interviewers.

After data collection and tabulation, $\bar{X}$ for the 475 incomes is calculated to be $10,500. The question that now arises is: Why didn't we get a $\bar{X}$ of $11,000, which corresponds to the registrar's μ figure? The reasons for this can be found in two types of errors encountered in collecting data. The first is *sampling error,* and the second is *measurement error.*

Sampling and Measurement Error

Sampling Error. Sampling error is due to chance or luck involved in sample selection. For example, we may by chance have selected a higher proportion of students having smaller incomes (as defined) in the sample than is true for the population. We should be *very* careful to note that this does not mean that a nonrandom sample was selected. In the case of a nonrandom sample the results would probably differ from μ by a far greater amount. It simply means that by chance the sample may differ from the population. The law of large numbers pertains here; that is, as we take a larger random sample, the sampling error should diminish. Note also, though, that the costs and time involved increase as the sample size increases. Sampling error will almost always be present if a sample is taken. The only way to eliminate it is to take a census, which, as we have noted, is often impossible.

Measurement Error. A potentially more dangerous and undetectable error is *measurement error.* Actually, this term covers a wide variety of errors that are generally human errors. For example, some respondents may not know their parents' exact income and report incorrect figures, some interviewers may write down wrong numbers, some numbers may be transposed in tabulation, typing errors may be made, or some people may lie. These types of errors are dangerous because they are often impossible to detect and their impact on the study is not known. The best way to handle these errors is to minimize them through thorough training of the data collectors, verification, and cross-checks. Very rarely, however, can errors be completely eliminated.

Assuming that the registrar's figure of $11,000 is accurate, the difference of $500 from our $10,500 can be attributed to sampling and measurement error. Each of these topics will be discussed in greater detail later. Once again, the student should be cautioned not to confuse the perhaps mislabeled term "sampling error" with poor sampling techniques. Sampling error refers to the chance variation of sample measures from the corresponding population measures.

We made an assumption earlier that the registrar of the university already had data pertaining to the 10,000 students. At this point, let us relax this assumption, as it was used only for examining sampling and measurement error. As we have all deduced by now, if we have data for the entire population, it will be unnecessary to take a sample. However, when we do not have population data, it may be very necessary to take a sample.

Uses of Sample Data

Description Versus Inference

One of the choices we shall have to make regarding the data is how we wish to use them. One choice we could make is simply to report the $10,500 as the $\bar{X}$ of the sample and leave it at that. This would be using the $\bar{X}$ to describe the sample. Statistics used in this fashion are called *descriptive* statistics.

A second choice could be made to use the statistic for inferential purposes. That is, we could use the $10,500 calculated from the sample of 475 incomes to make inferences about the population of 10,000 incomes.

Inferential Uses

We now realize that we do not have data for the 10,000 students, but we do have our measure $(\bar{X})$ for the 475 students. The question then becomes: What do we do with this measure of $10,500? Before proceeding with *how* we use this figure, let us think about *what* the measure really represents. The measure $\bar{X}$ can be used as an estimate of the population value μ. It is only an estimate because it will differ from μ, owing to sampling and measurement error. For now, let us concentrate on the sampling error component and later we shall look more fully at measurement error. The sample was taken so that we could give the governing board some data to use in retarding a tuition increase for our college or university. Recognizing that the $10,500 is only an estimate, we have to make a choice about how we want to provide the data to the board. There are two options that are commonly used. First, we can simply tell them that $10,500 is our estimate. This is called a *point estimate* because it is one single number.

A second option would be to provide them with an interval around the $10,500—say, $9,500 to $11,500. The purpose of this interval is to allow for the sampling error that we expect to be associated with our calculation. You will recall we are attempting to estimate μ. Therefore, it might be useful to the board if we could say that 95 per cent of the time when intervals are formulated this way, μ falls within them. Much of the important material we shall cover later deals with how to set up this interval and interpret its meaning. It appears to make more sense to provide an interval that allows for sampling error rather than giving one single estimate of $10,500, which we assume will be off by a certain margin.

Before moving to some more business examples, some important points need to be made regarding the previous example. It seems logical, and in fact it is true, that if this interval is established to allow for sampling error, one of the ways to make it narrower is to reduce the sampling error. As mentioned before, one way to do this is through increasing the sample size, because as the sample size n comes closer to the population size N, we will find that $\bar{X}$ tends to give a better estimate of μ. However, we must also note that when we increase the sample size, the costs of data collection and the time involved increase. What we have, then, is a balance or trade-off between the resources available and the accuracy needed.

A second important point involves how accurate $\bar{X}$ is to start with. We know that $\bar{X}$ can differ from μ because of the two types of error, measurement and chance. In addition, though, $\bar{X}$ can be thrown off by characteristics of the data themselves. For example, consider the following set of data which represent automobile mileage per gallon (mpg) of fuel.

Car	mpg
1	12
2	10
3	10
4	13
5	45
	90

$$\bar{X} = \frac{90}{5} = 18 \text{ mpg}$$

In this case the average mpg = 18 = $\bar{X}$. However, when we look at the data, 4 of the 5 cars got substantially less than 18 mpg. What has happened in this case is that the extremely high mpg of car 5 has pulled up the average for all cars substantially. This pehnomenon represents one of the most commonly used methods to "lie with numbers." We shall find later that there is a statistic that gives us an indication of how much the data vary about $\bar{X}$. Only when we have this statistic that shows the degree of variation can we know how accurate $\bar{X}$ is as an estimate. Once again, though, one way to lessen the problem of extremes influencing our average is to increase the sample size so that the extremes will tend to cancel one another out.

Other Uses of Sample Data

Another common use of the data might involve showing correlation. For example, we may collect data from colleges on tuition paid by students versus the need of the students for financial aid. We may be able to show that as tuition increases, the needs for grants, loans, and other forms of aid increase proportionately.

We might then want to imply causality. By this we mean that we would attempt to make statements to the board that universities and colleges adjust their tuition based on ability to pay. By doing this, we would be using statistical data to aid in presenting our case. More generally, from a business standpoint, sample data can be useful in making business decisions.

Other Business Examples

The processes of sampling, calculation of statistics, and estimation are commonly around us on an everyday basis. Perhaps two of the more widely known polls are the Harris political polls and the television ratings done by A. C. Nielsen. In the Harris poll, many questions are generally asked, but the sample size is surprisingly small, ranging from 1,500 to 2,000 respondents. From the results of this sample, projections are made regarding national sentiment on a topic. Such polls generally consist of random samples and estimates of the type we have discussed.

The A. C. Nielsen television-viewing survey is the basis for many advertising decisions as well as for cancellation or continuation of a program. This survey is of approximately 2,000 households and national projections are made from those estimates. Similar surveys are conducted regarding magazine readership and radio listening.

In manufacturing, samples and estimates are often made. For example, in quality control, where products are expected to meet a set standard or be within tolerance limits, pieces are often taken from the production line and estimates made regarding the quality of the product.

In accounting, especially with regard to auditing, it is often necessary to sample transactions to check the accuracy of accounting systems. Because of the large number of accounts and/or transactions, it would be impossible to take a full census, and sampling has thus proved to be very useful.

Politicians are beginning to poll their constituents fairly regularly with regard to one issue or another. Because of expenses and time constraints, these polls are most often samples of their constituents.

Governmental sampling is very important in our everyday lives. For example, the consumer price index (CPI) represents the level of prices for consumer products on a monthly basis. It would be impossible to examine the prices for all products in all stores nationwide. As a result, a "market basket" of products is examined for a sample of cities. This index is important to salary and wage adjustments, economic planning, and other functions. The index of unemployment is calculated similarly and is equally as useful.

Other governmental agencies often sample the opinions of people involved in an issue. State commerce commissions and other regulatory bodies will often sample to determine peoples' attitudes toward rate hikes, taxes, and environmental issues.

In closing, the student may wonder why we have paid so much attention in our opening chapter to the topic of sampling. First, we have seen that our basic definition of *statistic* is "a measure derived from a sample." Second, we have noted that samples are commonly used in business and government. As the great majority of us are, or will be, employed in business or government, it behooves us to learn all that we can about correct sampling and statistical methods.

Exercises

1.1 You have been assigned to select an unrestricted random sample from your class where $n = 5$.
   (a) State precisely how you would select this sample. Make your directions concise enough so that another person could follow them, and select the sample.
   (b) If the data you were collecting involved student age, as of his or her last birthday, how would you calculate $\bar{X}$ for the sample of 5?
   (c) How would you calculate μ for the entire class?
   (d) Do you expect your answers in (b) and (c) to correspond? Why?
1.2 In problem 1.1, what circumstances might suggest that a restricted random sample be taken rather than unrestricted?
1.3 How do the terms "statistic" and "parameter" differ?
1.4 It appears that the only way to completely eliminate sampling error and/or errors caused by extremes is to take a census. However, industry and government continue to sample. Why?
1.5 You have been assigned the task of taking a random sample in your town, $n = 250$. The data you are collecting involve family income before taxes for the past tax year. Your method of data collection is the personal interview.

(a) What types of measurement error could appear in this study?
(b) In what ways could you minimize error?

1.6 It is common in the manufacture of products, ranging from tires to drugs, to sample production to aid in detection of incorrect machine setups and/or human error. What problems regarding sampling error and measurement error might appear in these sampling procedures?

2

Learning Objectives

Reasons for Decisions

Definition of the Problem

Selecting Alternative Courses of Action

Testing Alternative Solutions
   Use of Secondary Information
   Observation
   Experimentation
   Models
   Surveys

Selecting the Best Alternative(s)

Implementation

A Business Example
   Bank-Problem Definition
   Selecting Alternative Bank Advertising Strategies
   Testing the Alternative Bank Advertising Strategies
   Selection and Implementation of the Advertising Alternative

Need for Computer Analysis

Additional Definitions

The Scientific Decision-Making Process

Learning Objectives

The material presented in this chapter helps to set the framework for statistics. Although some of the topics are conceptual in nature, the concepts are important in understanding statistics as they apply to business. The primary learning objectives of this chapter are:
1. To gain understanding of each step in the problem-solving process.
2. To recognize when a problem exists and whether quantitative methods apply.
3. To gain familiarity with some of the more common methods of data collection.
4. To become accustomed to the various formats in which numerical information can be presented.

Reasons for Decisions

Prior to a discussion of statistics and some useful decision-making strategies, we need to "set the scene." Persons engaged in business often take samples and calculate statistics. Why is this? From a somewhat mathematical standpoint the reason is that we want to estimate some parameter and/or describe some data. From a decision-making standpoint we are collecting and processing information in order to solve some problem. This chapter will be devoted to a discussion of the problem-solving process, some of the more common methods of collecting data, and the form that data should take once collected.

Definition of the Problem

Prior to defining any problem specifically, the decision maker must become "absorbed" in the details associated with the problem so as to collect all information that may bear on its definition. For example, if clerks in a retail store have been complaining about work conditions, the store manager should talk to key personnel before specifically defining the problem. If a particular grinding machine is turning out a product that is consistently defective, the plant manager should talk to the foreman, the quality engineer, and others prior to narrowing down the problem. As mentioned previously, the decision maker essentially has two types of information that can be collected. *Primary* information is data that are not currently available, whereas *secondary* data can be compiled from existing files, books, records, or other sources.

Once a problem area has been isolated, it is time for a more specific definition of the problem. A well-defined problem takes the form of an interrogative statement, or question, that asks about the relationship between two or more variables—for example: Does our new truck maintenance program lead to savings for the company as compared to the previous program? In this case the variables are the two maintenance programs. Or, is our health insurance program comparable to that offered by other firms of our size and type? Or, has our new bank advertising program led to increased public awareness of our bank? These problem statements meet two important criteria. First, they are clear, and, second, they suggest that alternative courses of action exist. If there is only one course of action, a problem does not exist. It is only when alternative courses of action are possible that a true problem has been defined which requires a decision.

The problem-definition stage is one of the most critical in solving a problem. It is at this point that many firms wander off into even deeper trouble. For example, a high turnover rate of employees at a hospital may be improperly diagnosed as a workload problem when in fact the real problem revolves around the compensation and benefit program. A far more dangerous approach is to proceed with decision making without *ever* correctly defining the problem. This can lead to incorrect decisions, loss of money, decrease in employee morale, and other undesirable circumstances.

Selecting Alternative Courses of Action

A correctly defined problem thus leads to alternative courses of action. In the first example given, we could continue with our current truck maintenance program or revert back to the old one, given that the benefits of conversion outweighed the costs. In the second example, we could adjust our health insurance program or leave it as it is. Finally, we could continue with our current bank advertising program or switch to another of a number of different advertising approaches. Once the alternative approaches are specified, they can be tested for effectiveness.

In all the problem situations discussed, multiple courses of action can be taken to solve the problem. The decision maker may have one single solution which he or she prefers; however, scientific decision making calls for testing this solution against other possible solutions. In general, the best solution will be the one that leads to the greatest payoff toward accomplishment of the goals of the enterprise.

Testing Alternative Solutions

When alternative solutions have been identified, we can then perform tests to evaluate or test each one. We should be careful to note that the term "test" does not necessarily mean that quantitative analysis has to be performed. Many problems can be solved through the application of logical, verbal decision rules. Although the subject of this text is statistics, it is important to recognize that quantitative analysis is not the only valid approach to solution of a problem. In most cases, however, the use of statistical techniques can aid us in arriving at a better solution. Generally the decision maker must collect information to use in determining whether one alternative is better than another. There are a number of approaches that can be used in collecting information. In this text we shall discuss the more commonly used methods and describe situations where each can be employed in the decision-making process.

Use of Secondary Information

In many cases, sometimes unknown to the decision maker, information already exists that can be of use in solving a problem. Perhaps someone has already conducted a similar study. There are many sources that offer secondary information; quite often information can be found in company files. For example, if we wanted to make a comparative study of our salespeople, much of the necessary sales and market analysis information may be available in our own records. In other cases, the information might be found in *The U.S. Census of Population, Historical Abstracts, Sales Management* magazine, or other sources.

When secondary information is not available, a manager must generally resort to the collection of primary data. There are a number of ways to collect this type of information. In many of these methods, statistics plays an important role.

Observation

One of the most common methods of collecting information is through simple *observation*. Through observation it is possible to record behavior as it occurs, thus eliminating errors that arise from the reporting of past behavior. Rather than asking people about their current behavior, it is often more accurate to observe it. For example, pupil-dilation cameras have often been used to record a participant's attentiveness to a new package design. In the case of the purchase of household products, pantry checks are often run where the participant will physically show an interviewer what brands of a canned good or household cleaner, for example, are on hand. Observation may be used as the only means of collecting data or in conjunction with other methods, such as experimentation.

Experimentation

Experimentation is defined as the controlled manipulation of one independent variable, or of a set of independent variables, and the measurement of the impact that this independent variable(s) has on a dependent variable. As an example, we may change our hospital compensation and benefit program (independent variable) and examine the impact of the change on the employee turnover rate (dependent variable). To measure the impact, we would collect data and use statistical methods to evaluate the effect of the change.

There are two types of experimental approaches. The first is the *laboratory experiment,* where the decision situation is isolated from its real-world setting and one or more independent variables are manipulated under controlled conditions. A second is the *field experiment,* which is run in a realistic setting with conditions as tightly controlled as possible. In both cases we would collect data and use statistical analyses.

The laboratory experiment is often not as realistic as it should be, but it enables the researcher to maintain tight control over outside influences. Consider, for example, the attempt to measure the reaction to advertising by checking eye-pupil dilation. A laboratory setting would allow for keeping other influences at a minimum, but might not really allow a realistic reaction to some advertisement or package design. The field experiment would be set in a realistic situation; however, it often suffers from outside influences. For example, assume that we experimented to see whether, by changing our program, our hospital compensation and benefit plan influenced employee turnover. While we were conducting our experiment, however, another hospital in town might also be altering its plan, which would introduce an outside influence and possibly leave our results somewhat inconclusive.

In the event that a controlled realistic setting is too difficult to achieve, reliance must be placed on methods other than those previously discussed. After the data are collected, it may be impossible to select an appropriate course of action. Further examination of data may be necessary through statistical analysis and/or model building.

Models

A *model* is a representation of a real-world situation. Basically it involves specifying the relationships between variables in order to explain or describe some system or process. There are generally three classes of models: physical, analogue, and symbolic. A *physical model* is a physical representation of some object or situation. Model cars, maps, and pictures are physical models. *Analogue models* do not physically depict an object or situation, although they represent a real-world situation. Flowcharts, organization charts, and demand curves are examples. A *symbolic model* can be either mathematical or verbal. A verbal model is a collection of statements describing a situation or the thoughts of management concerning a particular problem. It uses words to portray a real-world situation. Mathematical models also portray real-world situations, but differ in that they use the notation or language of mathematics. For example, $Y = 3 + 5X$ is a mathematical model.

It is important to recognize that regardless of which type of model is used, the purpose is to allow testing of alternative courses of action. The model will be manipulated under different decision situations, and outcomes will be examined to determine which alternative best meets the objectives.

Surveys

In some cases it is not appropriate to manipulate variables in order to see the impact on other variables. Instead, data may be collected on a number of points and treated somewhat more descriptively than in the case of an experiment. A *survey* is typically viewed as a fact-gathering process involved in discovering the relative incidence, or distribution, of economic, psychological, or sociological variables.

There are three fundamental types of survey: mail, telephone, and personal interview. The following brief discussion should prove helpful in understanding the advantages and disadvantages of each.

Mail Surveys. A point that is not generally recognized is that a mail survey generally provides for a more thorough distribution to the sample than does the personal interview. This means that the original distribution of mail questionnaires is more representative than the distribution of personal interviews. This advantage, however, is counteracted by people not responding to the mail questionnaires. This problem is called the *nonresponse problem* and is the major drawback of mail questionnaires.

Essentially, the nonresponse problem means that those people who respond to a mail questionnaire often are different from those people who do not respond. Typically, those who are very pro or con with regard to an issue respond, while others do not. Because the respondent can choose whether he or she wants to mail back the form, it is often said that the sample is "self-selecting." To return to the original point, although the mail survey may result in a good distribution of questionnaires originally, this advantage is negated by many potential respondents not responding. The final result is that the sample may not be representative of the population from which it was drawn and any statistics used to estimate parameters may be erroneous to a far greater extent than what would be due to sampling error.

On the positive side, if a good response rate can be obtained, the method allows the respondent to answer questions at his or her own pace and eliminates measuring error associated with an interviewer. In addition, even with increasing postage rates, mail surveys are typically the least expensive of the three types of surveys.

Telephone Surveys. Telephone interviews have two advantages over personal interviews and mail questionnaires. First, they are a rapid means of obtaining data. Second, the telephone directory, although sometimes incomplete, often offers a better list of the population than is available for mail or personal interview methods. The major disadvantage of telephone interviewing is that the interview must be restricted to a small number of questions that are nonpersonal in nature. If a small number of factual questions are to be asked, the telephone method can be very successful.

Personal Interview Surveys. A personal interview involves a face-to-face conversation between an interviewer and a respondent. The interviewer will record responses during the interview or shortly thereafter. Personal interviews lend themselves to conducting a more lengthy interview with probing questions where necessary. In addition, experienced interviewers can often detect potential misrepresentation of answers. The major disadvantage of this method is the expense. Because interviewers have to be paid on an hourly basis or by the interview, with travel expenses, training expenses, and meals, the cost can often run as high as $10 to $15 for a single completed interview of 15 to 20 minutes. These expenses can rapidly become prohibitive, and often sample size or randomness is sacrificed to monetary considerations at the expense of data accuracy.

Selecting the Best Alternative(s)

Once the problem has been defined and alternatives have been identified and tested through logic, experiments, models, or survey, the decision maker should be ready to select the alternative that best accomplishes the firm's objectives. Such objectives include sales volume, profits, growth, service, and/or market share. The task of making the decision comes at this point. It may seem somewhat simple, but the decision maker often has the difficult task of matching possible payoffs to the objectives of the business organization.

Implementation

Often the best course of action is selected but then is either partially implemented or not implemented at all. Once the appropriate action has been selected, it must be fully implemented before the problem can be solved. The implementation should then be followed up to verify that it is meeting expectations.

A Business Example

Statistical data can be collected in many forms. As a result, the following example is only one of the many, many situations where the decision maker is faced with collecting and processing information to answer a question or solve a problem. Many other situations will be examined later in the text. At this point, it will be very useful if we can examine how a problem-solution system was designed, the data collected, and alternatives selected.

One of the key stages in solving a problem is to define the problem correctly at the outset. A part of the definition process involves information for the topic area. Essentially, the decision setting for this example was as follows:

> American National Bank of Traversville had a long-standing reputation as one of the more successful banks in Traversville. One of the major problems they were facing was that, although they were successful, new residents and businesses in town seemed to be going to some of the newer banks rather than American National. In a college town of 80,000 people, this could make serious inroads into business over the long run. As the bank was already spending a considerable amount of money on advertising, they wondered whether their advertising campaign was truly effective. Being sound businesspeople, the managers of the bank realized that they were in a decision situation. The situation revolved around whether to change advertising strategies, and whether a change would result in greater awareness of the bank.

Bank-Problem Definition

After a series of meetings designed to allow discussion of the problem, personnel in the bank and its advertising agency were charged with the task of specifically defining the problem and working out a plan to solve it. Based upon the preliminary information, the problem was defined as:

> Will a different advertising campaign lead to greater awareness (recall) of our bank?

The agency had an idea for a campaign that would lend itself to the bank's objectives, but no one knew whether this campaign would be any better than the one they were currently using. After defining the problem, they began to specify alternative courses of action.

Selecting Alternative Bank Advertising Strategies

Because of the expense and timing of advertising, the bank managers and the agency decided that there were two feasible courses of action. The first was to remain with the old campaign, and the second was to adopt a new campaign that would involve the use of animated characters, community benefits, and developing a specific slogan for the bank. Prior to implementation of a new campaign, the decision makers wanted to establish a "benchmark" so they could measure whether the new program was working.

Testing the Alternative Bank Advertising Strategies

Because a bank's image is very important to management and to consumers, the decision was made to test the new campaign as it compared with the old one. After some deliberation, a combination of experimentation and surveys was selected as the means of data collection. We should note that only limited secondary information was available and that the major data would be primary data.

The implementation of the new campaign was delayed long enough to allow a survey while the old advertising program was still in effect, with a second survey planned 4 months after the new campaign had been launched. Essentially, this involved a combination of experiment and survey in that the advertising campaign implementation became the independent variable, awareness the dependent variable, and the two surveys the means of data collection.

A bank-preference mail survey was conducted through use of the bank-preference questionnaire shown in Figure 2-1. The questionnaire was sent to randomly selected customers drawn from bank files, as well as noncustomers drawn from the telephone directory. As management was concerned with how customers as well as noncustomers would react, it was important from the outset to collect data on both these groups. Data were collected regarding patronage patterns, attitudes, reasons for potentially switching banks, recall of advertising, media habits, and personal demographics. It was felt that the study could help not only in measuring advertising recall but also in finding what "types" of people recalled, and how to place television messages more effectively. The same questionnaire would be mailed out to two different randomly selected samples before and after introduction of the new campaign. To ensure adequate representation in the sample, repeat mailings and telephone call reminders were used to gain an adequate return rate of the questionnaires.

Although we shall thoroughly explore the use of statistical methods in the analysis of data later, we should pause to examine the nature of the data that we are collecting. The following definitions will prove useful for later work. For additional definitions pertaining to Figure 2-1, see the appendix at the end of the chapter.

Grouped Data. Through examination of questions 16, 17, and 20 in Figure 2-1, we can see that the respondent will indicate a response by simply checking a category. For example, when we get the data back on age and income we will not know the *exact* age or income but only a category. Data of this type are called *grouped data*.

Ungrouped Data. For question 19 we find out the *exact* number of years the respondent says that he or she has lived in town. Data of this type, where we retain the exact values, are called *ungrouped data*.

Dichotomous Data. The first parts of questions 1, 8, 9, and 14 are called dichotomous (two parts) because they simply ask for a yes or no

Bank Preference Study

Here are some questions about your banking preferences and attitudes. Please place your answers in the space provided.

1. Do you presently bank in Traversville?

    _____ Yes   _____ No   (If no, skip to question 9)

    If yes, which bank(s) do you use?   (Please check where appropriate)

    _____ American National Bank of Traversville
    _____ Dr. R. Lutz Memorial Bank
    _____ County Bank and Trust
    _____ National Bank
    _____ Commercial Bank
    _____ First National Bank
    _____ City Bank

2. Which of these banks would you call your "primary" bank, meaning at which bank do you do more business than any other?
    _____

    If you checked two or more banks in question number 1, which of these banks would you call your "secondary" bank?
    _____

3. Regarding your primary bank, would you rate its *"friendliness"* as

    _____ Extremely friendly
    _____ Friendly
    _____ Neither friendly nor unfriendly
    _____ Unfriendly
    _____ Extremely unfriendly

4. In terms of the *"service"* you get at your primary bank, would you rate it as having

    _____ Extremely good service
    _____ Good service
    _____ Neither good nor bad service
    _____ Bad service
    _____ Extremely bad service

5. How would you rate your primary bank in terms of *how much they care about you*?

    _____ Extremely good
    _____ Good
    _____ Neither good nor bad
    _____ Bad
    _____ Extremely bad

6. How would you rate the way your primary bank is managed?

    _____ Extremely good management
    _____ Good management
    _____ Neither good nor bad management
    _____ Bad management
    _____ Extremely bad management

7. Have you ever switched business from one bank to another?

    _____ Yes   _____ No

    If yes, why did you switch banks?_____
    _____
    _____

FIGURE 2-1. Bank Preference Study

8. Would you consider switching banks either again or for the first time?

　　　_____ Yes 　　_____ No   (If no, please skip to Question 9)

If yes, for which of the following reasons would you switch from your existing most used bank to another bank?

_____ Change in service charges at current bank
_____ Bank service package at another bank
_____ Closer location of another bank
_____ Friendlier people at another bank
_____ Additional banking services
_____ Less expensive checking
_____ New account premium offer
_____ Better parking facilities
_____ Other (please specify) _____

9. In terms of bank advertising, can you recall seeing or hearing any Traversville bank advertising in the last month (this includes newspapers, billboards, television, radio, etc.)?

　　　_____ Yes 　　_____ No   (If no, please skip to Question 11)

If yes, do you remember which particular bank sponsored this advertising?

　　　_____ Yes 　　_____ No   (If no, please skip to Question 11)

Please specify which bank(s).

_____ American National Bank of Traversville
_____ Dr. R. Lutz Memorial Bank
_____ County Bank and Trust
_____ National Bank
_____ Commercial Bank
_____ First National Bank
_____ City Bank

10. Do you recall where you saw or heard this advertising?

　　　_____ Yes 　　_____ No   (If no, please skip to Question 11)

If yes, please indicate.

_____ Radio
_____ Television
_____ Billboard
_____ Newspaper
_____ Other (please specify)

11. Are there any themes, slogans, or images that you can recall are associated with any banks in the Traversville area? Other examples are "real gusto" (Schlitz beer) and Ronald McDonald (McDonald's).

　　　_____ Yes 　　_____ No   (If no, please skip to Question 12)

Please specify which themes, slogans, or images you remember and the bank with which it is associated.

Theme, Slogan, or Image	Bank with which it is associated	Don't know
_____	_____	_____
_____	_____	_____
_____	_____	_____
_____	_____	_____

FIG. 2-1 (2)

12. Does your household do the majority of its banking by
    _____ Mail
    _____ In person in bank
    _____ In person — Drive-up

13. For times in a typical day of TV viewing for the adult members of your household, please indicate your television viewing frequencies.

    We (I) watch TV during this time.

	Never	Occasionally	Frequently	Almost Always	Always
7 AM — 9 AM	_____	_____	_____	_____	_____
9 AM — Noon	_____	_____	_____	_____	_____
Noon — 6 PM	_____	_____	_____	_____	_____
6 — 6:30 PM	_____	_____	_____	_____	_____
6:30 — 10 PM	_____	_____	_____	_____	_____
10 — 10:30 PM	_____	_____	_____	_____	_____
10:30 on	_____	_____	_____	_____	_____

14. Would you say adult members of your household watch one particular television channel for news shows more often than others?
    _____ Yes   _____ No   (If no, please skip to Question 15)

    If yes, please indicate which one.
    _____ Channel 12 (WDMA)
    _____ Channel 3 (WBBQ)
    _____ Channel 17 (WDFM)

15. We would like to collect a few items of information about your household for classification purposes only. What is the main occupation of the head of your household?
    _____

16. Which age category fits that in which your age at your last birthday is included?
    _____ 18-24
    _____ 25-34
    _____ 35-49
    _____ 50-64
    _____ 65 and over

17. Please check the educational category matching the last level the head of your household completed.
    _____ Grammar
    _____ Up to 2 years bf high school
    _____ High school graduate
    _____ Up to 2 years of college
    _____ College graduate
    _____ Graduate work

18. Your Sex
    _____ Female   _____ Male

19. To the nearest year, please indicate how many years you have lived in Traversville.
    _____

FIG. 2-1 (3)

20. For your household, please check which category best describes your pretax income for the past year.
   _____ $0-4,999
   _____ 5,000-9,999
   _____ 10,000-14,999
   _____ 15,000-19,999
   _____ 20,000-24,999
   _____ 25,000-29,999
   _____ Above $29,999 (Please indicate approximate amount)_____

21. Who makes most of the banking decisions within your household?
   _____ Husband
   _____ Wife
   _____ Both
   _____ Not married (Please specify who makes decisions)
   _____
   _____

THANK YOU!

FIG. 2-1 (4)

response. Any data that consists of two and only two responses or outcomes are *dichotomous*.

Discrete Data. Data in which there are breaks in the values that the variables can take on are known as *discrete data*. That is, if we collected data on the number of persons in the family we would get such answers as 1, 2, 3, etc., since we cannot get values in between. If we tabulated hat sizes stocked in a clothing store, we would find such sizes as $6\frac{1}{2}$, $6\frac{5}{8}$, $6\frac{3}{4}$, $6\frac{7}{8}$, 7, $7\frac{1}{8}$, and $7\frac{1}{4}$. These are discrete data since we cannot obtain values in between those given.

Continuous Data. In those cases where there are no breaks in the values that the variable can take on, the data are called *continuous*. In the bank-preference questionnaire, there are no true continuous variables. In a case in which miles per gallon (mpg) are reported, we may tabulate values for mpg such as:

$$18.730$$
$$11.487$$
$$15.625$$
$$\cdot$$
$$\cdot$$
$$17.617$$

These data do not have breaks, can take on any value between reasonable limits, and are therefore continuous. The same would be true for how fast a track team member runs the mile, grade-point averages, and times it takes to get to work or school. Continuous data are often later compiled into discrete classifications for ease in handling and presentation.

Selection and Implementation of the Advertising Alternative

Once collected and analyzed, the data indicated that the new campaign was working better than the old and was therefore the better alternative. The new campaign was selected over the old and was fully implemented as a long-run plan. We shall learn more about the collection and analysis of data later.

Need for Computer Analysis

As we look at the questionnaire shown in Figure 2-1, we can see that there are a great many statistical questions that could be asked. If we had collected 800 mail questionnaires in each of two mailings, there would be 1,600 total responses. To hand-tabulate the responses would be a long and difficult task; and if later statistical tests were to be run or cross tabulations (such as the ages of certain occupational classes) made, the task would be even more difficult. It is extremely common for data such as these to be processed on a computer. Essentially, the processing involves coding the data, keypunching and verifying, selecting from available computer programs, and running the program. The major asset of the computer is the speed and accuracy of its operation as well as the amount of data it can handle. Only 15 years ago many current studies and analyses would not have been possible because of the magnitude of the analysis necessary. Many of the common statistical procedures and computer programs will be demonstrated later.

Exercises

2.1 Why is it important for the decision maker to collect preliminary information prior to specific definition of the problem?
2.2 Define a problem that is pertinent to business decision making.
   (a) Are the terms in this problem specific and defined clearly to all participants?

(b) Show that your problem can be correctly worded as an interrogative statement.
(c) List the alternative courses of action.
2.3 Design a method that would help solve your problem.
2.4 Assume that you have to ask five questions of somebody regarding your problem. Write out these five questions in an unambiguous manner.
2.5 List decision situations where observation would clearly be more useful than a survey or experiment.
(a) Do the same for a survey and an experiment.
(b) Develop a list of the advantages and disadvantages of each data collection method.

Additional Definitions

The student should note that when data are collected the results can come back in many forms. The following are some additional definitions that may prove useful in future study.

Multiple-Response Data. As we examine questions 1, 8, 9, and 10 in Figure 2-1, we can see that the respondent can check multiple answers. One person may check one category while another checks four or five. Although this type of question is often used, it makes statistical analysis, other than percentage breakdowns, very difficult.

Attitude Scales. Questions 3–6 are simplified versions of attitude-type data. We can see that they range from good to bad or friendly to unfriendly. Although there is a continuum, say good or bad, along which attitudes are measured, the answers will still fall into one category or another.

Integer Data. Integer data represent a special case of discrete data where only whole numbers exist. For example, the answers to question 19 will be in whole numbers (or integers).

Open-End Responses. Questions 15 and 16 have open-end response categories. For example, in question 15 the respondent fills in his or her occupation, and, in question 16, the 65-and-over category has an open end in that we are not sure how far over 65 the respondent is when that category is checked.

Closed-End Responses. In question 20, because we have found the approximate amount of income in the above-$29,999 category, we have "closed" the response patterns. Questions such as 17 and 18 are also closed.

Nominal-Scaled Data. Data that imply no rank or continuum are nominally scaled. The yes–no answers imply no rank of importance or continuum and are nominal. The questions in 1, 8, 9, 10, 13, 14, 18, and 21 are also nominal in that there is no order within the data and no category is considered "better" or "higher" than another. Specifically, there is no origin, rank, or distance implied in the data.

Ordinal-Scaled Data. Data that are ordinarily scaled have a specified rank. For example, in question 3 *Extremely friendly* would be considered to be "better" than *Friendly*. In question 16, statements can be made about one respondent being *older* than another; in question 17 having *more* education than another; in question 20 making *more* income than another. We should carefully note, especially for the attitude questions, that no statement is made about how much more friendly, or how much better. It is simply stated that it was either more friendly or less friendly. For questions 16, 17, and 20 some differing assumptions can be made because the underlying data, age, years of education, and income are continuous. Methods for treating grouped data of this type will be discussed later.

Interval-Scaled Data. Data that are intervally scaled not only contain rank-order properties, but the notion of distance is also important. Popular examples are temperature, time, and scores. For example the base point on the Fahrenheit scale is 32° as it is the freezing point of water; and time has an origin at 12:00 noon and midnight. These origins are arbitrary in nature and could be changed. It does not make much sense to say that a score of 80 is twice as large as a score of 40 on a test, as we are then assuming that a score of 0 means no knowledge at all. However, we can say that one person scored 40 more points than another, or answered twice as many questions correctly.

Ratio-Scaled Data. Question 19 contains a ratio-scaled set of data. The origin of zero years has meaning in this question. For example, if someone has lived in Traversville for 10 years, this is twice as long as someone who has been there 5 years. Further, a person who has only been there 4 months has zero years of living in the town. Other popular examples are profit and income. Profit can take on negative values. The important point to note is that we can say that one of these types of measures is a multiple of another. For example, $100,000 profit is 10 times as much as $10,000 profit.

Learning Objectives

Need for Simplification in Data Handling and Display

Constructing a Frequency Distribution
    Examination of the Data Encountered
    The Array
    Class Limits
    Visual Presentation

Other Frequency Distributions
    Cumulative Distribution
    Relative Frequency Distribution

Measures of Central Tendency
    Arithmetic Mean
    Mean for Grouped Data
    Median
    Mode
    Characteristics of Measures of Central Tendency
    Representativeness
    Influence of Extreme Values
    Examples Where Measures Are Misleading

Measures of Dispersion
    Distance Measures
    Average Deviation
    Variance and Standard Deviation
    Variance and Standard Deviation for Grouped Data
    Coefficient of Variation
    Constant-Cause System
    Interpretation of Results: Possible Pitfalls

Frequency Distributions and Summary Measures

Learning Objectives

In business situations, or now as a student, it is important to know how to compile, interpret, and use information. Some of this information may have already been collected but need further interpretation. In other situations, the data must be collected first and then arranged such that it can be interpreted. One of our objectives in this chapter is to learn efficient ways of data compilation and presentation. In business, social, or scientific usage, it is important to use standard methods for presentation so that the data can be readily understood by others; often such information must be communicated to customers and suppliers, as well as to others in the organization. A further objective is to be able to understand and to use standard data displays prepared by others. In addition, the ability to compute and to use summary measures is a prerequisite to later chapters in this text and to other courses in marketing, finance, production, and management. Thus the primary learning objectives for this chapter are:

1. To learn efficient ways to compile data.
2. To understand the standard methods for presenting data in both tabular and graphic forms.
3. To understand data compiled and displayed by others.

Need for Simplification in Data Handling and Display

The successful use of statistical data depends to a great extent upon the manner in which they are arranged, displayed, and summarized. It is important that those persons for whom statistical data are intended be readily able to understand quantitative information and its meaning. Pictorial representation of data and related descriptive statements that interpret the results are vital inputs to many decision-making situations. Descriptive statements about a set of data can take either of two forms: verbal statements or summarized numerical values, such as the mean (or average). Standardized methods, developed to present and describe data, facilitate calculations and help communicate the conclusions to others for their use. The most widely used methods for carrying out calculations and presenting the results are discussed in this chapter.

The usability of information for descriptive purposes or in decision making depends heavily on the manner in which it is gathered and the form in which it is compiled and displayed. Mere lists of numbers or items are relatively useless, especially when the list is long. It is difficult to comprehend meaning from information in such a raw form. It is also difficult to make any valid general statements about the data or to draw conclusions that can be readily explained to someone else. In Chapter 2 we discussed the importance of collecting *good* information as the initial step in solving a problem.

Consider Table 3-1. The table shows a list of prices of single-family houses sold in the suburb of Abcon during 1977. The prices are listed in the order in which the sales occurred. A rough observation of the data would reveal that the highest-priced house was $64,000, whereas the lowest was $19,900. In the given form, usually called *raw data*, it is not

TABLE 3-1. Prices of 112 Single Family Houses Sold in the Suburb of Abcon During 1977

$24,000	$24,400	$41,500	$49,000
20,800	41,000	36,800	42,000
27,000	36,000	57,800	37,200
64,000	36,000	43,000	34,500
42,400	19,900	26,400	42,000
47,500	33,600	43,000	39,200
40,000	32,800	24,000	56,000
36,000	36,000	45,800	60,000
38,000	28,000	38,400	24,000
36,800	36,200	29,600	48,000
40,000	24,800	39,000	32,100
36,000	30,800	34,000	42,600
42,000	50,400	32,000	32,200
29,500	45,000	39,800	55,000
45,000	36,400	49,600	35,000
33,800	22,400	46,200	35,000
48,500	28,500	51,600	56,500
53,000	24,000	39,000	56,400
48,200	41,000	38,000	32,200
36,000	29,000	32,000	32,500
44,000	32,000	44,000	20,000
33,000	20,000	50,000	35,800
43,200	29,900	32,000	35,900
32,800	37,600	44,000	41,800
33,600	20,000	23,200	62,000
37,900	54,000	32,000	61,000
43,700	26,000	59,700	54,600
40,600	32,900	48,000	35,300

easy to make other valid conclusions. The greater the amount of data, the more difficult it becomes to reach valid conclusions from the raw data alone. For example, rather than having 112 houses, imagine that the list contained data from 11,200 houses sold in a larger community. Such a list would consist of pages and pages of numbers arranged in order by date sold. Without some kind of grouping, the numbers are almost meaningless except to show the total volume of houses sold.

Constructing a Frequency Distribution

Often we place data in the form of a frequency distribution to make the facts more presentable—that is, the data are put in a form that permits easy interpretation. We define a *frequency distribution* as a table that displays the data in groupings, along with the number of occurrences that fall into each group. This process of grouping the data into classes is often called *data classification*. A frequency distribution is shown in Table 3-2. It can be considered as a summarization of the data. We see that if only the frequency distribution is given we can no longer identify the individual values. To construct a frequency distribution (sometimes called a frequency table), the first step is to group the data into classes. Each class would include items from the raw-data list falling between the class limits. A class is formed by an upper and lower boundary. In the case of houses sold, all houses priced from $27,000 and up to (but not including) $35,000, for example, might be considered as one class. It is important that classes be contiguous and not overlapping so that any

TABLE. 3-2. Frequency Distribution of Home Prices Using 6 Classes and Intervals of $8,000

Home Prices	Interval Midpoint	Frequency (number of homes)
$19,000 and under $27,000	$23,000	15
$27,000 and under $35,000	31,000	28
$35,000 and under $43,000	39,000	35
$43,000 and under $51,000	47,000	20
$51,000 and under $59,000	55,000	10
$59,000 and under $67,000	63,000	4
		112

value in the set of data can fall into one and only one class. The question then arises as to how we choose the class size. Some guidelines for establishing class limits will be discussed after other aspects are considered.

To be able to record and classify data, it is necessary that the items under observation be measurable. In our case the price of each house is the characteristic measured. The price is considered the variable and can take on values that might be written out to 7 significant figures, such as $21,699.45. In practice, however, we find that offers and selling prices tend to be rounded off. In different situations, the data may be rounded off when they are collected if only a certain number of significant figures are considered important. We should note again before moving on that we arrange the data in classes to provide greater ease of interpretation and to highlight the pattern of variation. This pertains especially to situations where we have a large number of observations in our list.

Examination of the Data Encountered

Some data cannot be readily classified or formed into a frequency distribution. Let us look at other characteristics of houses besides price that might have been considered. For example, a house may or may not have a garage. This characteristic can be separated into only two categories, garage or no-garage. It is therefore referred to as an *attribute*. Another type of analysis might group houses by the number of bedrooms. Although such characteristics are not readily grouped into classes, they can be represented numerically. In the case of the garage, it is easy to represent the attributes numerically, since the case of no-garage would be represented by a 0 and a house with a garage by a 1. In the case of bedrooms, the bedroom count could be represented by the whole numbers 1, 2, 3, 4, or 5. The number of bedrooms is known as a *discrete variable*, where a discrete variable is defined as one that can take on only certain values within a specified interval. A *graph*, or *bar chart*, can be used to represent these values, as illustrated in Figure 3-1. Figure 3-1 shows that 86 of the 112 houses have garages and 26 do not. Another style of bar chart (often called a *line graph*) is used in Figure 3-2 to represent the number of houses having various numbers of bedrooms, as follows:

Number of Bedrooms	Number of Houses
1	0
2	11
3	54
4	28
5	19
	112

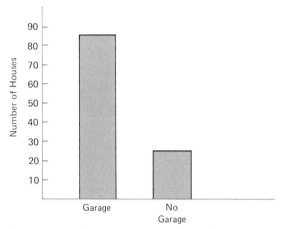

FIGURE 3-1. Portion of 112 Houses With and Without Garage

In these two figures, we can use wide bars, narrow bars, or even single lines, depending upon what we feel looks best and makes a clear presentation. In contrast to the discrete variable, measurements of a continuous variable can take on any intermediate value. Not all data can be compiled as a continuous variable. Obviously, a house cannot have 3.35 bedrooms so the number of bedrooms in a home is not a continuous variable. In measuring and compiling widths of lots, however, the data are continuous, since we could take measurements of 80.0 feet, 96.4 feet, 88.65 feet, or any other value, depending only on the preciseness of our measurement. The prices of the houses can be considered almost a

FIGURE 3-2. Bedroom Data for Group of 112 Houses

continuous variable because a price could be $24,694, or even $24,694.55. This is considered a continuous variable for practical purposes, even though it can be said that the measurement is to the nearest cent. In any case there will be a limited number of significant figures. We will see how to deal with this type of data as we learn how to prepare a frequency distribution.

The Array

Table 3-1 presented the home sale prices in the order in which the sales occurred during the year. When listed in that order it was not easy to make observations regarding the meaning of the data nor to readily draw any conclusions. We are now going to see how to prepare a frequency distribution to display the data and to help overcome these problems. The first step in the preparation of a frequency distribution is to reorganize the

TABLE 3-3. Home Prices in Abcon for 1977 Formed into an Array from Lowest to Highest Price

$19,900	$32,100	$37,200	$44,000
20,000	32,200	37,600	45,000
20,000	32,200	37,900	45,000
20,000	32,500	38,000	45,800
20,800	32,800	38,000	46,200
22,400	32,800	38,400	47,500
23,200	32,900	39,000	48,000
24,000	33,000	39,000	48,000
24,000	33,600	39,200	48,200
24,000	33,600	39,800	48,500
24,000	33,800	40,000	49,000
24,400	34,000	40,000	49,600
24,800	34,500	40,600	50,000
26,000	35,000	41,000	50,400
26,400	35,000	41,000	51,600
27,000	35,300	41,500	53,000
28,000	35,800	41,800	54,000
28,500	35,900	42,000	54,600
29,000	36,000	42,000	55,000
29,500	36,000	42,000	56,000
29,600	36,000	42,400	56,400
29,900	36,000	42,600	56,500
30,800	36,000	43,000	57,800
32,000	36,000	43,000	59,700
32,000	36,200	43,200	60,000
32,000	36,400	43,700	61,000
32,000	36,800	44,000	62,000
32,000	36,800	44,000	64,000

data so that values appear in sequence from the lowest to the highest. This has been done for the home prices in Table 3-3. This sequential arrangement is called an *array*.

Class Limits

The array is a preliminary arrangement of the data prior to the next step, classification of the data. *Classification* consists of arranging the data into groups, or *classes*. A class is merely a range of values, such as those prices from $19,000 and up to $27,000 as used in Table 3-2. In this case we have used a *class width* of $8,000 ($27,000-$19,000) to represent all home prices in this range. We could have also selected $19,000 to $22,000 (class width of $3,000) as in Table 3-6, or one of many other possibilities. It therefore becomes apparent that a decision must be made as to the number of classes to be used, which will determine the class width. This selection is not an exacting science and there is no one correct selection procedure. We want to avoid having the total graph either too wide or too narrow. A graph with either too many classes (too wide) or too few classes (too narrow) results in a pattern that cannot be easily interpreted. There are several guidelines to use in selecting an appropriate class width:

1. The number of classes should be from 6 to 16. The following formula is often used to obtain a first approximation of the number of classes to use:

$$\text{classes} = 1 + 3.3 \log N.$$

2. For any class to contain zero items is undesirable; however, in some instances it may be unavoidable.

3. All class intervals should be of equal width (such as $8,000 in Table 3-2). This provides a better pictorial representation and also facilitates later calculations. In the case of some variables, such as family income, however, equal intervals are often impossible because there would usually be a few very high incomes. If the frequency distribution is to be compared to other frequency distributions, it is especially important that each distribution have comparable class widths.

4. If possible, the class midpoints should be numbers that are easy to work with. If, in our example, the class interval $19,000 to $27,000 were selected, then the midpoint would be $23,000. We shall use midpoints in later calculations, and these calculations will be simplified considerably if the class midpoint is a round number such as $30,000 or even $29,000 rather than a number such as $28,750. The latter midpoint would result

if a class interval of $27,500 to $30,000 were selected and would be less desirable.

5. Open-end classes should be avoided. The class "$50,000 or over" would be considered open-end and would cause later calculations to be complex. Occasionally it will be necessary, however, to deal with a set of data where there are few very small or very large items and an open-end class may be unavoidable.

6. The class boundaries must be set such that any single item can fall into only one class. It should never be possible for an item of data to fit more than one class. For example, the following age classes would be ambiguous and should not be used:

$$
\begin{array}{c}
20-25 \\
25-30 \\
30-35 \\
35-40 \\
40-45
\end{array}
$$

Into which class would such ages as 25, 30, 35, or 40 fall? We would be uncertain where to place them. If these classes were used on a questionnaire, many people would have a tendency to check the "younger" bracket, which would provide a misrepresentation of the "true" data. The following would be a valid grouping:

$$
\begin{array}{c}
20 \text{ and under } 25 \\
25 \text{ and under } 30 \\
30 \text{ and under } 35 \\
35 \text{ and under } 40 \\
40 \text{ and under } 45
\end{array}
$$

When grouped in this manner, the 25 clearly falls into the second class.

With these guidelines in mind, we shall now examine the data more closely and establish suitable class limits for the home prices given in Table 3-3, using the six guidelines. The highest value (H) is $64,000 and the lowest ($L$) is $19,900, giving a difference of $44,100. Consider the first guideline requirement that the number of classes be from 6 to 16. This rough calculation can be represented as

$$\text{interval } (i) = \frac{H-L}{K} \quad \text{(where } K = \text{number of classes)}$$
$$= \frac{64,000 - 19,900}{K} = \frac{44,100}{K}$$

If 6 classes were used, each class width would be about $8,000, whereas if 16 classes were used, each class width would be about $3,000. This gives quite a range to select from. We have stated previously that there is no one correct K (number of classes); rather, we are trying to make a selection that presents the data in a manner that is easy to interpret. We shall consider the alternatives of 8 and 12 classes, as well as 6 and 16, to help us understand the process of class selection. If we use the formula presented earlier, Classes $= 1 + 3.3 \log 112 = 1 + 3.3(2.049) = 7.76$. By following the formula, we would select 8 classes; however, we might well decide to select a different number from what the formula indicates.

The data from the array in Table 3-3 have been compiled into four frequency distributions (Tables 3-2 and 3-4 through 3-6) to illustrate the effect of the class-interval selection upon the data presentation. We shall make some comparisons before a selection is made. Note that in our tabulations a value on the extreme upper limit of a class is tabulated into the class above; that is, a value of $26,000 does not fall into the class "$18,000 and under $26,000" but rather into the class "$26,000 and under $34,000." Thus the frequency distribution must be set up and the data tabulated such that there is no question as to which class the $26,000 is to be counted in. The midpoints of each class are also listed to help in the selection of a suitable class size. Note further that the beginning point for the lower limit of the first class is also selected according to the resulting interval midpoint. In Table 3-2, selection of the first-interval lower limit of $19,000 gives an interval midpoint of $23,000, which provides a midpoint at an even thousand.

In Table 3-4, the lower limit was selected as $18,000, giving a class midpoint of $21,000. Class limits in Tables 3-5 and 3-6 were chosen with the same considerations; however, when we got to 16 classes, the midpoints had to end in 500.

TABLE 3-4. Frequency Distribution of Home Prices Using 8 Classes and Intervals of $6,000

Home Prices	Interval Midpoint	Frequency (number of homes)
$18,000 and under $24,000	$21,000	7
$24,000 and under $30,000	27,000	15
$30,000 and under $36,000	33,000	24
$36,000 and under $42,000	39,000	27
$42,000 and under $48,000	45,000	17
$48,000 and under $54,000	51,000	10
$54,000 and under $60,000	57,000	8
$60,000 and under $66,000	63,000	4
		112

TABLE 3-5. Frequency Distribution of Home Prices Using 12 Classes and Intervals of $4,000

Home Prices	Interval Midpoint	Frequency (number of homes)
$18,000 and under $22,000	$20,000	5
$22,000 and under $26,000	24,000	8
$26,000 and under $30,000	28,000	5
$30,000 and under $34,000	32,000	23
$34,000 and under $38,000	36,000	20
$38,000 and under $42,000	40,000	17
$42,000 and under $46,000	44,000	12
$46,000 and under $50,000	48,000	6
$50,000 and under $54,000	52,000	5
$54,000 and under $58,000	56,000	7
$58,000 and under $62,000	60,000	1
$62,000 and under $66,000	64,000	3
		112

One of the primary objectives of this chapter is to provide an understanding of how to select suitable classes for the frequency distribution. Again, it is not an exact process, but by the end of this chapter the student should be able to select an appropriate number of classes for a group of

TABLE 3-6. Frequency Distribution of Home Prices Using 16 Classes and Intervals of $3,000

Home Prices	Interval Midpoint	Frequency (number of homes)
$19,000 and under $22,000	$20,500	5
$22,000 and under $25,000	23,500	8
$25,000 and under $28,000	26,500	3
$28,000 and under $31,000	29,500	4
$31,000 and under $34,000	32,500	21
$34,000 and under $37,000	35,500	18
$37,000 and under $40,000	38,500	8
$40,000 and under $43,000	41,500	16
$43,000 and under $46,000	44,500	7
$46,000 and under $49,000	47,500	5
$49,000 and under $52,000	50,500	5
$52,000 and under $55,000	53,500	3
$55,000 and under $58,000	56,500	5
$58,000 and under $61,000	59,500	1
$61,000 and under $64,000	62,500	1
$64,000 and under $67,000	65,500	2
		112

data. The four alternatives given in the home-price example will be closely examined as a learning illustration. Our intention is to help the reader develop his or her intuition for making a good single selection in the future without having to go through this trial-and-error process.

Visual Presentation

The next step will be to place the data from the frequency distribution into a visual presentation. The objective is to permit a quick interpretation of the total data. One of the most commonly used and easily interpreted forms of visual presentation is the histogram. The data for the 112 home prices will now be plotted in this form. Each *histogram* shown in Figures 3-3 through 3-6 uses a bar of length proportionate to the number of items in each class. All four histograms are plotted since we have yet to decide which is the best. Consider Figure 3-3, which represents the home prices from Table 3-2. The home-price class boundaries from Table 3-2 are set out on the X axis (or abscissa), and the Y axis (or ordinate) represents the frequency or number of homes in each class. It is common practice in constructing a histogram to plot the frequency on the vertical axis and the values on the horizontal axis. The bars touch because there can be no value between any two bars.

At this point, let us review the four frequency tables (Tables 3-2 and 3-4 through 3-6) and the four histograms in Figures 3-3 through 3-6 against the guidelines set forth and make a selection of one of the four as providing the best visual representation. The first three guidelines are met by all four histograms: The number of classes is from 6 to 16, there are no zero length bars, and all class intervals are equal. The midpoints of classes are shown in Tables 3-2 and 3-4 through 3-6, and all except Table 3-6 have midpoints of an even $1,000. The other criteria are also met by all four histograms, so that additional considerations must be brought in to provide guidance in making the selection.

Note that in Figures 3-5 and 3-6 several bars farther from the center are higher than adjacent bars closer to the center. These ups and downs are not desirable unless there is reason to believe that there is a valid reason causing this type of fluctuation. Note also that Figure 3-6 tends to have two peaks, which again is not desirable unless there is a known valid reason for the condition. Both these undesirable characteristics tend to occur if too many classes have been used. Figures 3-3 and 3-4 both form "smoother" histograms; however, Figure 3-3 has some rather large jumps between adjacent bars. Figure 3-4 seems to have the smoothest form and therefore would be the best choice for presenting the data. Although in most cases it is not usual practice to plot several histograms, it is done here to help show the types of characteristics that are good or poor in the histogram form of a visual presentation.

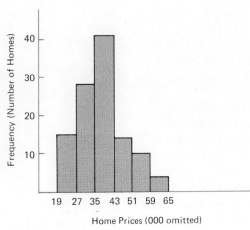

FIGURE 3-3. Histogram with Six Classes (From Table 3-2)

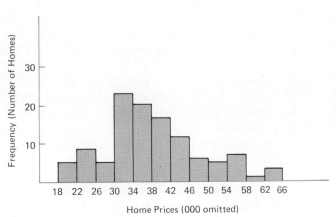

FIGURE 3-5. Histogram with Twelve Classes (From Table 3-5)

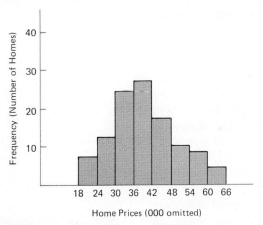

FIGURE 3-4. Histogram with Eight Classes (From Table 3-4)

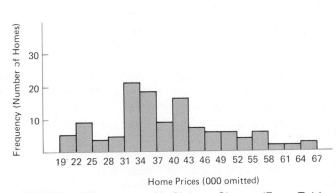

FIGURE 3-6. Histogram with Sixteen Classes (From Table 3-6)

The frequency polygon shown in Figure 3-7 is an alternative type of visual presentation of the home-price data. A polygon is a many-sided figure. A frequency polygon is constructed by marking the midpoints of the top of each bar on the histogram and connecting the midpoints by straight lines. The frequency polygon in Figure 3-7 represents the data from Table 3-4 and Figure 3-4, which we decided was the best of the four

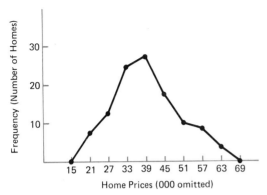

FIGURE 3-7. Frequency Polygon for Home Prices in Abcon for 1977

histograms shown. In order to close the polygon, an additional class is assumed at each end having a zero frequency. This type of visual presentation is often used together with (or in place of) the histogram. We must remember that only the midpoints used in the construction of the frequency polygon have meaning and that we cannot therefore interpolate between the points to obtain intervening values. Figure 3-8 shows the same data smoothed out to form a frequency curve, which is another form of presenting the same data.

If we desire to compare two or more frequency distributions, the frequency polygon or frequency curve often becomes the best visual method

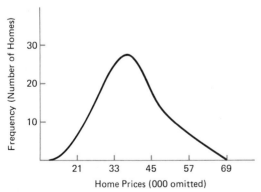

FIGURE 3-8. Frequency Curve for the Distribution of Home Prices in Abcon for 1977

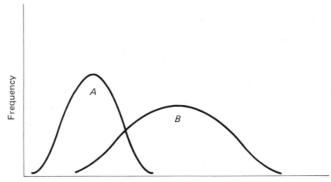

FIGURE 3-9. Comparison of Two Frequency Distributions by Use of Frequency Curves

for presenting data. As an example, Figure 3-9 shows two frequency curves. By observing the figure, we can see that variable A is more compact (narrower) than variable B but that in B the frequencies tend to be of smaller magnitude.

Before moving to other forms of frequency distributions, the student should examine each figure carefully and note that care was taken in labeling the axes, marking off the scale, and establishing a title for the figure. These are important factors for good visual presentation of data.

Other Frequency Distributions

The nature of the data or its intended use will often make a different type of visual presentation more useful. The objective is always to represent the data clearly so that it can be readily interpreted. A presentation subject to misinterpretation is often worse than no presentation at all. Cumulative and relative frequency distributions are two other useful forms of visual presentations.

Cumulative Distribution

As its name indicates, the *cumulative frequency distribution* cumulates the frequencies, starting at either the lowest or highest value. Consider the frequency polygon in Figure 3-7 as an example. This curve will be transformed into the cumulative distribution plotted in Figure 3-10. The data from Table 3-4 are repeated in the left portion of Table 3-7. These frequencies will now be tabulated to obtain the numbers in the center

TABLE 3-7. Frequency Distribution and Cumulative Frequency Distributions of Home Prices

Home Prices	Frequency	Cumulative Frequencies			
		Less than		More than	
		Less than $18,000	0		
$18,000 and under $24,000	7	Less than $24,000	7	More than $18,000	112
$24,000 and under $30,000	15	Less than $30,000	22	More than $24,000	105
$30,000 and under $36,000	24	Less than $36,000	46	More than $30,000	90
$36,000 and under $42,000	27	Less than $42,000	73	More than $36,000	66
$42,000 and under $48,000	17	Less than $48,000	90	More than $42,000	39
$48,000 and under $54,000	10	Less than $54,000	100	More than $48,000	22
$54,000 and under $60,000	8	Less than $60,000	108	More than $54,000	12
$60,000 and under $66,000	4	Less than $66,000	112	More than $60,000	4
	112			More than $66,000	0

portion of Table 3-7 and also to plot Figure 3-10. This is referred to as a "less than" cumulative frequency distribution, since the frequency tabulated represents a summation of all frequencies up to the specified value. The first item of the "less than" cumulative distribution is 7, just

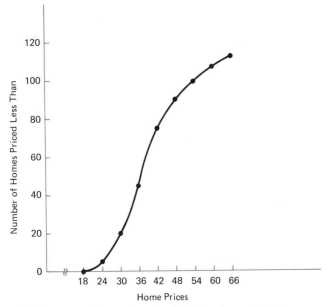

FIGURE 3-10. Ogive for the Distribution of 112 Home Prices Less Than a Specified Amount (thousands of $)

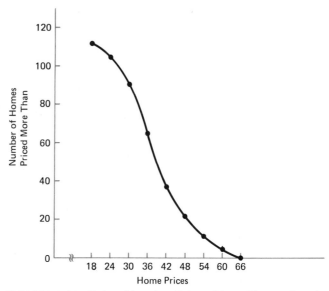

FIGURE 3-11. Ogive Showing Prices More Than a Specified Amount (thousands of $)

as in Table 3-4. The second point, representing values less than $30,000 is the 7 + 15, or 22. The third point in the center column of Table 3-7 is 7 plus 15 plus the 24 in the next class, or a total of 46. When these are plotted, as in Figure 3-10, the resulting curve is called an *ogive*. Any point on the curve represents the total number of homes priced less than that value.

We have drawn a "more than" ogive from the same data in Figure 3-11. The "more than" cumulative frequencies are tabulated in the right portion of Table 3-7. The 112 is interpreted to mean that all the prices are $18,000 or more. This appears in Figure 3-11 as the highest point at 18. The next point is interpreted to mean that 105 prices are $24,000 or more. This is obtained by subtracting the 7 from the 112. This process is continued to complete the tabulation and to prepare the plot in Figure 3-11. The choice between the "less than" or "more than" curves depends on the judgment of the preparer as to which will make the clearer presentation.

Relative Frequency Distribution

When a frequency distribution and histogram were prepared for the 112 home prices, the lengths of the bars were used to compare the frequency of prices in the various class groups. Each bar represented a

TABLE 3-8. Relative Frequency Distribution of Home Prices Equivalent to Table 3-4

Home Prices	Frequency, f	Relative Frequency, $f/\Sigma f$
$18,000 and under $24,000	7	.063
$24,000 and under $30,000	15	.134
$30,000 and under $36,000	24	.214
$36,000 and under $42,000	27	.241
$42,000 and under $48,000	17	.152
$48,000 and under $54,000	10	.089
$54,000 and under $60,000	8	.071
$60,000 and under $66,000	4	.036
	$\Sigma f = 112$	1.000

portion of the 112 houses in the suburb of Abcon. Sometimes we are more interested in the proportion or percentage of items in each group, rather than absolute quantities.

We can easily convert Table 3-4 to a *relative frequency distribution* by dividing each class frequency by the total of 112. The resulting relative frequencies are shown in Table 3-8. Since each class is represented as a proportion of the total, the relative frequency column totals unity (1.0).

Data in the form of relative frequencies would be useful in making year-to-year comparisons or comparisons with other communities. Assume that the number of houses sold in prior years in Abcon is as follows:

```
1975    76
1976    90
```

Suppose that we wanted to compare the proportion of houses over $50,000 in each of the 3 years, 1975, 1976, and 1977. With the difference in total houses sold each year, the quantity of units priced over $50,000 in each year does not provide a valid basis for year-to-year comparison. This is true because the total houses sold varies from year to year. The relative frequency would, however, provide the proportion, which could then be compared from year to year. A change from year to year might be the result of different buying preferences, inflation, or other factors.

In summary, several methods have been shown to compile data and present them in the form of a visual display. Other guidelines have been set forth in the tabulation of the data and in the preparation of displays. There are many ways to display the data, and the preparer will need to select one of these based on intended use.

Exercises

3.1 Tabulations were made of the number of orders shipped by the ABC Company for each of 30 days as follows:

47	62	98	74	89	41
94	69	77	104	72	63
87	91	64	79	66	77
76	83	56	107	59	82
78	54	106	96	84	82

Form the data into classes and determine the frequency and midpoint for each class.

3.2 A savings and loan institution held 60 mortgages which were placed before 1960. The interest rates were tabulated as follows:

Rates	Frequency
3.00–3.49	0
3.50–3.99	2
4.00–4.49	6
4.50–4.99	9
5.00–5.49	9
5.50–5.99	18
6.00–6.49	15
6.50–6.99	1
7.00–7.49	0
	60

(a) Draw a histogram to represent the data.
(b) Draw a frequency polygon and a smooth curve for the data.
(c) Prepare a cumulative "more than" frequency distribution.
(d) Plot an ogive for part (c).

3.3 Following are the amounts on customer meal checks at a small diner for one day:

$2.04	$2.60	$1.55	$2.58
1.83	2.84	1.78	3.04
.87	1.86	1.75	2.90
1.62	4.83	.86	2.10
5.85	1.10	1.85	2.16
2.04	4.08	3.62	1.16
2.64	2.00	.97	3.27
1.80	1.90	2.64	3.45
3.05	2.35	1.55	

(a) Tabulate into a frequency distribution using the following classes:

0– .99
1.00–1.99
etc.

(b) Draw a histogram and a frequency polygon for the data.
(c) Prepare a "less than" cumulative frequency distribution table and ogive for the meal charges.

3.4 Weekly earnings of 50 secretaries in a medium-sized town were tabulated as follows:

Weekly Earnings	Number of Secretaries
$50–74	2
75–99	5
100–124	10
125–149	17
150–174	10
175–199	4
200–224	2
	50

(a) What can you assume to be the actual class limits on each group?
(b) Plot a histogram for the data. What can you say about symmetry of the data?
(c) Prepare a cumulative frequency distribution and an ogive.

3.5 The following data show a tabulation of the weekly earnings of 428 skilled employees at the Jaston Assembly Company:

Weekly Earnings	Midpoint	Number of Employees	Per Cent of Employees
$215 and under $225	$220	0	0
225 and under 235	230	4	1
235 and under 245	240	46	11
245 and under 255	250	98	23
255 and under 265	260	126	29
265 and under 275	270	90	21
275 and under 285	280	50	12
285 and under 295	290	6	1
295 and under 305	300	8	2
		428	100

(a) Draw a histogram for the data.
(b) Draw a frequency polygon.
(c) Tabulate a "less than" cumulative frequency distribution.
(d) Plot a cumulative frequency curve showing employees earning less than a particular weekly earning.

3.6 Tire sales over 50 working days were tabulated as follows:

Daily Sales	Number of Days	Daily Sales	Number of Days
0	2	11	2
1	1	12	4
2	1	13	3
3	0	14	1
4	6	15	1
5	0	16	2
6	4	17	1
7	1	18	0
8	5	19	2
9	8	20	0
10	6		

The data were tabulated as follows:

Sales	Days
1–8	20
8–15	30
15–22	6

What are some poor practices in the tabulation of the data?

3.7 The number of policies sold during one year by each of 25 insurance salespeople are listed as follows?

205	154	169	134	88
195	127	78	143	221
156	54	233	192	208
241	170	118	92	126
99	286	184	168	159

Arrange the data into a frequency distribution, using a class width of 40, and calculate the relative frequencies in each class. Draw a histogram and a frequency polygon. Prepare a "more than" cumulative frequency distribution and draw the ogive.

3.8 A survey was made of a group of 40 new-car owners to determine how many items had required repair or adjustment in the first year of ownership. The following tabulation was obtained:

```
1  4  1  2  2
3  3  2  1  2
3  2  3  1  0
1  2  7  4  3
5  1  2  4  2
1  3  1  0  1
2  1  1  3  1
0  4  2  3  5
```

Tally the data into classes, considering each value as a single class.
(a) Prepare a frequency distribution.
(b) Prepare a relative frequency distribution.
(c) Draw a line graph for (a).
(d) Prepare a cumulative (less than) frequency distribution.
(e) Plot an ogive for (d).

3.9 A marketing research consultant conducted a survey of 40 persons who used a drugstore in one morning. The ages of the persons were recorded to the nearest year as follows:

```
53  40  43  52  19
58  37  38  52  20
61  24  31  39  31
64  13  16  33  32
40  42  10  21  39
34  38  21  10  39
21  28  33  49  47
12  13  39   7  48
```

(a) Set up a frequency distribution using ages 31.5 and up to 38.5 as one of the classes. Why was this grouping selected?
(b) Prepare a relative frequency distribution.
(c) Plot a histogram for (a).
(d) Prepare a "less than" cumulative frequency distribution.
(e) Plot the ogive for (d).
(f) Discuss the factors that you turned up in parts (a)–(e) and any limitations they may have on the usefulness of the tabulation or plot.

3.10 A test was conducted to measure the life of 80 experimental lamp bulbs giving the following results:

Length of Life (hours)	Number of Bulbs
2,000 and under 3,000	4
3,000 and under 4,000	16
4,000 and under 5,000	33
5,000 and under 6,000	20
6,000 and under 7,000	7
	80

(a) Plot a histogram and a frequency polygon and comment on the plots.
(b) Prepare a relative frequency distribution.
(c) Prepare a "more than" cumulative frequency distribution.
(d) Plot the ogive for (c) and comment on the results.

Measures of Central Tendency

In the first half of this chapter, we learned to summarize data into a meaningful presentation in the form of a frequency distribution. From the frequency distribution a graphical presentation was prepared which helped provide an understanding of the data pattern. The next step will be to calculate values from the data or frequency distribution that will permit us to define the whole data set in terms of summary values. Our objective is to formulate measures of the properties of the set of data, either for use in describing the data or for use in comparing it to other sets of data. As a further objective, we desire measures that are easy to calculate and are easily understood.

Measures of central tendency are used very widely to summarize and represent a set of data. Almost everyone is familiar with such terms as average, mean, or median. A *measure of central tendency* is defined as a single term that is considered most representative of the whole set of data. Following our definition and discussion of measures of central tendency, we shall then see how to measure and define variation or dispersion about the central value.

A summary measure is commonly used in describing the central tendency of a population or set of data. A teacher may say that the average grade on an exam was 74. In another instance, a car owner may state that his automobile averages 23.6 miles per gallon. In the first case the teacher was stating a mean value of all exams given; as you will recall from Chapter 1, this measure is called a *population parameter*. In the case of the car owner, he was expressing a mean value based on data which he had collected. Since his mileage data are only a portion of all data, they are sample data and the mean value would be referred to as a *statistic*. The population is the complete set of data, and a sample can be taken from the population.

There are several measures of central tendency from which we can select, and care must be taken to choose one that is well suited to the intended use. The advantages of each of several measures will be pointed out once they are defined and discussed.

Arithmetic Mean

The *arithmetic mean* is probably the most commonly used and readily understood measure of central tendency. Its many and varied uses include baseball batting averages, average miles per gallon, and other widely understood applications. A large segment of the U.S. population is familiar with one or more of these applications, and most people have learned how to calculate the mean. Consider a high school swim team with 6 boys of weights 124, 132, 143, 152, 155, and 158. The average, or arithmetic mean (μ), is calculated by adding these values and dividing by the number of boys, as follows:

$$\mu = \frac{124 + 132 + 143 + 152 + 155 + 158}{6} = \frac{864}{6} = 144$$

We have used the Greek letter μ to represent the mean, which, as noted in Chapter 1, is usual practice where we have the whole population (complete team). If it were a sample, we would, of course, use $\bar{X}$ instead of μ but perform the same calculation. If we let each individual value be represented by X_1, X_2, etc. and let N equal the number in the population, then the calculation of the mean can be symbolized by

$$\mu = \frac{X_1 + X_2 + X_3 + \cdots X_N}{N}$$

If we had a sample from the population, the lowercase n would be used instead of N. In either case the capital sigma is used as a summation sign to represent the sum of all the X's, so that the above expression can be written more conveniently as

$$\mu = \frac{\sum_{i=1}^{N} X_i}{N} = \frac{\Sigma X}{N} \quad \text{(for a population)}$$

The notation using *i* is mathematically correct; however, in basic statistical applications it is sufficient to drop the *i*'s and just say "the sum of the *X* over *N*." For sample data this would be

$$\bar{X} = \frac{\Sigma X}{n} \quad \text{(for a sample)}$$

There are means other than the arithmetic mean, but we shall not discuss them in this text. In this text the use of the term "mean" will refer to the arithmetic mean. The boys' weights above are referred to as raw data or ungrouped data and every value was added into the total. When the data are classified into a frequency distribution, such as the home prices in Table 3-4, they are referred to as grouped data and the calculation of the mean is handled differently.

Mean for Grouped Data

Earlier we took a set of data from home prices and compiled the data into a frequency distribution. The resulting classified data gave us a better picture of the data as a whole. In other situations the data might have been grouped into classes before we saw it, and we would not know what the original individual actual values were. In still other cases the available data might have been collected merely by tallying them into classes and the individual measurements never made. As an additional consideration, the calculation of the mean from raw data can be cumbersome and may be subject to error. The procedure that we shall now illustrate can be readily used to calculate the mean for a large amount of data after they have been grouped. Table 3-9 shows the calculations for the mean (μ) for the home prices originally given by Table 3-4. Let us assume that we had received the data already tabulated as in the first two columns of Table 3-9 and that we therefore do not know the individual prices. The next step would be to calculate the midpoint of each class interval, as shown in the table. For example, the 21 is the midpoint halfway between the 18 and 24. The formula for calculation of the mean (μ) of grouped data is

$$\mu = \frac{\Sigma fX}{N}$$

Note that the use of *f* and *X* are slightly different here, where

f = number of items in each class (frequency)
X = class midpoint
N = total number of items in the population

TABLE 3-9. Calculations of the Mean for Grouped Data

Home Prices (000 omitted)	Number of Homes, f	Interval Midpoint, X	fX
$18 and under $24	7	21	147
$24 and under $30	15	27	405
$30 and under $36	24	33	792
$36 and under $42	27	39	1,053
$42 and under $48	17	45	765
$48 and under $54	10	51	510
$54 and under $60	8	57	456
$60 and under $66	4	63	252
	112		4,380

$$\mu = \frac{\Sigma fX}{N} = \frac{4,380}{112} = 39.107 \text{ (or } \$39,107)$$

The last column in Table 3-9 shows the product of f (frequency) times midpoint (X) for each class interval. The formula in Table 3-9 illustrates the method of determining the mean for the grouped data. In this approach it might be considered that each class midpoint is weighted according to the number of cases in the class. Note also that the prices in Table 3-9 are expressed in thousands of dollars. In a calculation for the mean, we can sometimes simplify calculations by dropping a set number of decimal places and then reinstating them after the calculations are completed.

It should be recognized that the mean determined from the grouped data in Table 3-9 is an approximation to the mean that would be obtained by adding up the original 112 individual prices and dividing by 112. This difference exists because all the prices in each particular class have been represented by a single midpoint value. If we had selected a different class grouping, such as one of those in Tables 3-2, 3-5, or 3-6, the midpoints would have been different as would the frequencies, and the calculated mean for each would therefore have been slightly different. It must be recognized, however, that data often are really only part of a larger population. Also we need to recognize that many statistical measures are approximations anyway, and are thus subject to some inherent sampling errors. The science of statistics, however, has developed a set methodology that gives us valid and consistent methods of measurement, along with the methods needed to determine the likely errors that pertain to any case. Therefore, once the data have been grouped, it is important to make the calculations exactly by the formula.

Median

The median is also widely used as a value to represent central tendency for a set of data. The *median* is the middlemost value, or the value such that half the items lie above it and half lie below it. In some instances the median is preferred to the mean as a measure of central tendency.

If we have an odd number of items, the median will be the middle measurement. Consider a sample of 7 brands of cotton gloves priced at 95, 86, 65, 77, 92, 74, and 88 cents. To determine the median of these prices, the first step is to rearrange the data in ascending order as follows:

$$
\begin{array}{c}
65 \\
74 \\
77 \\
86 \\
88 \\
92 \\
95
\end{array}
$$

The median value is 86 because there are three items above 86 and three below it.

Suppose that we added one additional brand to our list with a price of 81 cents. The new sequence of 8 items is now:

$$
\begin{array}{c}
65 \\
74 \\
77 \\
81 \\
86 \\
88 \\
92 \\
95
\end{array}
$$

By definition, the median for an even number of items is midway between the two center items. In this case the two middle items are 81 and 86 so that the median (Md) is 83.5. This is calculated as

$$\text{Md} = \frac{81 + 86}{2} = 83.5$$

The median can have the same value as the mean, but usually it is different. Either the mean or median may be the better measure of central value for a particular set of data. Factors affecting the choice will be discussed later in the chapter.

The median for grouped data is not as easily obtained. The grouped data do not retain the individual values, and it therefore becomes necessary to estimate a median from the available grouped data. Look at Table 3-10 and assume that the original list of individual prices is not available. The median would have been between the 56th and 57th items since there is a total of 112 items. In grouped data, since we cannot identify individual items, the conventional formula uses the position $N/2$ as a basis. In this case, it is the 56th item (112/2). We know that the 56th value is not in the first, second, or third class because the cumulative frequencies to that point are only 46. There are 27 more in the fourth class so that the middle value is in the fourth class. The question is: How far do we have to go into the fourth class to reach our point? The answer is 10 more observations. As a result, we know that the median is greater than \$36,000 but less than \$42,000. Therefore, we go 10/27 of the way through this class. The formula to calculate the median for grouped data is

$$Md = L_{Md} + \frac{N/2 - \Sigma f_L}{f_{Md}} \times i$$

where
 Md = median
 L_{Md} = lower limit of class containing median
 Σf_L = the sum of the frequencies of all classes up to the median class
 f_{Md} = number of items in median class
 i = size of median class interval
 N = total number of data items

The median for the home prices in Table 3-10 is calculated by substituting the appropriate values in the formula given above.

$$\begin{aligned}Md &= 36{,}000 + \frac{112/2 - 46}{27} \times 6{,}000 \\ &= 36{,}000 + \frac{10}{27} \times 6{,}000 \\ &= 36{,}000 + 2{,}222 = \$38{,}222\end{aligned}$$

This formula applies to the calculation of the median for any set of grouped data.

Mode

The mode is the third of the often used measures of central tendency. The *mode* is the value that occurs most often. For example, in the series

TABLE 3-10. Determination of the Median for Grouped Data

Home Prices	Number of Homes
$18,000 and under $24,000	7
$24,000 and under $30,000	15
$30,000 and under $36,000	24
$36,000 and under $42,000	27
$42,000 and under $48,000	17
$48,000 and under $54,000	10
$54,000 and under $60,000	8
$60,000 and under $66,000	4
	112

$$Md = L_{Md} + \frac{N/2 - \Sigma f_L}{f_{Md}} \times i = 36,000 + \frac{112/2 - 46}{27} \times 6,000$$
$$= 36,000 + \frac{10}{27} \times 6,000 = 36,000 + 2,222 = \$38,222$$

of numbers 3, 4, 5, 5, 6, 7, 7, 7, 7, 8, 9, 9, the mode is 7 because it occurs most often. The dictionary definition of the term "mode" is "most usual, or conventional." In coat sizes, the size sold most frequently would be the mode, or the most typical size.

If the data are grouped, the calculation is different again. In the case of grouped data, the *modal class* is defined as the class with the greatest frequency. In Table 3-10, the modal class ($36,000 and under $42,000) contains 27 items, which is greater than any other class frequency. The modal class may or may not contain either the median or the mean. Although there are formulas for calculating the precise mode in the modal class, this precise number is not of sufficient practical use to warrant further discussion here. If a frequency distribution is represented by a smooth curve as in Figure 3-8, the mode is the value on the X axis below the highest point on the curve.

When some frequency distributions are represented by a smooth curve, they may have more than one peak point, as shown in Figure 3-12. This type of distribution is referred to as *bimodal* if there are two peaks. The term *multimodal* is used if there are more than two peaks. In the case shown there are two peaks and the distribution that appears to be bimodal. Before the bimodal shape is accepted as the true shape of the distribution, an attempt should be made to determine the reason for the existence of two peaks. Suppose that Figure 3-12 resulted from a tabulation of sales at a restaurant. Further analysis may then show that the sales in the bar area form one distribution and the dining room checks

FIGURE 3-12. Bimodal Distribution

form another distribution. If they are mixed together, we have combined two different distributions and a bimodal curve results. In cases such as this it is important to identify the two underlying distributions in order to properly analyze the data and arrive at valid conclusions. In the case of the restaurant we would probably decide to divide the sales checks into two groups and evaluate each distribution separately.

Characteristics of Measures of Central Tendency

Although the mean, the median, and the mode are all valid measures of central tendency, there are certain advantages to each. A particular set of data or particular intended use may dictate the selection of one or the other of these measures. The following are some helpful guidelines.

The mean, in addition to being the most widely used value, is also the most convenient to handle mathematically. If we have two or more mean values, they can be added and averaged to obtain the mean of the total. If the mean income for workers in one manufacturing plant is $220 per week and for a second plant with the same number of workers is $230, the mean of the combined total is $225. If the two plants had different numbers of employees, such as 2,000 for the first plant and 3,000 for the second, the mean would be

$$\mu = \frac{2{,}000(220) + 3{,}000(230)}{5{,}000}$$
$$= \frac{440 + 690}{5} = \frac{1{,}130}{5}$$
$$= 226$$

The means have been combined by weighting each value according to the number of workers that it represents. The median and the mode are not so readily combined.

Representativeness

When we calculate a measure of central tendency, our primary intent is to obtain a measure that represents the whole set of data. If the measure is to be compared to measures of other populations and possibly combined with them, the mean is the best measure. In later chapters we shall make estimates or inferences about a total population based on information from the sample. Standard computational procedures permit us to make these estimates and evaluations from sample means.

In the case of grouped data, the median tends to lose its value as a representative measure. We saw in our calculation of the median for grouped data that it was no longer a particular item from the original list. The mean is much more meaningful; however, the modal class that contains the most items is also a useful piece of information.

Figure 3-13 shows a histogram of suit sizes sold in a store. In this case we are working with discrete values since there are no fractional suit sizes. Also, the mean, if it turned out to be 40.4, would not really be very useful. In attempting to determine how many of each size to stock, a store would be more interested in the mode, as the most often used size, and in the relative frequencies of the other sizes that they would stock.

The mean is affected by each "individual" item in the list of data, and if any single item were to be changed, it would affect the mean. The median and the mode, in contrast, may not be affected if individual items are changed. In our example using prices of cotton gloves, the highest-priced item could be changed from $.95 to $2.00 without affecting the median. In Figure 3-13, suit sizes 42 and over could be eliminated entirely from the data without affecting the mode.

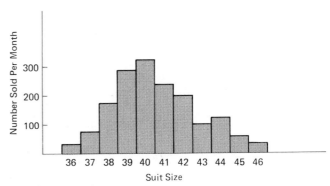

FIGURE 3-13. Distribution of Sizes of Mens Suits Sold in a Store

Influence of Extreme Values

Extremely high or low values in a set of data have a heavy influence on the mean. Consider a small tax consulting firm where the salaries are as follows:

President	$96,000
Consultant A	24,000
Consultant B	20,000
Consultant C	18,000
Consultant D	16,000
Consultant E	14,000
Associate A	10,000
Associate B	10,000
Associate C	8,000

The median value turns out to be $16,000 and is not influenced very much by the president's very high salary. This high salary does, however, have an appreciable effect on the mean of $24,000, which is much higher than the median thanks to the influence of the one high value. If a candidate for a position in this firm were told that the mean salary of the 9 employees

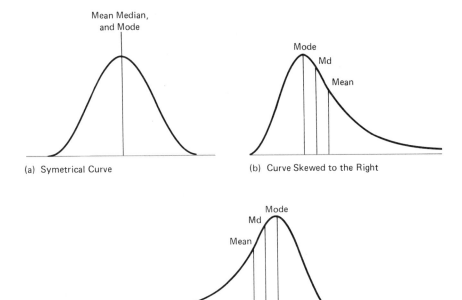

FIGURE 3-14. Relative Positions of Mean, Median, and Mode in Three Shapes of Distribution

was $24,000, the statement would be factually correct, but very misleading. We conclude, then, that the median usually provides a better measure of central tendency whenever there are any extreme values in only one direction.

The influence of extreme values on the mean can be examined further by observing the relative positions of the mean, the median, and the mode in three situations. The symmetrical curve in Figure 3-14a will have its mean, median, and mode at the same value. The distribution represented by the curve in Figure 3-14b is skewed to the right and the extreme values pull the mean to the right of both the median and the mode. The mode is farthest left of the three values at the highest point of the curve. A curve skewed to the right is said to have *positive skewness*. If the age of all licensed automobiles in the United States were plotted, the few very old autos still around primarily for their antique value would pull the right end of the distribution way out, resulting in a curve of this shape. The third curve (Figure 3-14c) is skewed to the left and is said to have *negative skewness*. In this case the mean is to the left of both the median and mode. Often, when the three values have been calculated for a set of data, the relationship of the mean, the median, and the mode will help a person visualize the shape of the distribution curve without actually preparing a frequency distribution or plotting the curve.

Examples Where Measures Are Misleading

A primary objective of statistics is to provide meaningful information to the user, and to provide it in such a manner that it is not readily misinterpreted. Unfortunately, however, there are those who misuse statistics for their own benefit. A statement used in advertising that more doctors use brand X than any other brand tested should raise questions. How many were tested? How many doctors were polled? Could the difference be due to chance or to errors? In addition to learning to present information such that it is not easily misunderstood, we must also be able to recognize the misuse of data by others. We previously cited the case where the statement "the mean salary of the people in this firm is $24,000" would be very misleading to a college graduate considering joining the firm, as only one person exceeded the $24,000 mean value. The median would be a better representation and would provide the candidate with a clearer understanding of his possibilities.

As another instance in which statistics may be misleading, consider a clerk who collected data on dress sizes and found that they came in discrete steps of half sizes. He calculated a mean of 12.3, a median of $12\frac{1}{2}$, and a mode of 12. These numbers would require much further analysis because, as most women would know, a dress size of 13 is not halfway between a 12 and 14, nor is a $14\frac{1}{2}$ halfway between a 14 and a 15. Sizes

5, 7, 9, 11, 13, and 15 are styled for younger girls; 6, 8, 10, 12, 14, and 16 are styled for the women of average build; and half sizes are styled for a mature figure. These should really be considered as three separate populations. This example stresses the importance of understanding as much as possible about the sources of data.

Consider another misuse of the mean when there are extreme values present. A survey of noise levels in a manufacturing plant resulted in an average well below the acceptable level to avoid ear damage. Further investigation, however, would have shown some areas to be quiet whereas in other areas the level was actually quite high. It might show that infrequent blasts created noise levels so high as to be dangerous. A person circulating through the plant without protection could receive permanent ear damage because of one peak incident. As another example, a hiker taking clothes for a July mean temperature of 72° in Vermont may be subjected to below-freezing temperature at higher altitudes on cooler days of his trip. All of us can imagine the results if a nonswimmer tried to walk across a lake which he was told averaged 3 feet deep.

The wrong basis of comparison can also lead to wrong conclusions. A manager of a retail department store chain brags that his average sales per store have increased over 8 per cent in the past year, compared to an average of 6 per cent per year for the past 5 years. Further investigation might have shown that the average increase registered by his competitors was 12 per cent. It is important to know all the standards with which to compare the data. Additional research might also reveal one of his stores with a decrease in sales for the year. An overall figure should not be allowed to cover up specific situations requiring action. Consider another example where the gas mileage of automobiles is being calculated. Four cars were tested with the results being 40 mpg, 13 mpg, 13 mpg, and 14 mpg. The average mpg is 20 with the median being 13.5 and the mode 13. If a consumer were not careful, he or she could be misled by the 20 mpg mean. These possible limitations in use of measures of central tendency lead us to the need for measuring variation from the mean. These will be called measures of dispersion.

Exercises

> **3.11** In a city block, the number of persons living in each house was tabulated as follows: 2, 4, 5, 3, 4, 3, 1, 2, 6, 7, 3, 3, 2, 5, 4, 6. Determine the mean, the median, and the mode. Which do you think is the best measure of central tendency?
>
> **3.12** The annual income of 13 salespersons is given as follows:

$ 9,000	$18,600
9,200	21,000
10,000	24,000
11,500	25,000
14,000	64,000
14,000	75,000
15,400	

Determine the mean, the median, and the mode and discuss the pros and cons of each measure.

3.13 A savings and loan institution still holds 60 mortgages which were placed before 1960. The interest rates have been tabulated as follows:

Rates	Frequency
3.00–3.49	0
3.50–3.99	2
4.00–4.49	6
4.50–4.99	9
5.00–5.49	9
5.50–5.99	18
6.00–6.49	15
6.50–6.99	1
7.00–7.49	0
	60

Calculate the mean and the median.

3.14 The following is a tabulation of scores obtained on an employment aptitude test by 1,000 applicants. Determine the mean value.

Interval	Frequency
92–100	60
83–91	140
74–82	160
65–73	120
56–64	140
47–55	80
38–46	119
29–37	81
20–28	50
11–19	32
2–10	18
	1,000

3.15 A survey of 100 students is made to determine the cost of food per week. The following results are obtained. Determine the mean, the median, and the modal class of this population.

Amount	Number of Students
$ 6.00 and under $10.00	2
$10.00 and under $14.00	23
$14.00 and under $18.00	47
$18.00 and under $22.00	23
$22.00 and under $26.00	5

3.16 The ages of homes in a small community were tabulated as follows:

Age (years)	Frequency
5 and less than 10	3
10 and less than 15	7
15 and less than 20	22
20 and less than 25	36
25 and less than 30	15
30 and less than 35	11
35 and less than 40	4
40 and less than 45	2

(a) Determine the midpoints and calculate the mean.
(b) Calculate the median.
(c) Discuss the shape of the distribution based on the calculations.

3.17 A trucking company kept records of 2 brands of tires used on its trucks as follows: (assume these are populations)

Life Mileage (in thousands)	Brand A (number of tires)	Brand B (number of tires)
20 and under 25	2	2
25 and under 30	6	4
30 and under 35	15	18
35 and under 40	10	32
40 and under 45	15	30
45 and under 50	17	10
50 and under 55	13	2
55 and under 60	9	2
60 and under 65	8	0
65 and under 70	2	0
70 and under 75	3	0
	100	100

(a) Calculate the mean for each brand.
(b) Calculate the median and the modal class for each brand.
(c) Plot the data in the form of 3 histograms. (brand A, brand B, & combined)
(d) Comment on the 2 distributions, the measures of central tendency, and the combined distribution.

3.18 The company in problem 3.17 used 100 tires of brand C. They took a sample of 20 tires and the mileage recorded on each was tabulated as follows (in thousands):

33	55	49	60
31	67	58	72
78	84	58	93
44	46	57	104
54	67	62	40

(a) Calculate the mean, the median, and the mode.
(b) Classify the data into classes using "30 and under 40" as one class, and calculate the mean, the median, and the modal class.
(c) Compare the results and give reasons for any differences.
(d) Is the distribution symmetrical or skewed?

3.19 An investor buys 100 shares of stock each year. The following were the prices paid per share for a 5-year period: $40, $24, $20, $30, $48. What is the mean price paid per share?

3.20 Calculate the mean, the median, and the modal class for the following discrete data. What can you say about symmetry?

Number of Defects per Item	Frequency
2–10	18
11–19	32
20–28	50
29–37	81
38–46	119
47–55	80
56–64	140
65–73	120
74–82	160
83–91	140
92–100	60
	1,000

Measures of Dispersion

We have seen how to calculate measures of central tendency to provide a single value to represent a set of data. These measures are not usually, by themselves, sufficient to present an adequate summary of the data under evaluation. An additional measure is dispersion (or variability, as it is sometimes called). A *measure of dispersion* describes the spread or scattering of the individual values around the central position. Consider the two distributions in Figure 3-15, which have identical means, medians, and modes. They differ considerably, however, with respect to dispersion, or spread. As an illustration of the significance of dispersions, consider the person who invests in common stocks rather than bonds because the average yield is higher. It is also important to understand that the wider variability of returns on stocks results in greater risks. Some stocks may provide yields much higher than the mean, whereas others may provide a lower yield or none at all. The person is taking the risk that although the average return may be higher, some stocks may provide little or no return.

Measures of dispersion can be grouped into categories. The first category measures the distance between values with the same scale as the original data and are called distance measures. Examples are the range and interfractile ranges, which are discussed in the next paragraph. The second category includes measurements in terms of deviations, such as the variance, standard deviation, and average deviation. Where the dispersion of one set of data is to be compared to the dispersion of another

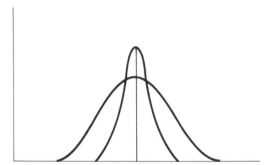

FIGURE 3-15. Two Distributions with Identical Means, Medians, and Modes but with Different Dispersions

set, a measure of relative dispersion is also available and will be discussed here.

Distance Measures

Distance measures of dispersion are expressed in the same units as the original data. The simplest of these is the *range,* or the difference between the highest and lowest values in the data. In the case of the home prices, the highest sales price was $64,000 and the lowest $19,900, so the difference or range equals $64,000 − $19,900, or $44,100. The range is expressed in dollars just as the original data are. The range is very easy to calculate and provides a rough feel for the spread of the data; however, it is not adequately precise for most analytical needs. Since it depends on two values only, the highest and the lowest, the range does not reflect the variability or clustering between all the other values that make up the available data. It is obviously affected by any extremely high or low values.

Although the value of the range depends solely on the two extreme values, other measures, called *interfractile ranges,* take additional points into consideration. Examples are quartiles and percentiles. Just as the median divides the data into two equal groups, the *quartiles* divide the data into four equal groups and *percentiles* would divide the data into 100 equal groups. Consider the following per-pound prices of 8 varieties of grass seed sold in a garden store. In this case we have an even number of items in the list. In any case the calculations are carried out in a fashion similar to those for the median:

$$.35$$
$$.39$$
$$.79 \leftarrow \text{first quartile} = \frac{.39 + .79}{2} = \frac{1.18}{2} = .59$$
$$.95$$
$$1.19 \leftarrow \text{median (second quartile)} = \frac{.95 + 1.19}{2} = \frac{2.14}{2} = 1.07$$
$$1.35$$
$$1.50 \leftarrow \text{third quartile} = \frac{1.35 + 1.50}{2} = \frac{2.85}{2} = 1.425$$
$$1.79$$

If there were 11 values (an odd number), the dividing points might be as follows:

$$.33$$
$$.35$$
$$.46 \leftarrow \text{first quartile point} = .46$$
$$.75$$
$$.79$$
$$.88 \leftarrow \text{median} = .88$$
$$.95$$
$$1.04$$
$$1.19 \leftarrow \text{third quartile point} = 1.19$$
$$1.33$$
$$1.54$$

The quartiles are shown with the data. The interquartile range is the range between the third quartile point and the first quartile point. In the first example, the interquartile range would be 1.425 − .59, or .835. In the second list, the interquartile range would be 1.19 − .46, or .73. The interquartile range represents the spread of the middle 50 per cent of the items in the data. As noted, the term "percentile" is used when the data are divided into 100 equal groups. We can determine the range that will include other percentages of the data, such as 90 per cent or 95 per cent, by measuring the spread over specified percentiles.

Average Deviation

The *average deviation* is an improvement over the range as a measure of dispersion in that it considers all elements in the set of data. The average deviation is calculated by first obtaining the mean. After the mean of the data has been calculated, the next step is to determine the difference between each value and the mean. Since we are concerned with the magnitude of the difference, the sign (+ or −) from the mean is ignored. If the sign were considered, by definition of a mean value, the differences would always add up to zero.

To illustrate the average deviation, the prices for dry-cleaning a suit at each of the 5 cleaning establishments in town were $1.30, $1.45, $1.60, $1.40, and $1.40. These are listed in Table 3-11. The steps in finding the average deviation are to calculate the mean, find the absolute difference of each item from the mean, and finally to take the average of these differences. The average deviation is defined by the following formula:

$$\text{average deviation} = \frac{\Sigma |X_i - \mu|}{N}$$

TABLE 3-11. Calculation of Average Deviation

X_i (cents)	$X_i - \mu$	$|X_i - \mu|$
135	−10	10
145	0	0
165	+20	20
140	−5	5
140	−5	5
$\Sigma X_i = 725$		$\Sigma |X_i - \mu| = 40$

$$\mu = \frac{725}{5} = 145$$

$$\text{average deviation} = \frac{\Sigma |X_i - \mu|}{N} = \frac{40}{5} = 8 \text{ cents}$$

We read this formula as "the average deviation equals the sum of the absolute values of each X from the mean divided by the number of X's." The calculations are shown in Table 3-11. If we had not used the absolute values (designated by two vertical lines), the 20 would be positive and the rest negative, to add up to zero. Since the data represent the complete population, or all the cleaning establishments, μ is used to represent the mean value. If it were a sample, $\bar{X}$ would be used instead of μ. From the results we could state that each price is different from the population mean by an average of 8 cents. The average deviation is expressed in cents as are the original data.

Sometimes the average deviation is measured about the median. In that case the median is used in place of the mean in the equation above.

Variance and Standard Deviation

The variance and standard deviation are the most widely used and important measures of dispersion. The *variance* is the square of the *standard deviation* and represents the average squared deviation from the population mean. It is calculated by taking the difference of each item from the mean, squaring each difference, and then taking the average of these squared differences. Where the complete population is used, the variance is represented by σ^2 (sigma squared) and the standard deviation by σ, and the calculation is as follows:

$$\sigma^2 = \frac{(X_1 - \mu)^2 + (X_2 - \mu)^2 + \cdots + (X_N - \mu)^2}{N}$$

$$\sigma^2 = \frac{\Sigma (X - \mu)^2}{N}$$

and

$$\sigma = \sqrt{\frac{\Sigma (X - \mu)^2}{N}}$$

The variance and standard deviation become larger as the variability, or spread within the data becomes greater. This is usually also true with range and other measures of dispersion. The following formulas for σ are mathematically equivalent to the formula above and are often more convenient to use in calculations.

$$\sigma = \sqrt{\frac{\Sigma X^2 - \frac{(\Sigma X)^2}{N}}{N}} = \sqrt{\frac{\Sigma X^2}{N} - \mu^2} = \sqrt{\frac{\Sigma X^2 - N\mu^2}{N}}$$

Table 3-12a and b shows the calculations for σ of the prices at 5 cleaning establishments by both methods. In this case we have all the establishments in town, so that it is a complete population.

TABLE 3-12a. Calculation of Standard Deviation for Population (Basic Method)

X (cents)	$X - \mu$	$(X - \mu)^2$
135	−10	100
145	0	0
165	+20	400
140	−5	25
140	−5	25
$\Sigma X = 725$		550

$$\mu = \frac{\Sigma X}{N} = \frac{725}{5} = 145$$

$$\sigma^2 = \frac{\Sigma (X - \mu)^2}{N} = \frac{550}{5} = 110$$

$$\sigma = 10.5$$

TABLE 3-12b. Calculation of Standard Deviation for Population (Alternative Method)

X (cents)	X^2
135	18,225
145	21,025
165	27,225
140	19,600
140	19,600
$\Sigma X = 725$	$\Sigma X^2 = 105,675$

$$\mu = \frac{\Sigma X}{N} = \frac{725}{5} = 145$$

$$\sigma = \sqrt{\frac{\Sigma X^2}{N} - \mu^2}$$

$$= \sqrt{\frac{105,675}{5} - (145)^2}$$

$$= \sqrt{21,135 - 21,025} = \sqrt{110}$$

$$= 10.5$$

If the data represent a sample from a population, rather than a complete population, it has been proven that the following formulas provide more valid calculations for the sample variance (s^2) and the sample standard deviation (s), based on a sample size of n:

$$s^2 = \frac{\Sigma (X - \bar{X})^2}{n - 1}$$

$$s = \sqrt{\frac{\Sigma (X - \bar{X})^2}{n - 1}}$$

The following formulas are mathematically equivalent to those above and often involve shorter calculations:

$$s = \sqrt{\frac{\Sigma X^2 - \frac{(\Sigma X)^2}{n}}{n - 1}} = \sqrt{\frac{n \Sigma X^2 - (\Sigma X)^2}{n(n - 1)}}$$

The only difference in the formulas is the use of $n - 1$ and $\bar{X}$ where we have sample data instead of N and μ where we have a complete population data.

In the prior illustration, there were only 5 dry-cleaning establishments in the town. Suppose that the town were larger and that we collected prices at a *sample* of 5 establishments. The calculations for this case are shown in Tables 3-13a and b, using the formulas above.

The use of the $n - 1$ is justified by statistical theory in that the estimate is based on data from a sample. As the sample size becomes larger and closer to the total population size, the resulting s will become closer to the value that we would obtain by use of the formula for σ.

An examination of the two formulas will show that individual deviations from the mean are squared in both. During calculations, the squar-

TABLE 3-13a. Calculation of the Standard Deviation for Sample Data (Basic Method)

X (cents)	$X - \bar{X}$	$(X - \bar{X})^2$
120	−13	169
150	17	289
130	−3	9
135	2	4
130	−3	9
$\Sigma X = 665$		$\Sigma (X - \bar{X})^2 = 480$

$$\bar{X} = \frac{\Sigma X}{n} = \frac{665}{5} = 133$$

$$s = \sqrt{\frac{\Sigma (X - \bar{X})^2}{n - 1}} = \sqrt{\frac{480}{4}} = \sqrt{120} = 10.95$$

TABLE 3-13b. Calculation of Standard Deviation for Sample Data (Alternative Method)

X (cents)	X²
120	14,400
150	22,500
130	16,900
135	18,225
130	16,900
ΣX = 665	ΣX² = 88,925

$$\bar{X} = \frac{\Sigma X}{n} = \frac{665}{5} = 133$$

$$s = \sqrt{\frac{\Sigma X^2 - \frac{(\Sigma X)^2}{n}}{n-1}}$$

$$= \sqrt{\frac{88,925 - \frac{(665)^2}{5}}{4}}$$

$$= \sqrt{\frac{88,925 - \frac{442,225}{5}}{4}}$$

$$= \sqrt{\frac{88,925 - 88,445}{4}}$$

$$= \sqrt{\frac{480}{4}} = \sqrt{120}$$

$$= 10.95$$

ing of these numbers will help in identifying the existence of extremes in the data. This is illustrated in Table 3-14, where the standard deviation is calculated for the salaries of the consultants in the tax firm previously discussed. The fact that the standard deviation comes out larger than the mean also indicates the possible presence of an extreme value(s).

Variance and Standard Deviation for Grouped Data

Tables 3-15a and b show the compilation of the daily sales of gasoline at a small service station for a period of 105 days. Calculations for the mean, the variance, and the standard deviation for grouped data are shown by the basic method and the alternative method. The first two columns provide the frequency distribution and the other columns provide the additional data necessary for the computation of $\bar{X}$ and s. The summation of the last column is used in the formula to solve for s. The data represent a sample, so $n - 1$ is used in the calculations.

TABLE 3-14. Calculation of Population Mean and Standard Deviation for Tax-Consulting-Firm Salaries

X	$X - \mu$	$(X - \mu)^2$
96,000	72,000	5,184,000,000
24,000	0	0
20,000	−4,000	16,000,000
18,000	−6,000	36,000,000
16,000	−8,000	64,000,000
14,000	−10,000	100,000,000
10,000	−14,000	196,000,000
10,000	−14,000	196,000,000
8,000	−16,000	256,000,000
216,000		6,048,000,000

$$\mu = \frac{216,000}{9} = 24,000$$

$$\sigma = \sqrt{\frac{\Sigma (X - \mu)^2}{N}} = \sqrt{\frac{6,048,000,000}{9}} = \sqrt{672,000,000} = \$25,923$$

TABLE 3-15a. Calculation of the Sample Mean and Standard Deviation for Grouped Data (Basic Method)

Number of Gallons Sold	Days, f	Class Midpoint, X	fX	Deviation, $X - \bar{X}$	$(X - \bar{X})^2$	$f(X - \bar{X})^2$
70 and under 80	2	75	150	−35.9	1,288.81	2,577.62
80 and under 90	6	85	510	−25.9	670.81	4,024.86
90 and under 100	16	95	1,520	−15.9	252.81	4,044.96
100 and under 110	28	105	2,940	−5.9	34.81	974.68
110 and under 120	24	115	2,760	4.1	16.81	403.44
120 and under 130	18	125	2,250	14.1	198.81	3,578.58
130 and under 140	8	135	1,080	24.1	580.81	4,646.48
140 and under 150	3	145	435	34.1	1,162.81	3,488.43
	105		11,645			23,739.05

$$\bar{X} = \frac{\Sigma fX}{n} = \frac{11,645}{105} = 110.9$$

$$s = \sqrt{\frac{\Sigma f(X - \bar{X})^2}{n - 1}} = \sqrt{\frac{23,739.05}{104}} = \sqrt{228.26} = 15.1$$

Column	Calculation or meaning
X	class midpoint
fX	multiply the frequency by the class midpoint
$X - \bar{X}$	subtract overall mean X from class midpoint
$(X - \bar{X})^2$	square the number in previous column
$f(X - \bar{X})^2$	multiply the number in previous column by the frequency f

Frequency Distributions and Summary Measures

TABLE 3-15b. Calculation of the Standard Deviation for Grouped Data (alternative method)

Number of Gallons Sold	Days, f	Class Midpoint X	$f \cdot X$	$f \cdot X^2$
70 and under 80	2	75	150	11,250
80 and under 90	6	85	510	43,350
90 and under 100	16	95	1,520	144,400
100 and under 110	28	105	2,940	308,700
110 and under 120	24	115	2,760	317,400
120 and under 130	18	125	2,250	281,250
130 and under 140	8	135	1,080	145,800
140 and under 150	3	145	435	63,075
	105		11,645	1,315,225

$$\bar{X} = \frac{\Sigma fX}{n} = \frac{11,645}{105} = 110.90$$

$$s = \sqrt{\frac{\Sigma fX^2 - \frac{(\Sigma fX)^2}{n}}{n-1}}$$

$$= \sqrt{\frac{1,315,225 - \frac{(11,645)^2}{105}}{104}}$$

$$= \sqrt{\frac{1,315,225 - 1,291,486}{104}}$$

$$= \sqrt{\frac{23,739}{104}} = \sqrt{228.26}$$

$$= 15.1$$

A short-cut, coded method for calculating the standard deviation for the same grouped data is illustrated in Table 3-16. By comparing Tables 3-15a and b and 3-16, it is apparent that the amount of calculations are reduced in Table 3-16. In order to use the short-cut, coded procedure, it is necessary that all class intervals be equal, and the symbol i is used to represent this interval. It is also necessary to take one of the class midpoints as an arbitrary assumed mean X_a. In Table 3-16 we take 105 as the assumed mean, which is the midpoint of the "100 and under 110" class. The value in the d column, then, represents the number of class intervals above or below this assumed mean. The next column (fd) is obtained by multiplying frequency by d. The last column (fd^2) is then d multiplied by fd. Note the identical answers by the two methods in Tables 3-15 and 3-16. The equations are mathematically equivalent; however, the calcula-

TABLE 3-16. Shortcut Coded Method for Calculation of Mean and Standard Deviation of Grouped Data

Number of Gallons Sold	Days, f	Deviation, d	fd	fd²
70 and under 80	2	−3	−6	18
80 and under 90	6	−2	−12	24
90 and under 100	16	−1	−16	16
100 and under 110	28	0	0	0
110 and under 120	24	1	24	24
120 and under 130	18	2	36	72
130 and under 140	8	3	24	72
140 and under 150	3	4	12	48
	105		62	274

$$\bar{X} = \bar{X}_a + \frac{i \, \Sigma fd}{n} = 105 + \frac{10(62)}{105} = 105 + 5.90 = 110.90$$

$$s = i\sqrt{\frac{\Sigma fd^2 - \frac{(\Sigma fd)^2}{n}}{n-1}} = 10\sqrt{\frac{274 - \frac{(62)^2}{105}}{104}} = 10\sqrt{\frac{274 - 36.61}{104}}$$

$$= 10\sqrt{\frac{237.39}{104}} = 10\sqrt{2.2826} = 15.1$$

tions are much easier in Table 3-16 owing to the smaller numbers involved. If the data in the frequency distribution have unequal-sized class intervals, we cannot use the shortcut coded method, and it is necessary to use the equation shown in Table 3-15 (a or b).

Coefficient of Variation

Often it is desired to compare two or more distributions by relating the dispersion of one distribution to the dispersion of another. Standard deviations of the two distributions cannot usually be compared directly if sample sizes or units of measure are different. Consider data collected from a sample of 50 of each of 2 brands of tires.

$$\bar{X}_1 = 50,000 \text{ miles} \qquad s_1 = 12,000 \text{ miles}$$
$$\bar{X}_2 = 30,000 \text{ miles} \qquad s_2 = 8,000 \text{ miles}$$

In order to obtain a true comparison, they first must be converted to similar terms. The *coefficient of variation* (CV) is used for this purpose where

$$CV = \frac{s}{\bar{X}} \times 100$$

The coefficients of variation are computed as follows:

$$CV_1 = \frac{s_1}{\bar{X}_1} = \frac{12{,}000}{50{,}000} \times 100 = 24\%$$

$$CV_2 = \frac{s_2}{\bar{X}_2} = \frac{8{,}000}{30{,}000} \times 100 = 26.7\%$$

When we compare the coefficients of variation, we see that brand 2 actually has a greater relative dispersion as compared to its mean than does brand 1.

The question often comes up when examining a standard deviation: What is too large an s? There are no set guidelines in statistics to answer this question, but a rough rule of thumb is that as CV becomes greater than 50 per cent, the standard deviation is becoming quite large compared to the mean. Much, however, depends upon the use to which we intend to put the data. For example, in our earlier example of tax consultant salaries, a high CV should be cause for alarm to an incoming employee. An examination of the salary figures would then show the one, single, very high salary. This would indicate to the candidate that others in the firm are not doing as well as the mean ($\bar{X}$) may tend to indicate.

In other situations where one is only trying to get a rough feel for an average, the coefficient of variation (CV) or standard deviation may not be as important. From a decision-making standpoint, it all boils down to how much it costs you to make an error through use of a mean that is not precise. The tax consultant candidate may become very unhappy if he accepts the position without understanding the true facts.

Constant-Cause System

In our discussion of dispersion we have talked about a set of data that we assume came from what we might call a *constant-cause system*. In other words, we are making the assumption that there has been no basic change in the population structure over the period in which the data were taken. Say that we are measuring parts turned out by a machine with a setting of 2.6 inches. The parts would then vary about this setting, depending upon the ability of the machine to hold tolerance, and we would probably obtain a symmetrical frequency distribution.

If an occurrence took place to cause the setting to change, then we would have variation due to a cause. After the change, the parts would really form a new distribution about the new setting. In a later chapter we shall discuss ways of detecting such a change. We must remember, however, that the techniques developed in this chapter apply to the constant-cause system.

Interpretation of Results: Possible Pitfalls

Having learned how to compute measures of central tendency and dispersion, it becomes important to interpret and use them correctly. The misinterpretation and resulting possible misuse of data and calculated measures often can cause worse consequences than having no data or measures to begin with.

Frequently the mean value of a set of data is presented without the measure of dispersion. One way of looking at the measure of dispersion is as an indicator for the validity of the mean or median as a measure of the total distribution. Low dispersion, lack of skewness, and symmetrical shape make the mean a more valid measure. Even if the opposite is true, the mean is still helpful, but more caution should be taken in its use. We will find this especially true later when we try to predict the mean of a population from a sample. In any case, however, there will normally be some error when the sample mean or dispersion is used to estimate the characteristics of the population. It becomes our responsibility to recognize this error and identify its possible extent to others who may rely on our data. One pitfall in interpretation stems from not considering the effect of extreme values, as in the tax-consulting-firm case. When the recent graduate, having accepted a position with a tax-consulting-firm on the basis of the fact that the average salary was $24,000, later learned that the president's salary of $96,000 was the only salary greater than $24,000, he realized that this extreme value had pulled the average up and caused him to be misled.

In addition to the mean and dispersion, the shape of the distribution is important. If the distribution is either bimodal or heavily skewed, the measures might sometimes be more misleading than helpful. If the bimodal nature of a set of data is not recognized and the cause determined, the mean can be very misleading and may result in improper decisions. Sometimes data take the form of a U distribution, as in Figure 3-16, and the mean falls in a position where it is not a representation of the majority of the data items.

One of the most frequent reasons for using statistical methods is to obtain information about the population from which the sample data came. An impression of the population distribution can be inferred from the sample mean and standard deviation calculated by methods studied in this chapter. Later we shall learn more precise methods of estimating the population parameters based on sample data. The histogram or frequency polygon gave us a general picture of how to visualize the population, but they are generally inadequate for most statistical use. Although the sample histogram may be a reasonable estimate, we do not know how good the fit really is. We are still in need of a way to determine how valid the estimates are in representing the actual population.

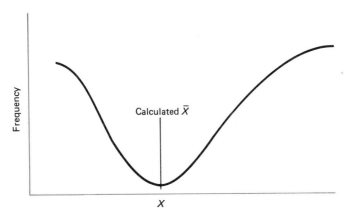

FIGURE 3-16. Example of U Distribution

The measures of central tendency and variability tend to create a mental picture of the set of data. The greatest disadvantage is that the measures are often used alone or used without the measures of their validity. All statistical conclusions have some uncertainty and it is important to recognize how much uncertainty is present in each case. When two or more persons discuss a set of data, it is easy to begin to use only the central measures and to forget the other important factors. The more information we can provide, the better will be the image picked up by others who are trying to understand what we are presenting. You can see that the use of a mean alone *can* be very misleading, as we also need an index of dispersion. It must also be remembered that smaller samples of data provide less reliable estimates of a population than do larger samples.

Exercises

3.21 Calculate the variance and standard deviation for the following population:

X	f
7	1
8	2
9	3
10	2
11	1

3.22 A sample of 20 families were surveyed to determine the number of cars owned by the household. Following are the results:

Number of Cars	Number of Households
0	4
1	7
2	5
3	3
4	1
	20

Determine the mean, the variance, and the standard deviation.

3.23 A retail store had 26 new families apply for charge accounts last month. In trying to determine its product line, the store manager compiled the customers as to number of children in the family. Compute the mean and the standard deviation, considering this to be a sample and using $n - 1$ to compute s.

Number of Children	Number of Families
0	7
1	0
2	9
3	6
4	4
5	0
	26

3.24 Compute the average deviation for the following figures (a) using the mean and (b) using the median.

1,068
1,027
1,010
  931
  852

3.25 The town of Lanvers tallied the yearly water bills for each of the 80 residents of the community as follows:

Water Bill	Number of Residents
40.00–49.99	0
50.00–59.99	2
60.00–69.99	2
70.00–79.99	5
80.00–89.99	9
90.00–99.99	14
100.00–109.99	19
110.00–119.99	15
120.00–129.99	10
130.00–139.99	3
140.00–149.99	1
	80

Compute the mean, the variance, and the standard deviation by the shortcut coded method.

3.26 A survey showed that the average per capita consumption of gasoline for private cars was 480 gallons per year. The population was estimated at 240 million persons. What is an estimate of total gasoline consumed by private cars for the period?

3.27 A retail establishment kept records on mileage of all 220 of their radial tires replaced by them last year. The data were tabulated as follows:

Miles	Number of Tires
24,000 and under 28,000	16
28,000 and under 32,000	30
32,000 and under 36,000	46
36,000 and under 40,000	50
40,000 and under 44,000	40
44,000 and under 48,000	26
48,000 and under 52,000	12
	220

Compute the mean, and the standard deviation using the formula in Table 3-16 (shortcut coded method) where $d = 0$ for the 32,000 to 36,000 class.

3.28 In problem 3.27, determine the median. What can you say about skewness of the distribution? Another brand K of radial tired had the following values:

$$\text{mean} = 39{,}400$$
$$\text{median} = 37{,}200$$
$$\text{standard deviation} = 8{,}600$$

Use this information to make a comparison between the two brands.

3.29 Bob earned the following for selling magazines on 8 different days: $8, $5, $11, $6, $6, $4, $7, and $9. (Assume that he works for 6 months, so that this is a sample.) Calculate:
(a) the mean
(b) the median
(c) the mode
(d) the average deviation about the mean
(e) the range
(f) the standard deviation
(g) the quartile deviation

3.30 Calculate the mean, the median, and the standard deviation of the following sample using the shortcut coded method:

Class Interval	Frequency
3–7	1
8–12	9
13–17	12
18–22	15
23–27	13

3.31 A company was interested in the distribution of hotel or motel rates that employees charged from their business trips by a sample of 112 expense reports. The rates, including taxes, were compiled as follows:

Charge	Number of Cases
$ 9.00 and up to $11.00	4
11.00 and up to 13.00	8
13.00 and up to 15.00	6
15.00 and up to 17.00	25
17.00 and up to 19.00	15
19.00 and up to 21.00	22
21.00 and up to 23.00	8
23.00 and up to 25.00	9
25.00 and up to 27.00	5
27.00 and up to 29.00	6
29.00 and up to 31.00	1
31.00 and up to 33.00	3
	112

Determine the mean and the standard deviation. (Use $20 as the assumed mean.)

3.32 A test conducted to measure the life of 80 experimental lamp bulbs gave the following results:

Length of Life (hours)	Number of Bulbs
2,000 and under 3,000	4
3,000 and under 4,000	16
4,000 and under 5,000	33
5,000 and under 6,000	20
6,000 and under 7,000	7
	80

Assume that this is a complete population. Determine the median, and the standard deviation (σ).

3.33 Calculate the mean, the variance, and the standard deviation for the following sample of discrete data using the shortcut coded method:

Number of Defects per Item	f
2–10	18
11–19	32
20–28	50
29–37	81
38–46	119
47–55	80
56–64	140
65–73	120
74–82	160
83–91	140
92–100	60
	1,000

3.34 Data have been collected on the life of two brands of light bulbs as follows:

$\bar{X}_1 = 800$ hours $\bar{X}_2 = 770$ hours
$s_1 = 100$ hours $s_2 = 60$ hours
$n_1 = 50$ $n_2 = 60$

Calculate the coefficient of variation and interpret the results.

3.35 In problem 3.34 assume all the data is the same except $s_2 = 30$ hours instead of 60. Does this change your conclusion?

4

Learning Objectives

Importance of Probability to Management

Probability As a Relative Frequency

Important Concepts Used in Probability
   Sets and Events
   Sample and Population
   Replacement
   Venn Diagrams
   Independence

Basis for Determining Probabilities
   Objective Probability
   Subjective Estimates

Types of Probability
   Marginal Probability
   Joint Probability
   Independent and Dependent Events
   Conditional Probability
   Example Illustrating Marginal, Joint, and Conditional Probabilities

The Laws of Probability
   Addition Laws
   Multiplication Laws

Probability Trees

Bayesian Concepts

Business Applications of Probability

Probability

Learning Objectives

In this chapter we introduce basic probability concepts that are useful in decision making and which can be applied to business situations. These probability concepts also serve as a background for the student going on to study more advanced courses in management. The understanding of probability is a necessary prerequisite to many advanced courses, both quantitative and nonquantitative. The primary learning objectives for this chapter are:

1. To understand when probability can be applied in decision-making situations.
2. To learn the basic terms used in probability.
3. To learn how to calculate probabilities, either objectively or without exact quantitative data.
4. To understand how to interpret results and apply them in decision making.

Importance of Probability to Management

Probability concepts have many exciting and useful applications in the management of business, social, or governmental enterprises. In the constantly changing areas of marketing, finance, and production, decisions frequently must be made under conditions of uncertainty or risk. Uncertainty exists where we cannot exactly determine all influences. Probability concepts are used to measure and evaluate uncertainty or risk, since these factors usually exist to some extent. Reaction of the market to a new product, for example, always has a considerable degree of uncertainty. Therefore, the introduction of a new product involves risks to the company. Poor decision making under these conditions can often have an unfortunate impact upon a firm. In addition, participants in wrong decisions in critical situations often find their future promotion prospects or job security threatened.

In many decision-making situations, it is possible to assess occurrences and outcomes in terms of probabilities. Probability concepts help to form a problem into a logical model. The person responsible for the decision can use these probabilities in conjunction with estimates of profits and costs as an aid in systematic problem resolution.

An understanding of probability can be very useful to a manager or anyone making business decisions. Forming strategies and making decisions is an important function at any level of management. Uncertainties are present in most decision-making situations, and the manager must determine the degree to which uncertainty can influence a decision or future event. Probabilities are used in measuring and evaluating these uncertainties and their potential effect on the firm. An understanding of the resulting risks directly affects the policies formulated by management.

Probability As a Relative Frequency

Probability is the chance or likelihood that an event will occur. It is represented by a number between zero and one. It can never be greater than 1.0 nor can it be negative. A probability of zero means that an event cannot occur. An event that is unlikely to happen will have a probability close to zero. As the event becomes more likely, the probability will increase. The chance of a head on the toss of a coin is 50–50, expressed as a probability of .5. The probability that our car will start in the morning may be .98, since it is very likely. When the probability is one (1.0), on the other hand, we say that the event is certain to occur.

If we are working with data that are already collected, a probability is defined as the relative frequency of occurrence. This is sometimes called an *experimental probability*, if we planned an experiment and collected the data. It can be written as

$$P(E) = \frac{\text{number of times that event occurred}}{\text{total number of trials}}$$

where $P(E)$ is the notation for the probability that event E will happen. The (E) following the P is often omitted. Consider an example where a wholesaler has collected data in order to predict the probability that orders for a product can be filled from inventory.

$$P(\text{filling order from inventory}) = \frac{\text{number of orders filled from inventory}}{\text{total number of orders processed}}$$

If a firm had 2,400 total orders over the past year and 1,920 of these were filled from inventory, the probability of filling an order from inventory would be expressed as

$$P = \frac{1{,}920}{2{,}400} = .80$$

In the inventory problem we used data to evaluate the probability. In other instances we might derive the relative frequency by logic. In tossing a coin, for example, we can reason that the probability of a head is one chance out of two (head or tail), giving a probability of 1/2, or .5. By similar reasoning the likelihood of a 6 on the roll of a die would be 1/6 or .167.

Important Concepts Used in Probability

Some of the important basic concepts related to probability will have been studied in algebra or other math courses. These concepts will be reviewed and expanded to some extent for use in this chapter. The following definitions and examples will set the scene for later topics dealing with probability logic, calculations, and interpretation.

Sets and Events

Any collection of objects or symbols can be considered a *set*. For example, if the possible results, or *outcomes*, in a vote for city manager were Markov (X), Houston (Y), and Harrison (Z), we could describe the set of possible outcomes as

$$\{\text{Markov, Houston, Harrison}\}$$

Or, using symbols, we could express the set as

$$\{X, Y, Z\}$$

This set of all possible outcomes is called the *sample space*. As another example, in the throw of a common 6-sided die, the set representing the sample space would be {1, 2, 3, 4, 5, 6}. We can say that there are 6 points in the sample space.

In the voting example, an *event* is any individual element in the set. For instance, the possible election of Harrison is an event. Statistically, this type of event would be called a *simple event*.

There are also complex events, which are made up of more than one simple event. Suppose that we might elect either a woman or a man from the set {woman, man}. At the same time a person could be in the set {college graduate, nongraduate}. If we were evaluating the likelihood of electing a woman college graduate, the candidate must be a woman as well as a college graduate, and we would call this a *complex event*. In this case the sample space would be

{woman and college graduate, man and college graduate, woman and nongraduate, man and nongraduate}

There are four possible outcomes in the sample space.

Events are said to be *mutually exclusive* if the occurrence of one precludes the occurrence of the other. As a simple example, if you select a single card from a regular deck of playing cards, the card can be either red or black, but not both. The events red and black are then mutually exclusive. Consider the 100 people who belong to a business club. The club members have been polled and asked whether or not they read *Space Set* magazine. Those polled could form the sets of events {man, woman} and {reader, nonreader}. The events man and woman are mutually exclusive, since only one can occur at a time. The events woman and reader are not mutually exclusive, however, because a person polled can have both characteristics. If a 6-sided die is cast, events {1, 2, 3, 4, 5, 6} are mutually exclusive, since only one side can turn up on a simple toss. It should be apparent, then, that if the occurrence of one event does not permit the occurrence of another, the events are mutually exclusive.

A set of events is said to be *collectively exhaustive* if the set contains all possible outcomes or occurrences. If the die is rolled, the collectively exhaustive set of events {1, 2, 3, 4, 5, 6} includes all possible events. In the reader poll, the set of events {man, woman, reader, nonreader} is collectively exhaustive.

Other examples of collectively exhaustive events include the "buy" or "not buy" decision when looking at a new car. These are the only 2 possible outcomes. You may wait and not buy at first and then return later and buy, but once again your initial decision is from the set {buy, not buy}. As another example, a firm specializing in measuring magazine readership may be interested in which sports magazines you read. Such magazines as *Sports Illustrated, Sport, Field and Stream,* plus all other sports magazines, would make up the collectively exhaustive set.

Sample and Population

For purposes of the current discussion, let us review some of the concepts presented in Chapters 1 and 2. If we were interested in determining attitudes of employees toward length of the lunch period at a company, the entire group of 6,000 employees would be viewed as the population. The *population,* then, is defined as the entire group of interest in a study. If we had the time, money, and facilities, we could question each employee, and the result would be called a census. A *census* is a poll of the entire population. Care should be taken to note that population does not necessarily refer to people; a population can consist of any group of items, such as dogs, automobiles, or ball bearings. If the ball bearings were from a production line, however, we may not be able to identify the population since some items making up the population may not be manufactured yet.

Recognizing that it is often too time-consuming, too costly, or flatly impossible to take a census of the entire population, we may often take a sample. A *sample* can be viewed as a subset of the population. For example, we may wish to select 500 out of the total 6,000 employees, obtain their views, and then make statements giving inferences about the entire population based upon the results from the sample.

Making statements about the entire body of employees based on the data from a sample of 500 may appear to be risky since errors can arise in any sampling process. To minimize the danger of error, it is common practice in scientific studies to make sure that we select random samples. When of adequate size, random samples generally provide reasonably accurate representations of the whole population. The simple *random sample* is one in which each member of the population, each employee in this case, has an equal chance (probability) of being selected in the sample. Where we are dealing with combinations, each combination must have an equal chance for selection in order to meet the requirements of a simple random sample.

In any type of scientific random sampling, it is very important that we *know* the chances or probability for selecting any member as part of the sample. This enables us to weigh the results and arrive at an accurate representation of the population. Sampling will be dealt with more extensively in Chapter 6.

Replacement

When the occurrence of an event depends upon a sampling or selection process, it is important to consider replacement and nonreplacement. Take an example where a complete deck of playing cards has a total of 52 cards and we are going to select 2 cards from the deck. If one card is drawn and then replaced in the deck before a second card is drawn, the population remains at 52 for the second draw. This is called *sampling with replacement*. If, however, the second card was taken without replacing the first card, the population would contain only 51 cards at the second draw. This is called *sampling without replacement*. The calculation of a probability is affected by whether or not items are replaced prior to subsequent events.

In some situations, such as with the deck of cards, replacement is possible. In other cases, however, replacement may not be desired, or in still others it might be impossible. As an example where replacement is impossible, a radial-ply tire has been taken from a lot of 500 off the production line and placed under test to determine its mileage before wear-out. The tire cannot be replaced because it is worn out during the test. Items subjected to various types of destructive tests cannot be replaced in the original population. On the other hand, a ball bearing taken from production and measured for diameter can easily be replaced.

The size of the population also has an effect on the probability. Where the population is large, such as 10,000, the probability of selecting a particular item is 1/10,000 (or .0001). After we select one item and do not replace it, the difference between 1/10,000 and 1/9,999 may not be considered significant. In cases of large populations, therefore, we may just perform calculations as if there were replacement and neglect the error, even though replacement did not take place.

Venn Diagrams

Diagrams used to portray a sample space or a relationship between events are called *Venn diagrams*. In Figure 4-1a the circle represents hearts (H) as a subset of the whole set (deck) of cards. The circle represents the event H and the remainder of the rectangle is the complement of H, often designated H', indicating "not hearts."

Two mutually exclusive events would be illustrated by Figure 4-1b. The example shows hearts (H) and spades (S), both of which are subsets of the deck. In this case hearts and spades are mutually exclusive. In Figure 4-1c, we show one circle to represent hearts (H) and another circle to represent aces (A). In this case there is an overlap or *intersection* of the circles, because the ace of hearts belong to both subsets. The subset "hearts and aces" is only one card in this case, and is represented by the small overlap area. Subset "hearts or aces" represents

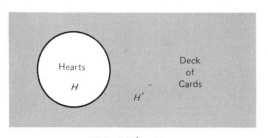

$P(H) + P(H') = 1.0$

(a) Hearts are a subset of the total card deck

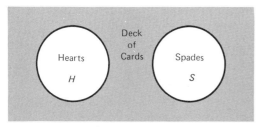

(b) H and S are Mutually Exclusive Events

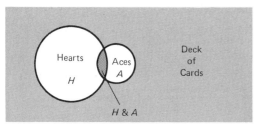

(c) In a deck of cards, Hearts and Aces have an intersection (Ace of Hearts)

FIGURE 4-1. Venn Diagrams Illustrating Various Relationships Involving Two Events

all cards that are either a heart or an ace or both. These diagrams are often useful in the portrayal of the logic of a problem where it is desired to determine the probability.

Independence

Probability has been discussed both in terms of the likelihood of occurrence of an event and also as the proportion of events of a particular type that are likely to occur. When a fair coin is tossed, the occurrence of a head on one toss does not affect subsequent events. This is known as independence. We have *independence* if the outcome of an event is not dependent upon the outcome of previous events. In the coin toss, even if we obtain 5 heads in a row, as long as the coin and flip were known to be fair, the probability of a head on the sixth toss will still be .5. The occurrence of 5 heads in a row has low likelihood and it may cause us to check further on the honesty of the coin. If the coin is still shown to be an honest one, however, the results of the sixth and any subsequent toss are independent of prior happenings, and the probability of a head remains at .5.

Consider a different situation where events would be dependent. Assume that data have shown that a machine in a manufacturing plant produces 90 per cent good parts and 10 per cent defectives over a period of time. We can then say that the probability of a defective part being selected at random from a production lot is .1. Further analysis of the data has shown, however, that bad parts frequently tend to be grouped because they are usually caused by an incorrect machine setting or similar problem. In other words, some bad parts might occur at random, whereas others occur because of a discrepancy in machine setting. If, for example, the data show that when a defect occurs, the probability of another defect occurring on the next item is .2, there would be dependence. The likelihood of a defect on any event is then said to be *dependent* on the result of the prior event, and it is not the same as the .1 for a random occurrence. In other words, having detected a defective part on the line leads us to think that the chances are higher (.2) for the next part to be defective as compared to the original chances of selecting a defective part (.1).

Consider another example where an insurance agent has determined that the probability of the home office canceling an automobile insurance policy for a customer is greater if the customer has had claims in the past year. We might then establish that the probability of the event "cancel" is dependent upon the event "claims in the past year." If the occurrence of one event can in no way be shown to affect the probability of occurrence or nonoccurrence of another, the events must be assumed to be independent. The evidence of dependence may come from compilation of current data, general knowledge of the situation, or collection of information through experimentation.

Basis for Determining Probabilities

In some situations where we work with probabilities, there are data available from which to make judgments. This information can be obtained either by a logical analysis of the set and events (as in the case of the die) or by conducting an experiment to provide data. In other cases someone else may have compiled the data for us. Our aim is to use the data to estimate probabilities. In these cases where we have or obtain sufficient information, we can say we are working with *objective probabilities*. Lacking sufficient data, we may be forced to work with *subjective probabilities*. It is important to recognize the source of data because it will help us in determining the degree to which we can rely on the results.

Objective Probability

If data are available that we can use to calculate probabilities, we can say that we are working with *objective probability*. In other cases the probability of an event's occurrence is known or can be determined by logic without experiment, because the process is already well understood. Dice, cards, or coins are familiar examples. In the case of a card deck, we can accurately estimate the probability of drawing an ace from a complete deck of cards without resorting to experimental trials; for example, $P(\text{ace}) = 4/52 = 1/13$. Even in more complicated situations, there are special probability distributions that we shall learn to use without recourse to an experiment. As an example, the probabilities of obtaining yes–no responses to a public poll are treated very differently than are the probabilities of obtaining a certain diameter of steel rod from a production process. Specific probability distributions and their usage will be discussed in Chapter 5, but it is important here to recognize these sources of information that can be used to calculate objective probabilities.

Subjective Estimates

There are many situations in which one cannot determine a relative frequency or find a suitable probability distribution. When a company is introducing a new product, the marketing manager may need to know the probability of the competition reacting in a certain way. Because this behavior is in the future and no historical data are available, information for objective estimates is lacking. Because of the uniqueness of the situation, a standard probability distribution would not apply. In these cases we may have to rely on subjective estimates. A *subjective probability* is a decision maker's estimate of the likelihood of an event's occurrence.

In these situations the decision maker might be required to estimate the probability of alternative competitive actions. It should not be inferred that this estimate is a wild guess, since it could be based upon sound judgment, collection of information, or past experience. Company management often has some previous experience in calculating the chances of the competition reacting in different ways. The combination of this prior experience plus information pertaining to the current situation can often yield a sound estimate for the probabilities of competitor actions. Once a subjective probability has been estimated, even though it may have less scientific basis, the value can be used in a manner similar to the way objective probabilities are used.

Exercises

4.1 Bill was born in September. What is the probability that it was on the 21st? What is the probability that it was on a Tuesday?

4.2 Mr. Jones had 4 screwdrivers, each a different size, in his workshop. He sent his son down to bring back a screwdriver. What is the probability that he brought back the smallest one?

4.3 June 11, 1976 fell on a Friday. What is the probability that June 12, 1976 was on Saturday?

4.4 The price of eggs this week is 72¢ per dozen. What is the probability that they will be higher a week from now? What is the probability that they will be higher 5 years from now? What is the set of possible occurrences in each case? Are these objective or subjective probabilities?

4.5 Two dice are thrown. What is the probability that the difference between the numbers is 3? List the sample space. What is the probability of a difference of 2?

4.6 A committee of persons engaged in real estate included 12 brokers. Of the 12 brokers, 3 sold both residential and commercial property, and 5 sold commercial property only. Draw a Venn diagram that shows how many brokers sold:
  (a) residential
  (b) commercial
  (c) residential only

4.7 The following are selected at random from a deck of 52 cards. Indicate whether in each case they are mutually exclusive, collectively exhaustive, both, or neither:
  (a) ten, diamond
  (b) nine, king, ace
  (c) red, black
  (d) club, red

4.8 Jack tells Bob that he will give him 2 to 1 odds that the Bears will win a certain game, and 3 to 1 odds that the Colts will win their game. What probability is Jack giving to each?

4.9 A jar contains 5 green chips, 5 white chips, and 5 black chips. Each color set is numbered from 1 to 5, and it is equally likely that any chip be selected. Draw a Venn diagram of the sample space, identify the following subsets, and find the probability for each for a single draw:
(a) green and even
(b) odd and black
(c) white and black
(d) green or white

Types of Probability

The 100 people who belong to our social club have been polled and asked whether or not they had read *Social Set* magazine last month. The collected data are given in Table 4-1 in the form of a contingency table. Note that the events listed horizontally in a contingency table (men and women in this case) are mutually exclusive and collectively exhaustive. This must also be true of the events listed vertically, in this case readers and nonreaders. The data in Table 4-1 will be used to illustrate three different types of probability: marginal, joint, and conditional.

Marginal Probability

In the club there were 54 readers out of the total of 100 persons polled. If we take one person at random from the set of 100, the marginal proba-

TABLE 4-1. Contingency Table Showing Reading Behavior of 100 Men and Women

	Men, M	Women, W	Totals
Readers, R	10	44	54
Nonreaders, NR	30	16	46
Totals	40	60	100

bility, or simple probability (the terms will be used here synonomously), of selecting a reader can be written as

$$P(R) = \frac{54}{100} = .54$$

A *marginal probability* is one that gives information about one variable. The term "marginal probability" is used because such probabilities are derived from the value located in the margin of the contingency table. The probability of a nonreader in Table 4-1 is a marginal probability, so $P(NR)$ is 46/100 or .46. A marginal probability gives information only about the basic category, in this case reader or nonreader, without regard to other categories. Probabilities related to the other basic category Men ($P = 40/100 = .4$) or Women ($P = 60/100 = .6$) are also marginal probabilities, since these probabilities are derived from values in the lower margin of Table 4-1. Note that the probabilities in either margin sum to 1.0.

Joint Probability

A *joint probability* represents the chance that two or more characteristics occur together. In this case, the probability of selecting a woman who is also a reader is a joint probability. A joint probability can be determined from a value found in the body of Table 4-1, as for example, the probability of selecting a woman who is also a reader is

$$P(W \ \& \ R) = \frac{44}{100} = .44$$

We have used the symbol "&" to represent "and." (We could have used $W \cap R$ instead.) Similarly, the probability of drawing a male nonreader, $P(M \ \& \ NR)$, is 30/100 or .30, since of the 100 members there are 30 who are both male and nonreaders. The joint probabilities are derived from values in the body of the table. By summing these values either across or vertically, we obtain the quantities used to determine the marginal probabilities. A joint probability refers to two or more characteristics together, while a marginal probability refers to the total for one characteristic only. In Table 4-1 we can calculate four joint probabilities. When we say two or more characteristics, we could have had a male reader who also was a cigarette smoker. If we were also investigating the variable "smoker" and "nonsmoker," our study would then involve three characteristics.

Independent and Dependent Events

Earlier in this chapter we defined independence and dependence. We said that two events are *independent* if the occurrence of one has no effect on the probability of occurrence of the other. In other words, we are saying that knowledge of what happened on one event does not help us in determining probabilities for the other event.

When the data are in the form of a contingency table, independence can be tested easily. Whenever the joint probability of two events is not equal to the product of the marginal probabilities, then the events are *dependent*. For example, consider in Table 4-1 the joint probability

$$P(W \& R) = .44$$

Compare this to the product of the marginal probabilities:

$$P(R) \times P(W) = .54 \times .60 = .324$$

Since these two values are unequal, this indicates that they are not independent (and, therefore, dependent). We can continue to check the other factors in Table 4-1 showing that all are dependent.

$P(M \& R) = .10$ $P(M) \times P(R) = .4 \times .54 = .216$
$P(M \& NR) = .30$ $P(M) \times P(NR) = .4 \times .46 = .184$
$P(W \& NR) = .16$ $P(W) \times P(NR) = .6 \times .46 = .276$

As an example where the comparison shows independence, look at Table 4-2.

$P(R \& W) = .2571$ $P(R) \times P(W) = .3 \times .857 = .2571$
$P(R \& S) = .0429$ $P(R) \times P(S) = .3 \times .143 = .0429$
$P(NR \& W) = .5999$ $P(NR) \times P(W) = .7 \times .857 = .5999$
$P(NR \& S) = .1001$ $P(NR) \times P(S) = .7 \times .143 = .1001$

TABLE 4-2. Contingency Table Where Independence Exists

	Weekday, *W*	Sunday, *S*	Totals
Rain, *R*	.2571	.0429	.3000
No Rain, *NR*	.5999	.1001	.7000
Totals	.8570	.1430	1.0000

In some cases independent events and dependent events can be found in the same table. After we have discussed conditional probabilities, another test for independence will be demonstrated.

Conditional Probability

A *conditional probability* is contingent upon prior knowledge. It is the probability that an event will happen, given that another event has happened; or, in other cases, it is the probability of an event, given that a certain condition exists. If we wish to know the probability that a person is a reader in Table 4-1, given that the person is a woman, this is a conditional probability. The probability of selecting a reader is conditioned upon the premise that the selection is being made from the women.

The conditional probability of R given W can be expressed as $P(R|W)$ where the vertical line is read "given that." It is calculated as follows:

$$P(R|W) = \frac{\text{joint probability of } R \text{ and } W}{\text{marginal probability of } W} \quad \text{or} \quad \frac{P(R \ \& \ W)}{P(W)}$$

With the data from Table 4-1, we are given that our selection is one of the 60 women and asking the probability that she is also a reader. The conditional probability in this example is

$$P(R|W) = \frac{.44}{.60} = .733$$

Another interpretation is that the .44/.60 represents the proportion of the women who are readers. Thus it may be somewhat helpful to simply view a conditional probability as referring to a subset of a marginal group, as for example, the proportion of readers who are male (10/54), proportion of males who are readers (10/40), and so forth.

We can now use the conditional probability to check for independence. If gender and readership are unrelated (independent), the conditional probability such as $P(M|NR)$ will equal the $P(M)$ because it did not make any difference if we knew readership when predicting gender. However, we would note in this case that

$$P(M|NR) = \frac{P(M \ \& \ NR)}{P(NR)} = \frac{.30}{.46} = .652$$

while
$$P(M) = .40$$

In this case knowledge of nonreadership influences our prediction as to whether the person will be a male and, therefore, dependence exists.

Using this same approach, we can calculate the probability of selecting a nonreader, given that he is a man:

$$P(NR|M) = \frac{P(NR \& M)}{P(M)} = \frac{.30}{.40} = .75$$

Note that $P(NR|M) \neq P(NR)$ so that these events also do not demonstrate independence. In Table 4-2,

$$P(R|W) = \frac{.2571}{.8570} = .3000$$

and
$$P(R) = .30$$

which demonstrates independence since they are equal.

Example Illustrating Marginal, Joint, and Conditional Probabilities

A record has been kept for 100 days showing the relationship between TUVW stock and the Dow Jones industrial average as shown in Table 4-3.

TABLE 4-3. Contingency Table for TUVW Stock and Dow Jones Industrial Index for Adjacent Days

TUVW Stock	Dow Jones Industrials Index			Totals
	Up, U	No Change, NC	Down, D	
Up, A	.12	.01	.17	.30
No Change, B	.24	.01	.25	.50
Down, C	.12	.01	.07	.20
Totals	.48	.03	.49	1.00

The marginal probabilities are as follows:

$$P(A) = .12 + .01 + .17 = .30$$
$$P(B) = .24 + .01 + .25 = .50$$
$$P(C) = .12 + .01 + .07 = .20$$
$$P(U) = .12 + .24 + .12 = .48$$
$$P(NC) = .01 + .01 + .01 = .03$$
$$P(D) = .17 + .25 + .07 = .49$$

The joint probabilities are:

$$P(A \& U) = .12 \qquad P(B \& D) = .25$$
$$P(A \& NC) = .01 \qquad P(C \& U) = .12$$
$$P(A \& D) = .17 \qquad P(C \& NC) = .01$$
$$P(B \& U) = .24 \qquad P(C \& D) = \underline{.07}$$
$$P(B \& NC) = .01 \qquad \text{Total} = 1.00$$

A test for dependence between A and U is as follows:

$$P(A \& U) = .12$$
$$P(A) \times P(U) = .30 \times .48 = .144$$

This test indicates dependence between A and U since the two values are not equal. We can check for dependence by another test, as follows:

$$P(A|U) = \frac{.12}{.48} = .25$$
$$P(A) = .12$$

Again, dependence is indicated between A and U by this test since the two are not equal. These two tests will now be used to check for dependence between B and U in Table 4-3:

$$P(B \& U) = .24$$
$$P(B) \times P(U) = .50 \times .48 = .240$$

This test shows independence between B and U since both are .24. Using the other method, we find

$$P(B|U) = \frac{.24}{.48} = .50$$
$$P(B) = .50$$

This again shows independence between B and U.

The conditional probabilities are:

$$P(A|U) = \frac{.12}{.48} = .25 \qquad P(U|A) = \frac{.12}{.30} = .40$$

$$P(A|NC) = \frac{.01}{.03} = .33 \qquad P(NC|A) = \frac{.01}{.30} = .033$$

$$P(A|D) = \frac{.17}{.49} = .35 \qquad P(D|A) = \frac{.17}{.30} = .57$$

$$P(B|U) = \frac{.24}{.48} = .50 \qquad P(U|B) = \frac{.24}{.50} = .48$$

$$P(B|NC) = \frac{.01}{.03} = .33 \qquad P(NC|B) = \frac{.01}{.50} = .02$$

$$P(B|D) = \frac{.25}{.49} = .51 \qquad P(D|B) = \frac{.25}{.50} = .50$$

$$P(C|U) = \frac{.12}{.48} = .25 \qquad P(U|C) = \frac{.12}{.20} = .60$$

$$P(C|NC) = \frac{.01}{.03} = .33 \qquad P(NC|C) = \frac{.01}{.20} = .05$$

$$P(C|D) = \frac{.07}{.49} = .14 \qquad P(D|C) = \frac{.07}{.20} = .35$$

We should note as a general rule that there are twice as many conditional probabilities as there are joint probabilities.

Exercises

4.10 The characteristics of a sample of students at a midwestern university are as follows:

Males	100
Females	90
Male undergraduates	65
Male graduates	35
Female undergraduates	65
Female graduates	25

(a) Set up a contingency table to show the relationships.
(b) What are the marginal probabilities?
(c) What are the joint probabilities?
(d) What are the conditional probabilities?
(e) Are sex and type of student independent of one another?

4.11 An office has 13 men and 7 women. Of these, 10 men and 4 women drive to work and the rest ride the bus.
  (a) What is the probability that a person selected at random rides the bus?
  (b) What is the probability that a woman selected at random rides the bus?

4.12 An auditor selects accounts for inspection at random from the file of accounts receivable. Accounts that are less than 6 months old are termed new accounts. The accounts are classified as follows:

	Current, C	Delinquent, D
New accounts, N	108	12
Old accounts, O	792	88

Let N denote the selection of a new account; O, the selection of an old account; C, the selection of an account that is current; and D, the selection of an account that is delinquent. Calculate the following probabilities:
  (a) $P(N|C)$
  (b) $P(D \& O)$
  (c) $P(N \text{ or } D)$
  (d) $P(N|C \text{ or } D)$

4.13 A person who sells farm implements classifies tractor sales over a year according to the customers' methods of payment, listing the proportion of the total sales for each payment method:

	Method of Payment	
Type of Tractor	Cash	Credit
New	.04	.20
Used	.32	.44

  (a) What is the simple probability of a new-tractor purchase if the salesperson selects a buyer at random?

(b) What is the conditional probability that a buyer of a used tractor will pay in cash?
(c) What is the joint probability of selling a used tractor on credit?
(d) Are the type of tractor and the method of payment statistically independent?

4.14 If 45 per cent of the freshman class at a local university have portable electric typewriters and 25 per cent have calculators, and if one third of those who have electric typewriters have calculators, what proportion of those who have calculators also have electric typewriters?

4.15 A newspaper girl is considering using a list of addresses for door-to-door advertising. She knows that 35 per cent take the daily paper only, 15 per cent take just the *Sunday* paper, and 20 per cent take both. If a subscriber takes the daily paper, what is the probability that he or she also subscribes to the Sunday paper?

4.16 The following contingency table shows the relationship between height and weight for men:

Weight (lb)	Short, S	Medium, M	Tall, T	Totals
Up to 140, A	.04	.16	.00	.20
140 and under 160, B	.01	.22	.02	.25
160 and under 180, C	.02	.17	.09	.28
180 and over, D	.01	.14	.12	.27
Totals	.08	.69	.23	1.00

(a) What is the probability that a man weights under 160 lb?
(b) What is the probability that a man weights under 160 lb and is not tall?
(c) If a man weighs under 140 lb, what is the probability that he is tall?
(d) Calculate $P(T|180 \text{ or over})$.
(e) What is the probability that a man is of medium height, given that he weighs under 180 lb?
(f) What is the probability that a man is at least 180 lb or tall?

4.17 A local newspaper carries the probability of rain for each day. The following compilation has been made showing the prediction versus the actual results.

	Prediction Probability			
Actual	A (0–29%)	B (30–69%)	C (70–100%)	Totals
Rain, R		.14	.11	.37
No Rain, NR	.32	.24	.07	
Totals	.44			

Complete the missing items and determine the following:
(a) Calculate $P(R|C)$.
(b) Calculate the probability of rain if the prediction is 29 per cent or less.
(c) What is the probability of rain if the prediction is not known?
(d) If it is raining on a particular day, what is the probability that the prediction for rain was 70 per cent or over?
(e) If the prediction is C, what is the probability that it will not rain?
(f) Check the prediction and actual for independence.

The Laws of Probability

Business decisions frequently involve the likelihood of occurrence of events that are related to each other. The manner in which the events are related determines the procedures for handling the numerical probability calculations. When considering the probability of two or more events together, the individual event probabilities are either added or multiplied, depending on their relationship. We shall now examine various interrelationships so that usage laws (rules) can be established.

Addition Laws

The law of addition governs cases where we desire to calculate the probability that a number of events will occur, either together or separately. The *law of addition for events that are mutually exclusive* states that the probability of one or the other events occurring is found by summing the probabilities of the events occurring individually. This is expressed as

$$P(A \text{ or } B) = P(A) + P(B)$$

If a coin is tossed,

$$P(\text{head or tail}) = P(\text{head}) + P(\text{tail}) = .5 + .5 = 1.0$$

From this addition rule, it can be noted as in the above example that the sum of the probabilities for the complete set of all mutually exclusive and collectively exhaustive events equals one. If we wanted the probability of drawing a king, queen, or jack from a deck of cards, the probabilities would be added since they are mutually exclusive; however, since the events (K, Q, or J) are not collectively exhaustive, the probabilities would not total one. In this example,

$$P(K, Q, \text{ or } J) = P(K) + P(Q) + P(J) = \frac{1}{13} + \frac{1}{13} + \frac{1}{13} = \frac{3}{13}$$

If the events are not mutually exclusive, the law is different. There will be an overlap between them, since both can occur at the same time. In Figure 4-2, besides the independent occurrences of A and B, there is

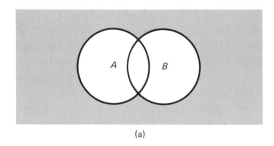

(a)

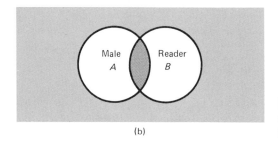

(b)

FIGURE 4-2. Venn Diagram of Nonmutually Exclusive Events

also a possible simultaneous occurrence of A and B. This simultaneous occurrence would not be possible for mutually exclusive events. This leads to the *law of addition for events that are not mutually exclusive,* which states that the probability of either one or the other or both event's occurrence is found by summing the probabilities of the events occurring individually and then subtracting the probability of their joint occurrence, which has been counted twice. As an example for two events,

$$P(A \text{ or } B) = P(A) + P(B) - P(A \& B)$$

From our previous example, if we wanted to determine the probability of selecting a woman or a reader, we would find it by placing the values from Table 4-1 in the equation above as follows:

$$\begin{aligned} P(W \text{ or } R) &= P(W) + P(R) - P(W \& R) \\ &= .60 \quad + .54 \quad - .44 \\ &= .70 \end{aligned}$$

We can see that, when we took the probability of a female, the female readers were included. Likewise, when we took the probability of a reader, female readers were included here also. In other words, the 44 female readers were double-counted. As a result, the formula must subtract out the 44 that have been double-counted, which is accomplished by the last term in the equation. From this we should also note that

$$P(W \& R) = P(R \& W)$$

which will be useful to us in solving certain probability problems.

Consider next the males, readers, and male readers in the sample of 100 persons. In Table 4-1 and Figure 4-2b we can see that the probability of a man or a reader is

$$P(M) + P(R) - P(M \& R) = .40 + .54 - .10 = .84$$

When we include the probability of M, or $P(M)$, moving from left to right in the figure, we have included the overlap area, and when we include $P(R)$, moving from right to left in the figure, we have added in the overlap area once again. Therefore, we must subtract the overlap to avoid double-counting. The overlap in the magazine-reading example is $P(M \& R)$, which represents the probability of one respondent being both a man and a reader.

Going back to the deck of cards, what is the probability of a jack or a heart?

$$P(J \text{ or } H) = P(J) + P(H) - P(J \& H)$$
$$= \frac{4}{52} + \frac{13}{52} - \frac{1}{52} = \frac{16}{52}$$

Again, the jack of hearts was counted in $P(J)$ and also in $P(H)$ and had to be subtracted out once.

Multiplication Laws

The *law of multiplication* is a restatement of the definition of conditional probability. For dependent events, identified as cases where $P(A|B) \neq P(A)$, conditional probabilities are often used to derive joint probabilities. From the section on conditional probabilities, we learned that

$$P(A|B) = \frac{P(A \& B)}{P(B)}$$

Solving for $P(A \& B)$, we find the law of multiplication expressed as a formula:

$$P(A \& B) = P(A|B) \times P(B)$$

In the reading-behavior example of Table 4-1.

$$P(NR) = .46 \quad \text{and} \quad P(M|NR) = \frac{.30}{.46} = .652$$

Therefore,

$$P(M \& NR) = P(M|NR) \times P(NR) = .652 \times .46 = .30$$

Solving for $P(NR \& M)$ in the same way, we calculate that:

$$P(NR \& M) = P(NR|M) \times P(M) = .75 \times .40 = .30$$

Therefore, we can see that order does not affect the result. This is because in each case the same cell in the probability table is involved and the joint probabilities are equal. Knowledge of this procedure for determining the joint probability will be important in many of our later applications.

The special multiplication law deals with independent events. When we are dealing with independent events, $P(A|B) = P(A)$. Therefore, when events are independent, the multiplication law expressed as a formula becomes

$$P(A \ \& \ B) = P(A) \times P(B)$$

In other words, the *special multiplication law* states that the probability of the joint occurrence of independent events is the product of the marginal probabilities of the events.

Consider the case where 2 dice are thrown. What is the probability of both turning up a six? The events are independent

$$P(\text{six} \ \& \ \text{six}) = P(\text{six}) \times P(\text{six}) = \frac{1}{6} \times \frac{1}{6} = \frac{1}{36}$$

This rule also applies to more than two independent events. Consider the throw of 3 dice:

$$P(\text{six} \ \& \ \text{six} \ \& \ \text{six}) = P(\text{six}) \times P(\text{six}) \times P(\text{six}) = \frac{1}{6} \times \frac{1}{6} \times \frac{1}{6} = \frac{1}{216}$$

As a check for independence in a contingency table, the multiplication rule for independent events discussed above was used. For example, if reader and woman are independent events,

$$P(R \ \& \ W) = P(R) \times P(W)$$

However, from Table 4-1 we know that

$$P(R \ \& \ W) = .44$$

while

$$P(R) \times P(W) = .54 \times .60 = .324$$

Since the .44 is not the same as the .324, we conclude that readership and being a woman are related (or dependent). We can also show that all other joint probabilities are unequal to the product of the two respective marginal probabilities. We conclude that readership and sex are not independent events in this example.

A summary of the various formulas used to this point can be found in

TABLE 4-4. Summary of Probability Formulas

Mutually Exclusive Events
$$P(A \& B) = 0$$
$$P(A \text{ or } B) = P(A) + P(B)$$
Independent Events
$$P(A \& B) = P(A) \times P(B)$$
Dependent Events
$$P(A \text{ or } B) = P(A) + P(B) - P(A \& B)$$
$$P(A \& B) = P(A|B) \times P(B) \quad \text{or}$$
$$ = P(B|A) \times P(A)$$
$$P(A|B) = \frac{P(A \& B)}{P(B)}$$
$$P(B|A) = \frac{P(B \& A)}{P(A)}$$

Table 4-4. When we work probability problems, it is usually very helpful to set them up using a table similar to Table 4-1. If some values are missing, we can sum across or vertically to fill in the missing numbers. Once we have inserted all the known information into appropriate positions in the table, computations become much more apparent.

Exercises

4.18 A sample of 420 bottles from a production process are checked with the following results:

Number of Flaws	Frequency
0	10
1	100
2	200
3	100
4	10
	420

A bottle is selected at random. Determine the probability that the number of flaws is:
(a) at least 2
(b) 3 or fewer
(c) 1 or more
(d) more than 2
(e) less than 4
(f) 4 or less

4.19 A company has 10 persons who have applied for a position as follows:

Name	College Grad	Age	Married or Single	Previous Experience
Mr. A	No	28	S	Yes
Mr. B	No	42	S	No
Mr. C	Yes	54	S	Yes
Ms. D	Yes	28	M	No
Mr. E	No	35	M	Yes
Mr. F	Yes	44	M	No
Miss G	Yes	33	S	Yes
Mr. H	No	39	S	Yes
Mr. I	No	35	M	Yes
Mrs. J	No	37	M	Yes

Determine the probability of the following when a person is selected at random from the group:
(a) college graduate
(b) no experience
(c) married
(d) 35 or under
(e) same age as another applicant

Prepare a contingency table for education and experience and determine:
(f) the number of experienced college graduates
(g) the percentage of experienced applicants who have graduated from college
(h) the probability that an experienced person is not a college graduate

4.20 A sales manager's records at a store that specializes in selling used television sets indicate that 30 per cent of the TV sales are of color sets; that 40 per cent are of the larger (23-inch and over) tube sizes; and that of those who purchase sets with larger tubes, 70 per cent buy color sets. Indicate which of the following four statements are correct and support your answers. If a purchase record is selected at random:
(a) The probability is .60 that the set will have a smaller tube (less than 23 inches) and .70 that it will be a black-and-white rather than a color set.
(b) The probability is .28 that the set will be a color model with a larger tube.
(c) The probability is .72 that the set will be a small-tube set, a black-and-white set, or a set with both these characteristics.

(d) The probability is .93 that a person who selects a small-tube set will buy black-and-white instead of color.

4.21 A retail store found that its sales for a day (good or bad) were independent of the weather (clear or raining). The probability of a good sales day is .4 and the probability of clear weather is .6.
 (a) Find the probability of clear weather and bad sales.
 (b) If sales were bad, what is the probability that it was a clear day?

4.22 Suppose for a moment that you are the manager of the computer-manufacturing division of QMB. You have just had a phone call from a supplier of transistors, who says that he has cleaned his warehouse and inadvertently mixed several cases of transistors of various types in one large bin. He is willing to sell this mixed bin at a discount; however, some of the transistors are of a type that you no longer use. Your job is to ascertain the various probabilities of extracting usable transistors from this bin, given the following data. The mixed bin contains the following transistors, which have been coded by color:

300	green with no markings
400	green with a white star
500	green with white rings
500	blue with no markings
200	blue with a white star
100	blue with white rings

 (a) What is the probability that a transistor is green, given that it has white rings?
 (b) What is the probability that the transistor has a white star, given that it is blue?
 (c) What is the name of the type of probability you have calculated in parts (a) and (b)?
 (d) If the only usable transistors are green (any green), what is the probability of selecting a usable transistor from the bin in a random draw?

4.23 A clock manufacturer uses either a black or white face. He paints the case either red or green. A poll of his customers showed that 20 per cent preferred a black face *and* a red case, whereas half his total clients desired a black face *or* a red case. He has already ordered 30 per cent black faces. What per cent should have a red case? Use a contingency table.

4.24 The probability that an automobile salesman will use sales strategy A is .6 if the customer is a woman and .25 if it is a man. The probability that he will use demonstrator car D, given strategy A, is .5 and is .3 in other cases. Twenty per cent of the customers are women. What is the marginal probability that car D is used as a demonstrator?

Hint: First prepare the contingency table for strategy versus sex. Then prepare a second separate contingency table.

4.25 A salesman, from his records, figures that the probability of making a sale on the first call is .40. If he makes subsequent calls on the same person who has not yet bought, the probability falls by .1 each call. What is the probability of selling on the third call? If the salesman selects a potential new customer, what is the highest probability of selling to him in a sequence of 4 calls?

4.26 Mr. Jack, a salesman, is taking a 1,000-mile automobile trip. He has taken the trip before and determined that the probability of any tire going flat is .01.
(a) What is the probability that he will have no flat tires?
(b) Assuming that he carries a spare, what is the probability that the left front tire will go flat and then the right front tire?
(c) What is the probability that he will have one or more flats?

Probability Trees

Probability trees can be utilized to portray certain problems in such a way as to help establish the logic. As an initial example, consider three flips of a coin. The potential outcomes and the associated probability for each outcome can be represented in the form of a probability tree, as in Figure 4-3. The tree makes a rather straightforward portrayal of the problem. Since the events are independent, the probability of a head or a tail remains constant from one flip to the next.

In Figure 4-3, the sequence of events moves from left to right with an event (flip of the coin) represented by a circle. The two branches coming out from each circle represent the two possible outcomes, head or tail. The probability of .5 represents the likelihood of taking that branch that

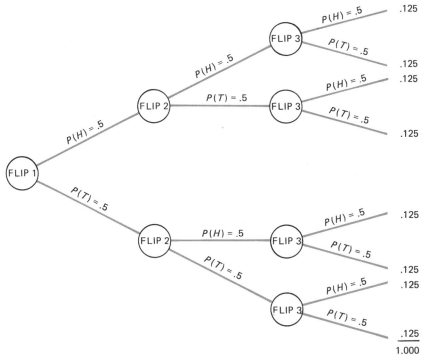

FIGURE 4-3. Probability Tree

is equivalent to the probability of a head (H) or a tail (T). The value .125 at the end of a final branch represents the product of the probabilities (since they are independent) from the original circle along a path containing three events. For example, using the multiplication rule for independent events,

$$P(H, H, H) = P(H) \times P(H) \times P(H) = .5 \times .5 \times .5 = .125$$

In many situations the use of a tree simplifies the understanding of a problem and helps to identify the type of probability—marginal, joint, or conditional—with which we are dealing. As the number of trials increases, the tree would contain more branches and become more and more cumbersome to work with. A practical business problem is analyzed using a probability tree in the following example.

Imagine that our firm, the Lester Company, is in the process of developing a new toothpaste, Glip. We currently have 30 percent of the toothpaste market, and our major competitor has 70 per cent. Given certain technical breakthroughs in the chemical formulation of the product, our laboratory researchers estimate that they are 80 per cent sure of being successful in developing this new toothpaste. If the new product breakthrough is made, Glip will be used as a new entrant in the market, which consists of our current products as well as the products of our competitor.

As a marketing manager you are interested in the chances that our company, Lester, will gain at least 50 per cent of the market share. This problem can best be depicted on a probability tree. Figure 4-4 shows the procedure for solving this problem. A_1 and A_2 are alternative development results for Lester, Z_1 and Z_2 are the competitive reactions by Meggett Company and R_1 through R_7 are market reactions. In this case, each A or Z is represented by a probability since marketing action is dependent upon the success of the development.

In our decision as to whether or not to market Glip, we need to be very careful to make estimates of how our competitor, Meggett Company, will react. We estimate that there is a .60 chance that Meggett will introduce a new product to counteract a new Lester product. This is shown by the

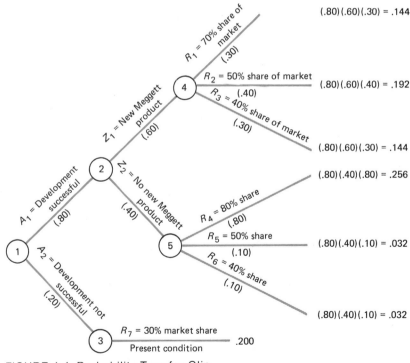

FIGURE 4-4. Probability Tree for Glip

first two legs of the upper branch in Figure 4-4. If this happens, there is a .30 chance that Lester will have a 70 per cent market share, a .40 chance that Lester will have a 50 per cent market share, and a .30 chance that Lester will have a 40 per cent market share. These alternatives are illustrated by the branches from circle 4 in Figure 4-4. The marketing research department also estimates that if Meggett is unable to react with a new product, Lester will have a .80 chance of obtaining an 80 per cent market share, a .10 chance of obtaining a 50 per cent market share, and a .10 chance of obtaining a 40 per cent market share. These alternatives are represented by the branches from circle 5. In the event that the development is not successful, Lester will maintain its current 30 per cent market share, as illustrated by the lower branch of the tree.

To find the chance of gaining at least 50 per cent market share, we are interested in each combination of events that leads to a 50 per cent or greater market share. Market reactions R_1, R_2, R_4, and R_5 qualify. Therefore, combinations A_1, Z_1, R_1; A_1, Z_1, R_2; A_1, Z_2, R_4; and A_1, Z_2, R_5 should be included. The solution—in other words, the probability of any one of these 4 combinations—is obtained by adding values from the tree as follows:

$$P = .144 + .192 + .256 + .032 = .624$$

The probability that Lester will achieve at least a 50 per cent market share is .624.

Probability trees are very useful in management decision making. The tree graphically portrays a situation which, because of the interaction of events, may be quite complex. In addition to simplifying the problem, trees can often make a decision process more scientific by forcing concise definition of alternative courses of action as well as the specification of potential outcomes for each decision. In other types of problems to be dealt with in Chapter 12, a branch point can represent possible alternative management decisions.

Bayesian Concepts

In many decision situations it becomes necessary to compare the costs of making an erroneous decision with the expense of collecting new information. Bayesian decision concepts provide a means to handle such problems. These concepts are a logical extension of probability concepts for situations where there is only partial information. New information costs money, and the business manager must decide whether the value of the information makes up for the cost of securing it. Bayesian decision theory provides a scientific approach to selecting the best course of action when the outcome depends upon an unknown factor not under the control of the decision maker, but where it is possible, however, to derive information about the unknown factor by judgment or experimentation.

Many practical problems deal with the necessity to revise current information, based on the occurrence of a new event or the availability of new information. The methods that are available to handle these types of problems can also be used to determine probabilities of causes, based on observed results. These concepts are often referred to as *Bayesian statistics,* named after Thomas Bayes, the English mathematician who developed their use.

As an example, assume that information has been collected about readers of *College Set* magazine at South State University. At South State 40 per cent of the students are men and 60 per cent are women.

Of the men, 20 per cent are readers and 80 per cent are nonreaders. Of the women, 30 per cent are readers and 70 per cent are nonreaders. The joint probability of a reader and a man is

$$P(R \ \& \ M) = P(R|M) \times P(M)$$
$$= .2 \times .4 = .08$$

Likewise for a woman

$$P(R \ \& \ W) = P(R|W) \times P(W)$$
$$= .3 \times .6 = .18$$

We can use this information to form a contingency table as shown in Table 4-5.

TABLE 4-5. Contingency Table for *College Set* Magazine

	Men, *M*	Women, *W*	Totals
Readers, *R*	.08	.18	.26
Nonreaders, *NR*	.32	.42	.74
Totals	.40	.60	1.00

We desire to know what proportion of our readers are male, which is not stated in the given data. This might also be expressed as the probability of a male, given that the person is a reader. We obtain this by dividing the readers who are male by the total readers:

$$P(M|R) = \frac{.08}{.08 + .18} = \frac{.08}{.26} = .3076$$

This is expressed by the equation

$$P(M|R) = \frac{P(R|M) \times P(M)}{P(R|M) \times P(M) + P(R|W) \times P(W)}$$

It could also be expressed as

$$P(M|R) = \frac{P(R \& M)}{P(R \& M) + P(R \& W)} = \frac{P(R \& M)}{P(R)}$$

This relationship is known as Bayes' theorem, and in this case it permits us to assess the makeup of our reader population from the available data. We can quickly see from our previous work in the chapter that the end result is nothing more than a conditional probability in the reverse direction. It is found by dividing a joint probability by a marginal probability. Thus we can solve this type of problem either by the contingency table or Bayes' rule.

Consider a different problem where it is necessary to determine the validity of a production test to evaluate product quality. A solid-state device is being manufactured for communication sets. From data collected, owing to uncontrollable aspects of the process, only 80 per cent of the devices work properly when installed in the set. This is caused by difficulties in matching characteristics of the device to the overall set. A new test was developed and of all devices installed in sets and working well, 90 per cent had passed the new test. Of those installed and not working (NA), 10 per cent had passed the test. Assume that a device was tested prior to installation in a set and that the test instrument indicated that it was good. What is the probability that the item will work when placed in the set?

In this problem, we started with a population where 80 per cent of the items were usable in the end product; therefore, the probability of getting a usable item was .8. We then collect additional information showing that of all items found to be workable when placed in a set, 90 per cent had passed the test. It is desired to use this new information to revise the original .8 probability. Let the events be represented by:

$A =$ the event that the item is actually usable
$A' = NA =$ the event that the item is not actually usable
$B =$ the event that the test indicates a good item
$B' = NB =$ the event that the test indicates not a good item

These data can be placed in a joint probability table as follows:

	Test Results		
Usability in Set	Pass, B	Fail, NB	Totals
Usable, A	$P(B\|A) \times P(A)$ $.9 \times .8 = .72$	$P(NB\|A) \times P(A)$ $.1 \times .8 = .08$	$.8 = P(A)$
Not Usable, NA	$P(B\|NA) \times P(NA)$ $.1 \times .2 = .02$	$P(NB\|NA) \times P(NA)$ Subtract to get $.2 - .02 = .18$	$.2 = P(NA)$
Totals	.74	.26	1.0

$P(B|A) = .9$, the probability that the item passed the test, given that it was usable in a set (given)
$P(NB|A) = 1 - P(B|A) = 1 - .9 = .1$
$P(B|NA) =$ probability that an unusable item had passed the test (given as .1)
$P(A|B) =$ probability that the item is usable, given that it has passed the test; this is the value we are searching for

Our next step will be to solve for all the items that pass the test. Then we will determine what part of these are expected to be usable. The items that pass the test are in two groups where

$P(B|A) \times P(A) =$ the joint probability of those that are usable and that also pass the test
$P(B|NA) \times P(NA) =$ the joint probability of those that are not usable and that pass the test

We then solve for the proportion of those passing the test that are usable:

$$P(A|B) = \frac{P(B|A) \times P(A)}{P(B|A) \times P(A) + P(B|NA) \times P(NA)}$$

We can see that although we are looking for the $P(A|B)$, we are utilizing information pertaining to $P(B|A)$ to derive it. This is the real contribution that Bayes' theorem provides in that we can use the available "reverse" of what we are looking for to find the value we need. This illustrates a situation where information is provided without complete reliability and it is desired to assess the validity of the measuring device in identifying acceptable items. Inserting the numbers into the equation, we get

$$P(A|B) = \frac{(.9) \times (.8)}{(.9)(.8) + (.1)(.2)} = \frac{.72}{.72 + .02} = \frac{.72}{.74} = .973$$

The probabilities we have been working with can be specified by certain terms. In the last example, the decision maker knew that 80 per cent of the items produced were usable. This is called a *prior probability* because it represents the chances of a usable item before the test results are known. After the test was conducted, we were able to say that there was a probability of .973 that the passing items were usable. This is a *posterior probability* since it applies after the new information was obtained. The new information might either raise or lower the prior probability.

In a marketing problem, prior probability may represent the estimated probability of success of a new product before other information is gathered. The posterior probability could then represent the probability of success of the new product after added survey or experimental information would be collected.

Sometimes the availability of time or the importance of the decision calls for the decision maker to delay his decision pending the collection and analysis of additional information. The firm must pay for this information, and it will almost never be entirely accurate. The decision maker will have to weigh the cost of the information against its value to the firm, or in other words against the risks or potential results of a decision based only on the prior analysis. This kind of analysis, known as *preposterior*, involves a decision whether or not to collect the additional information in the first place.

Bayesian concepts will be dealt with at greater length in Chapter 12 as they apply to decision making under uncertainty and to the evaluation of whether the collection of new information is worth the estimated cost of its collection.

Bayesian concepts can also apply to a set of decision-making steps that deal with progressive improvements in the information needed to make a decision. In this way it is not merely a question of whether or not to gather the additional information but rather a question as to when the information collected becomes sufficient.

This discussion was designed to familiarize the reader with the basic fundamentals of the Bayesian approach. A more detailed treatment of Bayesian decision making is given in Chapter 12.

Business Applications of Probability

In any business decision-making situation, the degree of validity of a decision and the degree of risk associated with a decision are related to the preciseness with which the influencing factors can be described. Those influencing factors which can be stated in quantitative figures are more precise and easier to evaluate than those which must be considered in only a subjective manner. Bringing quantitative probability figures into a decision-making situation generally results in a decision having greater validity and less undefined risk.

Probability is important not only to management; it is also important in the day-to-day work of any person associated with business. The potential effects of possible events are always present in the work of the accountant, salesperson, market researcher, or engineer. Our aim is to bring these unknowns within manageable bounds. Seldom, however, can we obtain sufficient information to remove all the unknowns. The gathering of information can be costly and managers cannot wait until all unknowns are eliminated. They must always proceed to carry out their responsibilities with varied degrees of uncertainties present.

A person with a good insight into these uncertainties can use them successfully to improve company performance. Sales personnel, for example, can plan their work more efficiently if they are able to recognize those contacts which have a greater probability of resulting in a sale. As another example, a company plan has greater value if it discusses the alternatives in terms of probabilities of certain events taking place. As a final example, an investor should certainly understand the probabilities associated with the success or failure of alternative investments before making a decision. In all these decision-making situations, probability comes into the picture whenever a logical analysis of the influencing factors is needed.

Exercises

4.27 An egg farm grades the egg into grades A, B, and C. In the long run, 20 per cent are grade A, 50 per cent grade B, and 30 per cent grade C. It has been found that 3 per cent of grade A are cracked, 1 per cent of grade B, and 2 per cent of grade C are cracked. An egg is selected at random before grading and found to be cracked. What is the probability that it is grade A?

4.28 A person chosen at random from the population of a remote Pacific island is given a tuberculin skin test and the test shows positive. The probability that a person with tuberculosis will have a positive skin test is .98. The probability that a person without the disease will get a positive test is .05. From data available, 1 per cent of the population in this area has tuberculosis. What is the probability that the particular person tested positive has tuberculosis?

4.29 A bank, in reviewing its credit accounts, has found that 75 per cent own their own home. Also 90 per cent of all persons applying for credit are acceptable credit risks. Further, 80 per cent of the good credit risks own their own homes. In this area, 75 per cent of the families own their own home. Given that a person does not own his home, what is the probability that he is a poor credit risk?

4.30 Three parts, A, B, and C, are produced independently and then assembled together. An inspector then inspects the item and has built up the following records, where S is satisfactory and D is defective (i.e., AS denotes that A is satisfactory. AD denotes that part A is defective, and so forth).

$P(AS) = .6$ $P(BS|AD) = .6$
$P(CS|AS \text{ and } BS) = .9$ $P(CS|AS \text{ and } BD) = .8$
$P(CS|AD \text{ and } BS) = .7$ $P(CS|AD \text{ and } BD) = .6$
$P(BS|AS) = .8$

(Note: Interpret $P(CS|AS \text{ and } BS)$ to be the probability that C is satisfactory, given that A and B are both satisfactory.)
Prepare a tree diagram and determine the following probabilities for a set comprised of three parts.
(a) All 3 defective
(b) No defectives
(c) Exactly 2 satisfactory
(d) C being satisfactory.

4.31 Our investment firm is attempting to obtain a large loan—considered as event A. The new sales manager is given the names of three companies B_1, B_2, and B_3 who might take the loan. The probability of his approaching either of the three first is 1/3. From past experience we know that

$$P(A|B_1) = .70$$
$$P(A|B_2) = .85$$
$$P(A|B_3) = .75$$

The sales manager returned and stated that he had placed the loan at the first company contacted. What is the probability that he went to (a) B_1? (b) B_2? (c) B_3? Draw a tree diagram to represent the problem.

4.32 Jack has 3 disks as follows:

A: red on one side and green on the other
B: red on one side and green on the other
C: green on both sides

The 3 disks are placed in a box. We leave the room. Jack takes a disk from the box at random and flips it in the air. When we return to the room, we see a green side up. Draw a tree to illustrate the problem. What is the probability that the side facing down is also green?

4.33 Mr. Jack interviewed four candidates (*A*, *B*, *C*, and *D*) for the position of secretary. After interviewing, he felt that each person had an equal probability of being hired. Each was then given a test, which indicated the following probabilities of success in the position:

A: .6
B: .8
C: .7
D: .9

By the time he had evaluated the test he found that 3 candidates had accepted positions elsewhere, so he hired the fourth. After 3 months he determined that she was handling the position successfully. Prepare a tree and determine the probability that *D* was hired, assuming the above probabilities are valid.

4.34 Albert, Bob, and Carl were tossing coins, with the odd man to buy coffee. It appeared that each had an equal chance to lose. Someone got the only head and paid for the coffee. It was then noticed that Bob's coin had 2 heads. Plot a tree and determine the probability that Bob bought the coffee. Was the game fair?

4.35 Three lots of material were set aside, each from a different supplier, A, B, and C. Another inspector came by and took a part from one of the boxes at random. The part was found to be defective. Past records show that lots from each supplier were of the following quality:

A: 10% defective
B: 20% defective
C: 27% defective

Plot the probability tree and determine the probability that the part was selected from C's box.

Learning Objectives

Introduction

Expected Value

Normal Distribution
  Standard Deviation Units

Bernoulli Processes

Binomial Distribution
  Review of Counting Methods
  Calculating Probabilities with the Binomial
  Binomial Expansion
  Summary of Binomial Distribution Assumptions
  Tables of the Binomial Distribution
  Binomial Distribution: Mean and Variance
  Normal Approximation to the Binomial

Poisson Distribution
  Applications of the Poisson Distribution
  Poisson Calculations
  Poisson Distribution As an Approximation to the Binomial

Learning Objectives

The primary learning objectives for this chapter are:

1. To understand the common types of distributions and their characteristics.
2. To learn to calculate and use the mean and standard deviation for each type of distribution.
3. To acquire the ability to recognize which type of distribution is best suited for various practical situations.
4. To understand how to use probability distributions to solve practical problems.

Introduction

Probability Distributions

We have studied ways in which data can be arranged, compiled, and then portrayed in a frequency distribution. In this chapter we examine how expected outcomes from probabilities can also be portrayed and evaluated by use of a distribution. In this case they are called *probability distributions*.

Probability distributions take on different forms, and will be classified into categories. The form of the probability distribution depends on the inherent characteristics of the item being measured. These characteristics include the methods by which the data are measured, the number of items considered, and whether the data are considered continuous or discrete. As an example, the expected number of yes votes in a survey of 5 people would form one type of distribution, whereas measurements of miles per gallon of an automobile would form a different type of distribution. The first is a discrete variable, and the latter is a continuous variable. In addition, the yes–no vote can take on only one of two possible values, whereas the miles per gallon can take on a wide number of possible values.

Probability distributions can be constructed either from data which have been collected or else from a logical analysis of the factors affecting the probabilities. To illustrate this point, consider a pair of dice. The sum of the two dice could turn up any value from 2 to 12. By use of logic we can count the sides, figure the possibilities, and calculate the probabilities, as shown in Table 5-1. We

TABLE 5-1. Probability Distribution for a Throw of 2 Dice

Sum of 2 Dice	Frequency	Probability
2	1	1/36
3	2	2/36
4	3	3/36
5	4	4/36
6	5	5/36
7	6	6/36
8	5	5/36
9	4	4/36
10	3	3/36
11	2	2/36
12	1	1/36
	36	36/36

can obtain the value of 2 in only one way (one and one) but 7 can occur in several ways (4 + 3, 3 + 4, 2 + 5, 5 + 2, 6 + 1, 1 + 6). The frequency for each possibility is shown. The alternative method is to toss the pair of dice many times and collect the data. After a large number of trials, the data collected would form a frequency distribution. In either case, we can arrive at a probability distribution.

At this point we can define a probability distribution. The *probability distribution* of a random variable shows a probability for each possible value that the variable can assume. The sum of these probabilities must be equal to 1.0. A *random variable* as used here is defined as a variable where the outcome is determined by chance. The tabulation in Table 5-1 meets these requirements of a probability distribution. In the case of a die we have a *discrete random variable* since it can take on only certain values. If ball bearings were being manufactured on a production line, where the intended mean diameter was 1.000 inch, the ball bearings might take on a number of values such as .998, .999, 1.000, and 1.001. If we use a precision instrument, we can obtain values in between these (such as .9995). Where the value can assume any value on a numerical scale, it is defined as a *continuous random variable*.

When attempting to estimate the shape of a distribution from collected data, we may proceed as we did in Chapter 3 to form a frequency distribution. Using this frequency distribution as a starting point, we would then attempt to associate relative frequencies (or probabilities) with the various quantities. Using the house price data presented in Table 3-9 as an example, we might attempt to determine the probability that a house selected at random would have a price between $27,500 and $32,500.

Since there are 17 in this group and 112 in the total, the probability would be about 17/112 or .15. In the case of the dice we could use either the logical or experimental approach. In most practical problems we would choose one or the other, depending on the nature of the process, the costs to collect data, and the time available.

If we were making a logical analysis of the process, we would first attempt to determine the nature of the process and the type of distribution we would expect it to form. The normal, binomial, and Poisson are distributions that are very common in everyday situations. Each will require a different method of analysis. Criteria for selecting the proper distribution and performing the analysis will be studied in this chapter. Previous to our further discussion of probability distributions, however, we want to present the concept of expected value.

Expected Value

Expected value is an important measure used in the evaluation of a set of data or a distribution. It is sometimes called *mathematical expectation*. The *expected value* of a random variable is a long-run average. Whether a distribution consists of continuous or discrete data, either the average or the expected value may turn out to be a fraction. For example, a valid statement might be that the average (or expected) number of automobiles sold per day by a dealer was 2.60. This is a valid statement even though we know that he cannot sell a fraction of a car.

Consider a game in which a person tosses a die and receives in dollars the amount of the die face, either 1, 2, 3, 4, 5, or 6. We might ask, "What should a person pay to play this game?" The expected value of the win or the mathematical expectation of the game are other ways to define this value. Using the symbol $E(X)$ to represent the expected value of the random variable X, in our example,

$$E(X) = \Sigma\, XP(X)$$
$$= 1\left(\frac{1}{6}\right) + 2\left(\frac{1}{6}\right) + 3\left(\frac{1}{6}\right) + 4\left(\frac{1}{6}\right) + 5\left(\frac{1}{6}\right) + 6\left(\frac{1}{6}\right) = 3.5$$

If a person played this game for a period of time, he could be expected to have an average win of $3.50, even though he could not win exactly 3.5 on any one play.

In the example above, each outcome was equally likely and was therefore weighted with equal value. As another illustration suppose that records of a bakery show the following for number of cakes sold per day:

Cakes Sold (X)	Number of Days	Probability = P(X)
7	20	.2
8	40	.4
9	30	.3
10	10	.1
	100	1.0

In this case we would calculate the expected number of cakes sold per day by weighting each value by the probability or frequency as follows:

$$E(X) = \Sigma\, XP(X) = 7(.2) + 8(.4) + 9(.3) + 10(.1)$$
$$= 1.4 + 3.2 + 2.7 + 1.0 = 8.3$$

This again is our long-run average. Past data are used to calculate the expected daily sales and to predict what we might expect in the future.

Normal Distribution

The first of the distributions to be studied is the *normal distribution*. Many measurable elements in the business or scientific world and in our day-to-day life take the form of a normal distribution. This distribution is sometimes called the *normal curve*. A typical curve is shown in Figure 5-1, although it can be either more flat or peaked as in Figure 5-2. Examples of data that tend to form normal distributions are test scores, many manufacturing characteristics such as diameters of ball bearings, and human characteristics such as heights of boys age 12. Although data that we might collect will seldom exactly fit a normal curve, where the normal curve is a good representation we say that the data are *normally distributed* or came from a population that is normally distributed. In addition, the normal distribution often represents the distributions of sampling statistics; this important aspect will be dealt with in Chapter 6.

A normal distribution has the following characteristics:

1. The normal curve involves a continuous variable, as previously discussed, rather than discrete data. The variable being measured theoretically can take any value along the X axis.

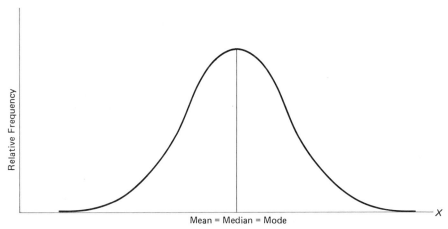
FIGURE 5-1. The Normal Curve (or Normal Distribution)

2. The curve is bell-shaped and symmetric about the mean, as labeled in Figure 5-1. The mean is also equal to the median and the mode of the normal distribution.

3. The normal curve is asymptotic in that its tails gradually approach but theoretically never reach the X axis.

4. The dispersion of the data affects the shape of the normal curve. Three normal curves with three dispersions are shown in Figure 5-2, with the right curve being more "peaked" than the others.* Two parameters, the mean μ and standard deviation σ, completely specify a particular normal distribution. A distribution with a smaller standard deviation, would be narrower, whereas a larger σ would define a flatter curve.

5. The area under the normal curve represents probability, and each half of the curve contains 50 per cent of the possible outcomes. In other words, the entire area under the curve represents 100 per cent of the possible outcomes; in terms of probability, the total area under the curve is 1.0. Any point along the X axis would actually represent zero area. Therefore, we shall talk in terms of areas between points along the X axis, or beyond certain points on the X axis. For our purposes here, the probability of a particular value is considered to be zero.

* The equation for the normal curve is

$$y = \frac{1}{\sigma\sqrt{2\pi}} e^{-1/2(X-\mu)^2/\sigma^2}$$

where μ is the *mean*, σ is the *standard deviation*, π (the Greek letter pi, which is the ratio of the circumference of a circle to its diameter) equals 3.14159, and e equals 2.71828.

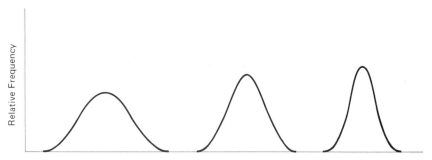

FIGURE 5-2. Three Normal Curves with Varied Degrees of Peakedness

Standard Deviation Units

A table showing areas under the normal curve appears in Appendix A. The total area under the curve equals unity, and, therefore, the area under a portion of the curve can be expressed as a fraction or part of unity. The two values that are most often used when working with this curve are the mean and the standard deviation from the mean. The mean is either known or can be calculated from data. The data can have various units of measure, such as dollars, inches, pounds, or other units. Thus, to provide a standard unit whereby a single table can be used, the standard deviation unit (SDU) is introduced. The deviation $X - \mu$, indicating the distance of a point X from the mean μ, is divided by the standard deviation σ to provide the figures in the first column of Appendix A. The resulting value is called a *normal deviate,* or *z,* and is written as

$$z = \frac{X - \mu}{\sigma}$$

In Figure 5-3, since $\mu = 110$ and $X = 185$, if $\sigma = 60$, then

$$z = \frac{X - \mu}{\sigma} = \frac{75}{60} = 1.25 \text{ SDU}$$

From Appendix A, the area under the normal curve between the mean and 1.25 SDU is .3944. Although the table lists probability values only up to .4990, the normal curve is symmetric and, therefore, a z of $+1.25$ will represent the same area as a z of -1.25. As a result, the area between $\mu \pm 1.25$ SDU $= 2(.3944) = .7888$. We might say that for an item taken at random from the population, the probability that it lies in the range $\mu \pm 1.25$ SDU is .7888. We could also say that 78.88 per cent of the items fall in this range.

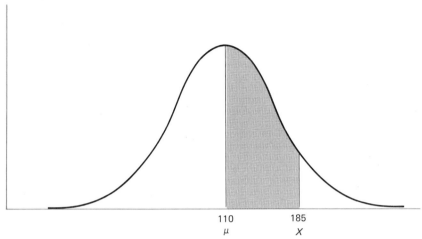

FIGURE 5-3. Area Under the Normal Curve Between μ and X

EXAMPLE 1

A teacher has 200 test scores that are adequately described by a normal distribution with a mean of 74 and a standard deviation of 6. How many test scores would we expect to be between 74 and 83 (as illustrated in Figure 5-4)?

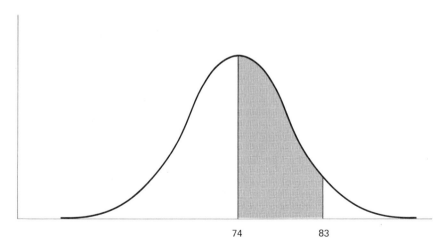

Shaded area = 1.5 SDUs = .4332.

FIGURE 5-4. Test Scores for Example 1

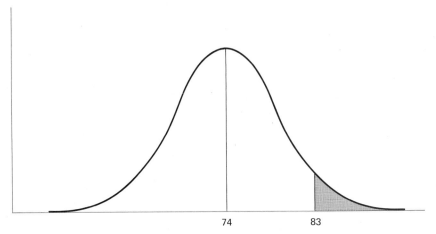

Shaded area = .5000 − .4332 = .0668.
FIGURE 5-5. Test Scores for Example 2

$$z = \frac{83 - 74}{6} = \frac{9}{6} = 1.5 \text{ SDU}$$

From Appendix A we see that 1.5 SDU is equivalent to .4332. Therefore, 43.3 per cent of the scores, or .4332 × 200, or approximately 87 of the test scores fall between 74 and 83.

EXAMPLE 2

How many of the test scores in Example 1 are above 83? Since the total area under one half the normal curve is .5000, the area beyond 83, as illustrated in Figure 5-5, is .5000 minus the portion of the area from 74 to 83, or .5000 − .4332 = .0668. Thus 6.68 per cent, or .0668 × 200, or approximately 13 of the 200 test scores are above 83.

EXAMPLE 3

Information from both sides of the curve can be used to determine the number of tests between 70 and 80. Each side of the curve is dealt with separately, as illustrated in Figure 5-6.

$$\frac{80 - 74}{6} = 1.0 \text{ SDU} \quad \text{representing an area of .3413}$$
$$\frac{70 - 74}{6} = -.67 \text{ SDU} \quad \text{representing an area of .2486}$$

Thus the area between 70 and 80 = .3413 + .2486 = .5899.

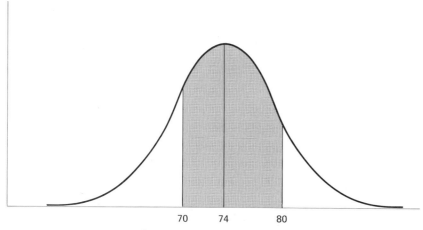
Shaded area = .3413 + .2486 = .5899.
FIGURE 5-6. Test Scores for Example 3

In the calculation, each side of the curve was considered separately. Although we can add or subtract areas under the normal curve as shown in Examples 2 and 3, we *cannot* add or subtract standard deviation units. The area represented by a SDU differs depending upon its position on the curve. In this case the number of tests between 70 and 80 is calculated as .5899 × 200, or approximately 118.

EXAMPLE 4

If we want to know the number of scores between 80 and 85, that value is represented by the area under the curve shown in Figure 5-7. To solve this problem, we subtract areas under the normal curve. The area between 74 and 80 is subtracted from that between 74 and 85; the remainder is the area between 80 and 85:

$$\frac{85-74}{6} = \frac{11}{6} = 1.83 \text{ SDU (Area} = .4664)$$
$$\frac{80-74}{6} = 1.0 \text{ SDU (Area} = .3413)$$

Thus the area between 80 and 85 = .4664 − .3413 = .1251. There are approximately 25 tests (200 × .1251) with scores between 80 and 85.

EXAMPLE 5

A machine produces parts with a mean diameter of 2 centimeters and a standard deviation of .0050 centimeter. What percentage of the parts

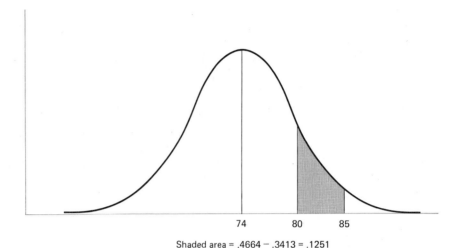

Shaded area = .4664 − .3413 = .1251

FIGURE 5-7. Proportion of Test Scores Between 80 and 85

will fall outside the specified tolerance limits of 1.987 to 2.014? To solve this probelm, it is necessary to determine the area in each tail of the curve separately.

$$z_1 = \frac{1.987 - 2.000}{.0050} = \frac{-.013}{.005} = -2.6$$

Area$_1$ = .5000 − .4953 = .0047   (from Appendix A)

$$z_2 = \frac{2.014 - 2.000}{.0050} = \frac{.014}{.005} = 2.8$$

Area$_2$ = .5000 − .4974 = .0026   (from Appendix A)

Thus the total area or probability = .0047 + .0026 = .0073. That is, .73 per cent of the parts will be outside the tolerance limits, which is less than 1 per cent.

The uses of the normal distribution demonstrated in these examples apply in many real situations. To employ them effectively, the decision maker must know as much as possible about the characteristics of the relevant data so that any limitations on the use of the normal distribution can be identified. These can then be considered when the results are used to make decisions.

We can see that as the distance from the mean exceeds 3.0 SDU, the area gets closer and closer to .5000. The curve never fully reaches the

X axis, however. In most instances when z gets to values beyond the table, we can assume that the area is .5. In some cases we may obtain results that indicate a very remote chance of an event happening, such as when $z = 4.0$.

Exercises

5.1 A fair die is thrown 15 times. What is the expected average of the numbers to appear face up? What is the expected total for 15 throws?

5.2 A trip is planned to a potential customer. The trip costs $300 and there is a .2 probability that the customer will make a purchase resulting in a profit of $4,600. What is the expected value of the trip? What other factor should be considered in deciding whether to make the trip?

5.3 The charge accounts at a certain department store have an average balance of $140 and a standard deviation of $30. If the account balances are normally distributed:
 (a) What proportion of the accounts is over $160?
 (b) What proportion of the accounts is between $100 and $150?
 (c) What proportion of the accounts is between $50 and $100?

5.4 A company gives aptitude tests to potential employees. The scores are normally distributed with a mean of 250 and a standard deviation of 25. What is the probability that a score selected at random will be:
 (a) Between 250 and 256?
 (b) Between 250 and 280?
 (c) Between 240 and 260?

5.5 A machine cuts metal rods for use in an assembly. The mean length is 9.15 inches with a standard deviation of .05 inch.
 (a) What is the probability that a rod will exceed 9.30 inches?
 (b) If the measuring device can measure only to the nearest .01 inch, what is the probability that an item will measure 9.10 inches? (*Hint:* Calculate the area from 9.095 to 9.105.)

5.6 A chemical company fills jugs of antifreeze with an average of 1 gallon per jug and a standard deviation of .04 gallon. Assuming a normal distribution, above what value will 90 per cent of the jugs be?

5.7 In problem 5.6, if the chemical manufacturer wants to assure that no more than 1 per cent of the jugs will be below .9488 gallon, how should he set the mean?

5.8 The chemical manufacturer filled 6,000 jugs per day. He is considering purchasing a new machine having a standard deviation of .02 gallon. If he still wants to allow only 1 per cent of the jugs to be below .9488 gallon, what is his expected daily savings in gallons if he buys the new machine? (Refer to problem 5.6.)

5.9 The chemical manufacturer decided that he wanted a probability of .99 that all jugs would contain at least 1 gallon. With the standard deviation of .04, where should he set the mean? (Refer to problem 5.6.)

5.10 Considering 250 working days per year and an antifreeze cost of $2.00 per gallon, the manufacturer wanted to retain the 99 per cent probability of no jug below 1.0 gallon. Should he pay $38,000 for the machine with a standard deviation of .02 to replace the .04 machine, if the machine has an expected 2-year life? (Refer to problem 5.9.)

5.11 The mean height of men at an Air Force base is 68 inches with a standard deviation of 2 inches.
   (a) What is the probability that a man is over 72.5 inches?
   (b) The shortest 15 per cent of the men are to be screened to perform a special task where the men must work in a small compartment. What is the maximum height allowable?

5.12 Several sets of brake linings are tested to determine the life. If 5.16 per cent of the sets last over 50,000 miles and the mean is 36,000, what is the standard deviation? Assume a normal distribution.

5.13 Another set of brake linings is checked. If 7.93 per cent last over 54,000 miles and 33 per cent last less than 30,000 miles, what is the estimated mean and standard deviation?

5.14 A shipment of 1,000 parts contains 93 per cent good parts and 7 per cent defectives. It costs $1 per part to test and sort the parts. If all parts are used, and the assembled unit is tested afterward, it costs $5 per unit to replace defectives. Which is the most economical approach?

5.15 A lot contains 1,000 switches, each with a cost of $10 if good. It is estimated that 10 per cent of the items are shorted and would cost $5 each to fix. It is estimated that another 20 per cent need touch up for paint damage at a cost of $3 each. What is the expected price that should be paid for the lot? It costs $1 to check each item to see if it is good.

5.16 A hot dog vendor at the stadium has kept track of sales as follows:

Sales per Day (dozen)	Number of Days
20	12
21	22
22	16
	50

What are the expected daily sales?

Bernoulli Processes

The next two distributions, the binomial and the Poisson, differ from the normal in that they are discrete, rather than continuous distributions. They are also similar in that they are both distributions of *Bernoulli processes*. A Bernoulli process is defined as one in which:

1. There are two and only two possible outcomes for each event.
2. The probability of each outcome is known and does not change over time or over a number of trials.
3. The probability of either outcome is independent of previous outcomes.

Some events are easily identified as Bernoulli processes, but others depend on a strict definition of the possible outcomes. For example, we easily recognize the toss of a coin as meeting all three conditions for a Bernoulli process: the coin will yield either "heads" or "tails" ("not-heads"), the probability of heads does not change from one toss to the next, and the probability of heads on one toss is independent of the results on a previous toss. Other events must be carefully defined to meet Bernoulli criteria. For example, the roll of a die may produce any one of six outcomes, but may be defined as a "five" or "not-five." Similarly, a card drawn from a deck could be defined as "Ace" or "not-Ace," but care must be taken to insure that independence is achieved by replacing the drawn card after each draw.

In some cases when we run trials, it is important to replace the previous item before the next trial. Consider a group of 100 parts which are 20 per cent defective. The probability of drawing a defective is 20/100, or .20. If we take one part and it is good, the probability of a defective on the second draw is 20/99. To assure that our concepts will continue to apply exactly, we must replace the items after each trial so that the probability remains constant. This is called *replacement*.

Binomial Distribution

When the Bernoulli process consists of a distinct number of trials, n, we say that it is a binomial process and the distribution of probabilities associated with it is called a Binomial distribution.

Let us consider a binomial distribution formed from a process that displays all the required characteristics. We shall consider dogs which have litters of 3 puppies. Assuming that the occurrences of male and female puppies are independent, equally likely events ($P = .5$), we could tabulate all the possible combinations as in Table 5-2. The probability of 3 successive males, for example, would be $.5 \times .5 \times .5$ or .125. The table then represents the probability distribution of the various numbers of females in the litters.

Table 5-2 shows the possible outcomes and expresses the relationship between the variable R (the number of females) and the probabilities of each value of R. The column labeled R in this case can be called a *sample space* since it represents all possible situations or events. Although the litters in our sample always consist of exactly 3 pups, there are 4 (or $n + 1$) events in this sample space, since there can be 0, 1, 2, or 3 females. The possibility of *zero* females represents the fourth possibility.

TABLE 5-2. Probabilities of Number of Females in a Litter of 3 Puppies

Possible Outcomes				Probability,
First Pup	Second Pup	Third Pup	Total Females, R	$P(R)$
M	M	M	0	1/8 = .125
M	M	F	1	
F	M	M	1	3/8 = .375
M	F	M	1	
M	F	F	2	
F	M	F	2	3/8 = .375
F	F	M	2	
F	F	F	3	1/8 = .125
				8/8 1.000

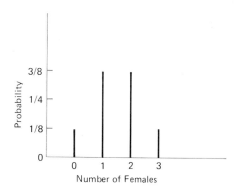

FIGURE 5-8. Probability Distribution for Number of Females in Litter of 3 Puppies

Since we are dealing with independent events, the occurrence of a male or a female pup is considered a *random variable*. It is not restricted and is free to take on any value (0, 1, 2, or 3) for each event. The variable is random because we cannot establish ahead of time which event will take place as each birth occurs. The events in Table 5-2 are mutually exclusive and collectively exhaustive. The sum of the probabilities for all events is unity.

These values can now be plotted in the form of a distribution as in Figure 5-8. This distribution portrays the probability that each of the events will occur, and is therefore called a probability distribution for this particular binomial situation.

Review of Counting Methods

Certain counting techniques are very helpful in understanding the binomial distribution as well as other types of probability problems. The following rules permit us to quickly determine numbers of events:

1. If event A can happen in n ways, and event B in m ways, then the two events can happen together in $n \cdot m$ ways.
Example: We have 6 colors of wrapping paper and 4 colors of ribbon. Packages can be wrapped in

$$6 \times 4 = 24 \text{ different ways}$$

2. A factorial (!) representation is used to abbreviate a product as follows:

$$n! = n(n-1)(n-2) \ldots (2)(1)$$

Also, by definition $0! = 1$.

Example: The factorial

$$6! = 6(5)(4)(3)(2)(1) = 720$$

3. The number of permutations of *r* items taken from *n* distinct items is given as

$$_nP_r = \frac{n!}{(n-r)!}$$

A permutation ($_nP_r$) considers which *r* objects are selected and also the sequence.

Example: How many 3-digit numbers can be formed from the digits 1, 2, 3, 4, and 5?

$$_nP_r = {_5P_3} = \frac{5!}{(5-3)!} = 5 \times 4 \times 3 = 60$$

4. The number of ways to sequence *n* items is *n*!. This is the same as

$$_nP_n = \frac{n!}{(n-n)!} = \frac{n!}{0!} = n!$$

Example: How many ways can *A, B, C, D,* and *E* be sequenced?

$$5! = 5 \cdot 4 \cdot 3 \cdot 2 \cdot 1 = 120$$

This can be illustrated by considering the manner in which each sequential item is selected. The first item can be selected in *n* ways (in our example, 5). There are then 4 items from which to select the next in sequence. This continues such that there is only one item left for the last selection.

5. The number of combinations of *r* items taken from *n* distinct items is

$$_nC_r = \frac{n!}{r!(n-r)!}$$

A combination differs from a permutation in that the combination does not consider sequence.

Example: How many combinations of 3 letters can be taken from *A, B, C, D,* and *E*? The following permutations are all one combination since they contain the same items:

```
ABD    BDA
ADB    DAB
BAD    DBA
```

In this case we have an r of 3 so that we divide the number of permutations by $3 \times 2 \times 1$ to obtain the number of combinations, or:

$$_5C_3 = \frac{5!}{3!(5-3)!} = 10$$

6. The number of permutations of n items where r are of one kind and $n - r$ are of another kind is

$$\frac{n!}{r!(n-r)!}$$

Example: Determine the number of possible permutations of $A, A, A, B,$ and B. Then,

$$n = 5$$
$$r = 3 \quad (A, A, A)$$
$$n - r = 2 \quad (B, B)$$

Then,

$$\frac{n!}{r!(n-r)!} = \frac{5!}{3!(2)!} = \frac{5 \cdot 4 \cdot 3 \cdot 2 \cdot 1}{3 \cdot 2 \cdot 1 \times 2 \cdot 1} = 10$$

These are illustrated by:

```
AAABB    ABBAA
AABAB    BAAAB
AABBA    BABAA
ABAAB    BAABA
ABABA    BBAAA
```

Calculating Probabilities with the Binomial

The binomial distribution applies when we deal with independent events that have stable probabilities, and where there are only two possible outcomes in any trial. In the dog-litter example, we were calculating the probabilities of outcomes of 0, 1, 2, and 3 females from litters having 3 pups. The probability of a female birth was assumed to be stable at .5; however, if it were .47 or some other number, we could still use a binomial because there are only 2 possible outcomes, male and female.

Take another example in which a town survey shows that 60 per cent of the people are in favor (yes answer) of a proposed school bond issue and 40 per cent are opposed (no). Assuming that the survey responses are independent, we can evaluate the probabilities of various sample results through use of the binomial distribution. In a random sample of 4 people, the probability of 4 yes answers $P(4 \text{ yes})$ is calculated as follows, using the multiplication rule from Chapter 4:

$$P(4 \text{ yes}) = .6 \times .6 \times .6 \times .6 = .1296$$

The probability of finding 4 not in favor of the bond issue $P(4 \text{ no})$ can be found in the same manner:

$$P(4 \text{ no}) = .4 \times .4 \times .4 \times .4 = .0256$$

There are many other possible arrangements of yes–no answers (such as 3 yes and 1 no), so it becomes necessary to perform additional calculations. In our illustration, there are 16 possible outcomes:

YYYY	NNNN	NNYY	YNYN
YYYN	NNNY	YYNN	NYNY
YYNY	NNYN	NYYY	YNNY
YNYY	NYNN	YNNN	NYYN

It is easy to count the number of cases for 0, 1, 2, 3, and 4 yes answers. One case has none and one case has 4 Y's. Three cases have one Y and three cases have 3 Y's. Six cases have 2 Y's. In order to count possible combinations, we could make up a list like this each time we encountered a problem where the binomial could be applied. As a shortcut, however, we use the formula, since the enumeration can become cumbersome if there are a larger number of trials.

The following formula is used to calculate the number of combinations of successes (or yes answers in this case) in n trials.

$$_nC_r = \frac{n!}{r!(n-r)!}$$

where r stands for the number of successes, $n - r$ the number of nonsuccesses, and C the number of possible combinations. The number of possible combinations of 3 yes answers out of 4 trials is calculated as follows:

$$_4C_3 = \frac{4!}{3!1!} = \frac{4 \times 3 \times 2 \times 1}{3 \times 2 \times 1 \times 1} = 4$$

The result can be verified by counting the number of outcomes in the list having 3 Y's. By definition, zero factorial (0!) equals one, so when $n - r$ equals zero, calculations are still possible.

Proceeding with our development of the binomial distribution, let p represent the probability of a yes answer and $1-p$, or q, represent the probability of a no answer. Consider the following sequence:

Yes, Yes, Yes, No

The probability of this sequence would be

$$p \cdot p \cdot p \cdot q = p^3 q = p^r q^{(n-r)}$$

where $n = 4$ and $r = 3$. The following sequence would also be given by the same $p^3 q$:

$$p \cdot p \cdot q \cdot p$$

Since there are $_nC_r$ of these groups of 3 yeses and 1 no, the probability of 3 yes answers in 4 trials becomes $_nC_r p^r q^{(n-r)}$. In our example this would be

$$_4C_3(.6)^3(.4)^1 = 4(.6)^3(.4)^1 = .3456$$

Binomial Expansion

The *binomial expansion* can also be used to calculate probabilities by listing all possible outcomes and combinations of outcomes in a situation. In our example, the binomial expansion would be

$$_4C_4 p^4 q^{(4-4)} + {_4C_3} p^3 q^{(4-3)} + {_4C_2} p^2 q^{(4-2)} + {_4C_1} p^1 q^{(4-1)} + {_4C_0} p^0 q^{(4-0)}$$

When terms are combined, the result is

$$p^4 + {_4C_3} p^3 q + {_4C_2} p^2 q^2 + {_4C_1} pq^3 + q^4$$

The formula developed above,

$$_nC_r = \frac{n!}{r!(n-r)!}$$

can be used to calculate each individual combination of n events taken r at a time and to evaluate each term in the binomial expansion.

There are several general statements which apply when dealing with the binomial probability model. Use of these rules allows short cuts in determining the terms in the binomial expansion.

1. There are always $n+1$ terms in the binomial expansion, where n is the number of trials.
2. In the sequence of terms, the exponents of p are $n, n-1, n-2, \ldots,$ 1, and zero.
3. The exponents of p and q in each term always add up to n.
4. The expansion is symmetrical in that the coefficients of terms at similar distance from the ends are equal. For example, the coefficient of the second term from the beginning is the same as the coefficient of the second term from the end (that is, $_4C_3 = {_4C_1}$). The coefficients become greater going to the middle of the series and then decrease. The graphical portrayal is symmetrical only if $p = q$.
5. The coefficient of each term can be calculated as follows:

(a) Number each term in the expansion from 1 to $n+1$. In our example, there are 5 terms:

$$p^4 + p^3q + p^2q^2 + pq^3 + q^4$$
$$1 \quad\ \ 2 \quad\ \ \ 3 \quad\ \ \ 4 \quad\ 5$$

(b) Multiply the coefficient of each term by the exponent of p and divide by the number of the term to obtain the coefficient of the next term. The coefficient of the second term in our example is calculated as

$$\frac{1 \times 4}{1} \quad \text{or} \quad 4$$

which forms the second term, $4p^3q$. The coefficient of the third term then becomes

$$\frac{3 \times 4}{2} \quad \text{or} \quad 6$$

(c) Continue this operation to calculate all terms in the expansion:

$$p^4 + 4p^3q + 6p^2q^2 + 4pq^3 + q^4$$

Substituting $p = .6$ and $q = .4$ from our problem, we have

$$(.6)^4 + 4(.6)^3(.4) + 6(.6)^2(.4)^2 + 4(.6)(.4)^3 + (.4)^4$$

This completed binomial expansion allows calculation of any probabilities or combination of probabilities in our polling example.

In practice, two types of solutions are generally sought in a binomial problem. The first is the *individual-term* probability. For example, the probability of 2 yes answers and 2 no answers would be a single term. The third term in the expansion corresponds to this:

$$P(2 \text{ yes \& } 2 \text{ no}) = 6p^2q^2 = 6(.6)^2(.4)^2 = .3456$$

The *cumulative* probability is the second type, which is often very useful. If we asked, "What is the probability of 2 or more yes answers?" this would be represented by the first 3 terms in the expansion. Since 4, 3, and 2 yes answers satisfy this question, add these 3 terms:

$$\begin{aligned} P(2 \text{ or more } Y\text{'s}) &= p^4 + 4p^3q + 6p^2q^2 \\ &= (.6)^4 + 4(.6)^3(.4) + 6(.6)^2(.4)^2 \\ &= .1296 + .3456 + .3456 = .8208 \end{aligned}$$

The expansion represents all possible combinations, so it is collectively exhaustive and the sum of all terms in the expansion equals 1.0. We could, therefore, solve this problem by calculating and summing the last two terms—which represent the probability of less than 2 yes answers—and subtracting the resulting sum from 1.0:

$$P(1 \text{ or } 0 \text{ yes}) = 4(.6)(.4)^3 + (.4)^4 = .1536 + .0256 = .1792$$
$$P(2 \text{ or more yes}) = 1.0 - .1792 = .8208$$

We can now plot Figure 5-9, which shows the probability of occurrence for each of the possible number of yes answers as follows:

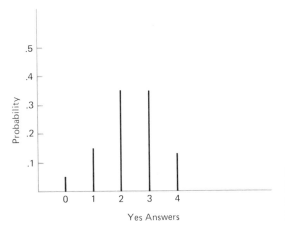

FIGURE 5-9. Probability Distribution for Number of Yes Answers in Survey Example

Number of Yes Answers	Probability
0	.0256
1	.1536
2	.3456
3	.3456
4	.1296
	1.0000

Summary of Binomial Distribution Assumptions

There are three characteristics that must be met before we can be sure that use of the binomial distribution is valid. In some cases, of course, we may allow slight deviation from the rules and still use the binomial as an approximation.

1. There must be exactly 2 possible outcomes on each trial of the experiment, such as success or failure, good or bad, pass or fail, etc. The probability of one attribute (success) is represented by p, and the probability of the other (failure) is represented by $1 - p$, or q. In our example, the probability of a yes answer p is .6, the probability of a no answer $1 - p$, or q, was $1.0 - .6$, or .4.

2. The trials and events are independent. As an example, when dealing with the sex of children at birth, the sex of any child is not influenced by the sex of prior children. As another example, if we collect opinions from 5 randomly selected people, the responses are assumed to be independent of the answers from prior persons polled.

3. The probability of each of the two outcomes (such as success or failure) stays constant over the time period involved. For instance, opinions do not change because of influences, nor are processes adjusted to affect good or defective items produced.

As noted previously, a process that meets these requirements is often called a Bernoulli process, after James Bernoulli, a Swiss mathematician who made significant contributions to the field of probability.

Tables of the Binomial Distribution

Performing evaluations using $_nC_r p^r q^{(n-r)}$ or terms from the binomial expansion becomes almost impossible when the number of trials n becomes very large. Even in small problems it becomes cumbersome. The binomial distribution tables in Appendixes C and D help considerably in reducing the computational effort required. Appendix C lists the individual-term probabilities, and Appendix D lists the cumulative probabilities. The cumulative table lists the probability of r or more successes for a given value of r, for selected values of n and p.

EXAMPLE 1

In a mail-order company, it is known that 40 per cent of the orders received contain errors in the price owing to use of outdated lists. In a sample of 10 orders, what is the probability of getting exactly 6 orders needing a price correction? From the binomial table in Appendix C,

$$P(r = 6 | n = 10, p = .40) = .1115$$

What is the probability of 4 or more orders needing to be repriced? This probability is found in the cumulative binomial probability table in Appendix D:

$$P(r \geq 4 | n = 10, p = .40) = .6177$$

What is the probability of 3 or fewer orders with errors? This cannot be found directly from the table; however, from our previous discussion on the binomial expansion, the probability of 3 or fewer orders with errors is the same as 1 minus the probability of 4 or more orders with errors:

$$P(r \leq 3 | n = 10, p = .40) = 1 - p(r \geq 4 | n = 10, p = .40)$$
$$= 1 - .6177 = .3823$$

EXAMPLE 2

In a complex process for making special TV picture tubes, the probability of producing a correct tube is .80 and the probability of producing a defective tube is .20. We desire to ascertain the probability of selecting 8 correct tubes in a sample of 12.

When working with problems such as this one, it is important to ascertain that the assumptions of the binomial distribution apply. First, there are only 2 possible outcomes: correct and defective. Second, it complies fully with the requirement that the trials be independent. If the population is not large, each time we select a part we would note whether it is defective and then return it to the batch. When taking a small sample from a large batch of items, failure to replace each part may not have much impact. The impact would also be small if the sample were taken from a continuous process or production line. In these cases we can consider the population to be infinite. When the sample is more than five per cent of the population, however, we should be very careful to sample with replacement; if this is not done, each trial will influence probabilities on later trials. As the third requirement, the probability of selecting a correct part remains constant. In our situation, the assumptions appear to be complied with.

From Appendix C, the probability of 8 good parts in the sample of 12 is

$$P(r = 8 | n = 12, p = .80) = .1329$$

The probability of 8 *or more* good parts is found in Appendix D:

$$P(r \geq 8 | n = 12, p = .80) = .9274$$

The probability of 5 or fewer good parts could also be found as

$$1 - P(r \geq 6 | n = 12, p = .80) = 1 - .9961 = .0039$$

Any problem that involves 2 possible outcomes and which meets the necessary assumptions may be solved as a binomial distribution. Other examples where the binomial could be applied are the number of people passing or failing a test, or number of persons reading or not reading a particular journal.

Often in real situations, items are not replaced as the sampling proceeds. This is called sampling without replacement, and since it violates the assumptions of the binomial distribution, another distribution, called the *hypergeometric distribution,* should really be used. In practical situations, however, if the sample is less than 20 per cent of the population, the binomial is an adequate approximation of the hypergeometric, so it will not be dealt with here.

Binomial Distribution: Mean and Variance

In our discussions of the normal distribution and of frequency distributions in general, we determined measures of the mean value and of the dispersion for the distribution. For the binomial distribution, we can also calculate a mean or expected value as well as a variance and standard deviation to measure dispersion.

In the polling example used previously, we can determine the expected number of yes answers as follows:

Number of Yes Answers (r) in a Sample of 4	Probability, $P(r)$	$r P(r)$
0	.0256	0
1	.1536	.1536
2	.3456	.6912
3	.3456	1.0368
4	.1296	.5184
		$\Sigma \, r P(r) = 2.4000$

We can say that the mean or expected number of yes answers in a sample of 4 would be 2.4. Where μ is the mean, n is the sample size, and p is the probability (in this case .6) for a yes answer, the mean is given by the formula

$$\mu = np$$
$$= 4(.6) = 2.4$$

The variance and standard deviation of a binomial distribution are defined as follows where q equals $1 - p$:

$$\sigma^2 = np(1-p) = npq$$
$$\sigma = \sqrt{np(1-p)} = \sqrt{npq}$$

For the polling example where $n = 4$, $p = .6$, and $q = .4$,

$$\sigma^2 = 4(.6)(.4) = .96$$
$$\sigma = \sqrt{4(.6)(.4)} = .9798$$

We can interpret μ as the expected number of yes answers in the sample of 4, where σ is a measure of the variation of the results about the expected value. The use of this concept will be covered more extensively when sampling distributions are discussed.

Normal Approximation to the Binomial

When working with the binomial tables, we noted that the sample size n only went as high as 25 in the table. As the sample size n becomes larger, the form of the binomial distribution comes closer and closer to the form of the normal curve. In Figure 5-9 we plotted the probabilities of various numbers of yes answers in a sample of 4 ($n = 4$) where the probability of an individual yes answer was .6 ($p = .6$). Figure 5-10 shows how the binomial distribution becomes more like the normal as the sample size increases, using the illustration where $p = .5$ and n is 4 and 8. The theory behind this will be discussed later in Chapter 6. Where n is greater than 25, the normal distribution can be used to approximate the binomial quite well. If p is very large or very small, however, the error in making the approximation may be significant. A good rule to follow is that both np and $n(1 - p)$ should exceed 5 in order for the normal to provide a reasonable approximation to the binomial.

Usage is illustrated by the following example, where n is selected as 25 so that the value from the binomial table can be compared to the value approximated by using the normal. As an example, we ask for the probability of obtaining 9 or more damaged items out of a sample of 25, given that the total population has 40 per cent damaged items.

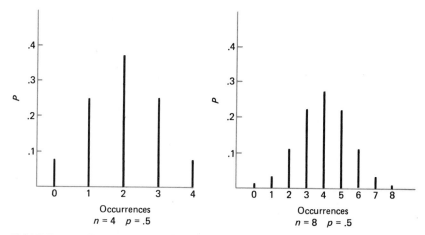

FIGURE 5-10. Two Binomial Distributions

$$n = 25$$
$$r \geq 9$$
$$p = .4$$

From the table, the probability of 9 or more damaged items is .7265. In order to use the normal as an approximation, we need a mean and standard deviation. Using the formulas for the binomial mean and standard deviation, we obtain

$$\mu = np = 25(.4) = 10$$

and

$$\sigma = \sqrt{npq} = \sqrt{25(.4)(.6)} = \sqrt{6} = 2.45$$

We can then say that when np and $n(1-p)$ both exceed 5, the binomial variable will be adequately approximated by a normal distribution with $\mu = np$ and $\sigma = \sqrt{npq}$. It is now necessary to determine the relationship of the binomial discrete elements to the continuous normal curve. To investigate the relationship, let us again consider the bond issue where 40 per cent of the people are opposed (no votes). We shall take a sample of 15 ($n = 15$) and consider it as a binomial variable with $p = 0.4$. (We could have obtained a similar situation by considering the yes votes, using a $p = 0.6$.) Using the $p = 0.4$, we can calculate the mean and standard deviation of the binomial as follows:

$$\mu = np = 15(0.4) = 6$$

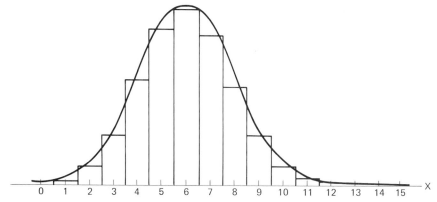

FIGURE 5-11. Normal Curve Superimposed upon the Binomial with $n = 15$ and $p = 0.4$

$$\sigma = \sqrt{npq} = \sqrt{15(.4)(.6)} = \sqrt{3.6} = 1.897$$

Figure 5-11 shows the relative frequency or probability of obtaining various numbers of "no" answers out of the sample of 15. Superimposed upon this binomial distribution is a normal curve also having a mean of 6 and a standard deviation of 1.897.

In order to illustrate the use of the normal to approximate the binomial, let us consider two examples.

EXAMPLE 3

What is the probability that we will obtain exactly 3 no answers?

The probability of the binomial variable X assuming the value of 3 would be equivalent to the area of the rectangle between x values of 2.5 and 3.5. In Figure 5-12 the area under the normal curve between 2.5 and 3.5 has been shaded in, and it is this area which will be used as the approximation to the area of the bar. The area under the normal curve is obtained as follows:

$$z_1 = \frac{2.5 - 6}{1.897} = -1.85$$
$$z_2 = \frac{3.5 - 6}{1.897} = -1.32$$

The corresponding areas under the normal curve from Appendix A are subtracted:

For z_1 area = .4678
For z_2 area = .4066
                       .0612

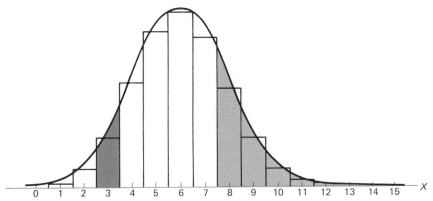
FIGURE 5-12. Normal Approximations to the Binomial for Two Examples

This value is close to the value of .0634 obtained from the binomial table (Appendix C) for $n = 15$ and $p = 0.4$.

EXAMPLE 4

What is the probability that we will obtain 8 or more "no" answers?

This probability is approximated by the shaded area in Figure 5-12 above 7.5 under the normal curve. It is calculated as follows:

$$z = \frac{7.5 - 6}{1.897} = .79$$
$$\text{corresponding area} = .5000 - .2852 = .2148$$

The exact binomial probability from Appendix C of obtaining 8 or more where $n = 15$ and $p = 0.4$ is .2131, which is close to the value obtained.

Exercises

> **5.17** A salesman plans his work so that he makes 10 business calls in a day. The probability that he will make a sale at a randomly selected establishment is 0.2. What is the probability that a salesman will make no sales in a day? What is the probability of exactly 3 sales? What is the probability of one or more sales in a day?
>
> **5.18** In 85 per cent of the homes in a particular community, someone is at home in the evening. A salesman calls at 12 homes per evening. What is the probability that he will find someone at home in exactly 7 homes?
>
> **5.19** Suppose that 50 per cent of the citizens in city A favor Allyne

for mayor. A sample of 5 persons is selected at random.
(a) What is the probability that all 5 favor Allyne?
(b) What is the probability that exactly 3 favor Allyne?

5.20 Determine the mean, variance, and standard deviation for each of the following, assuming a binomial distribution:
(a) $n = 100, p = .1$
(b) $n = 100, p = .8$
(c) $n = 25, p = .5$

5.21 A large box is filled with items of which 30 per cent are oversized. A random sample of 50 is taken. Assuming that the binomial distribution applies, determine the probability for each of the following, where r is the number of oversized items. Use normal approximation.
(a) $r < 18$ (d) $r \geq 20$
(b) $r = 15$ (e) $17 \leq r < 23$
(c) $r > 25$

5.22 Twenty pennies are tossed into the air. Determine the probability that:
(a) Exactly half will land heads.
(b) Less than 75 per cent will land heads.

5.23 The probability of the success of an experiment is .2. By use of a tree diagram, work out the probabilities of 0, 1, 2, and 3 successes in 3 tries.

5.24 A salesman has a history of successful sales on one third of his calls. If he makes 4 calls, what is the probability that he will fail on the first 2 and succeed on the next 2 calls?

5.25 In the previous problem, what is the probability that the salesman will obtain 2 sales in the 4 calls?

5.26 A manufacturer uses a sampling plan where a sample of 10 is taken and the lot is accepted if 2 or fewer defectives are found. If a lot is actually 5 per cent defective, what is the probability of rejecting it?

5.27 Use the normal curve to approximate the binomial for the following:

$$P(r \leq 8 | n = 16, p = .5)$$

Compare the result to that found in the binomial table. Determine:

$$P(r = 8 | n = 16, p = .5)$$

and compare it to the value given in table.

5.28 A batch of 60 items is 20 per cent defective. What is the probability of selecting 3 defectives in 3 trials with replacement? Compare this to the probability if the items are not replaced.

5.29 A company uses a sampling plan to accept or reject lots of switches received from a supplier. The plan calls for taking a sample of 6 items and rejecting if more than one defective is found in the 6.
(a) If a shipment of 24 items, 30 per cent of which are defective, is received, what is the probability of acceptance?
(b) What assumption must be made to use the binomial table?

5.30 In problem 5.29, if the plan called for taking a sample of 48 from a lot of 2,400 (also 30 per cent defective) and rejecting if more than 8 defectives were found, what would be the probability of rejection?

Poisson Distribution

In some Bernoulli processes, there is no clear enumeration of trials. For example, the occurrence of a telephone call may represent one possible outcome in a Bernoulli process and the number of telephone calls in some period can be counted readily. However, one cannot count the number of opposite outcomes — the absence of a phone call. In this case the Bernoulli process is a Poisson process and the number of phone calls is distributed according to a Poisson distribution. The Poisson distribution is applicable to situations where there are a number of occurrences per unit of time or space, whereas the binomial is applied to cases where there are a number of occurrences in a given number of trials. In the Poisson process, there is no sample size or specific number of trials, so the mean number of occurrences m serves as the only parameter. The Poisson is also used as an approximation to the binomial when the probability is very small and n is large.

Applications of the Poisson Distribution

The Poisson can be applied to a variety of practical problems. It is widely used to represent probabilities in queuing, or waiting-line, problems. Consider a manager of a supermarket who, knowing that he may lose business if customers have to stand in long lines to check out, may wish to estimate the probabilities of various line lengths. This will enable him to determine how many checkout lanes to have and how to staff them. It will also permit him to compare the cost of additional service facilities as compared to the cost of lost time waiting in line. The Poisson is also useful for certain problems involving defects in a manufactured product, the movement of goods or materials, machinery breakdowns, and inventory control.

In addition, the Poisson distribution is applicable to many problems that deal with a rate or frequency. Examples are aircraft arrivals per hour, material defects per square foot, or errors per page of a form or document. In these cases, we see that we count occurrences but we cannot count nonoccurrences. We cannot, for example, count nonarrivals or nondefects. Thus there is actually no total sample size. In order for the Poisson to apply, the value of p (representing rate) must be very low and be constant. The events must also be independent. In cases where errors on a machine occurred successively and were caused by a particular factor, such as the machine being out of adjustment, then the Poisson distribution would not apply.

Poisson Calculations

The Poisson is applicable to cases where the mean numbers of occurrences n is constant. The probability of X successes in n trials is approximated by the formula

$$P(X) = \frac{e^{-m}m^X}{X!}$$

where e is a constant that is the base of natural logarithms (2.71828) and m is the mean number of occurrences. The Poisson has the unique characteristic that the variance is also equal to m, so the standard deviation would be $\sqrt{m}$.

Consider an example of the use of the Poisson distribution. Cars arrive at a toll gate on the average of 2.5 per minute. Determine the probability that 5 or more cars will arrive in a one-minute period. In this case, $m = 2.5$. From Appendix F,

$$P(X \geq 5) = .1088$$

From Appendix E, we determine that the probability of exactly 5 cars in a one-minute interval is

$$P(X = 5) = .0668$$

We also see from the table that the probability of 11 cars in any one-minute period is essentially zero.

Consider another example where the Poisson is applicable. A clerk fills out shipping documents for a distributor. On the average he makes one typing error for a set of documents. If 3 shipments are made on Tuesday, what is the probability that there are exactly 2 errors on the 3 sets of documents? In this case the average is 1 for a single set of documents; however, the question relates to 3 sets of documents. The average number of errors in 3 documents $= m = 3 \times 1.0 = 3.0$:

$$P(X = 2) = .2240 \quad \text{(from Appendix E)}$$

This is the probability that there will be exactly 2 errors in the 3 sets of documents.

Poisson Distribution As an Approximation to the Binomial

We saw earlier that when n is larger than 25, we use the normal curve to approximate the binomial. It is also true, however, that in those cases where n is large and p is small, the shape of the Poisson distribution serves as a better approximation to the binomial. A good rule to follow is to use the Poisson as an approximation to the binomial if p is less than .05 (or greater than .95) and where n is also greater than 20.

As an illustration, assume that a machine is known to turn out parts which are 1 per cent defective. The quality control engineer takes a sample of 50 parts and desires to know the probability of getting exactly 0, 1, 2, 3, . . . defectives in the sample of 50 items. This would form a binomial distribution; however, the tables do not go beyond 25 for n and therefore we shall use the Poisson as an approximation. In this case, $n = 50$ and $p = .01$, so that $np = .5$. The following values are taken from the Poisson table for $m = np = .5$:

Number of Defectives	Probability
0	.6065
1	.3033
2	.0758
3	.0126
4	.0016
5	.0002
6	.0000
	1.0000

The probabilities are shown to 4 decimal places and add up to unity. There is a small probability that we will get 6 or even 7 or more defectives; however, we consider it to be zero when using 4 significant figures. We could, however, calculate the value of these small probabilities from the Poisson formula. In this example where we could not readily calculate the binomial, the Poisson is a better approximation to the binomial than is the normal curve.

As another example, suppose that a new type of flashcube is known to flash an average of 97 out of 100 times. In an experiment of 100 tries, we wish to know the probability of 0, 1, 2, and 3 failures to flash. In this case m is $100 \times .03$, or 3.00. From the formula or from Appendix E, which lists the individual-term Poisson probabilities, we find that

$$P(X = 0 | m = 3) = \frac{e^{-3} 3^0}{0!} = .0498$$

$$P(X = 1 | m = 3) = \frac{e^{-3} 3^1}{1!} = .1494$$

$$P(X = 2 | m = 3) = \frac{e^{-3} 3^2}{2!} = .2240$$

$$P(X = 3 | m = 3) = \frac{e^{-3} 3^3}{3!} = .2240$$

The last number represents the probability of exactly 3 occurrences. What if we wanted to determine the probability of X or fewer occurrences? We could obtain the probability of 3 or fewer occurrences by summing the values we have just calculated:

$$P(X \leq 3 | m = 3) = .0498 + .1494 + .2240 + .2240 = .6472$$

Appendix F lists the cumulative Poisson probabilities for X or more occurrences for various values of m. Using the individual and cumulative Poisson tables in Appendixes E and F is very similar to using the binomial tables. For instance,

$$P(X = 2 | m = 3) = .2240$$
$$P(X \geq 3 | m = 3) = .5768$$
$$P(X < 3 | m = 3) = 1 - P(X \geq 3) = 1 - .5768 = .4232$$

Exercises

5.31 The average number of customers arriving at a cafeteria during the lunch period is 2 per minute. Assume the Poisson distribution and evaluate the probability that exactly 3 customers will arrive in a one-minute interval. What is the probability that there will be no arrivals in the one-minute period? What is the probability of more than 3 arrivals?

5.32 On large sheets of steel there is an average of one flaw per each 20 sq-ft area. What is the probability that a 30 sq-ft piece will contain no flaws? What is the probability that it will contain 2 or more flaws?

5.33 A binomial population contains 3 per cent defectives. Use the Poisson approximation to determine the probability of exactly 3 defectives in a sample of 100. What is the probability of 1 or fewer defectives?

5.34 The demand for items from an inventory are often in the form of a Poisson distribution. Assume that the demand for cases of brand A soap from a wholesaler averaged 4 per week. Prepare a probability distribution for the weekly demand. If the wholesaler replenishes his stock to a level of 6 at the beginning of each week, what is the probability that the demand for a particular week will exceed the supply?

5.35 Ships arrive at a port with a mean of 2 ships per day. On a particular day, what is the probability of no arrivals? If the port has facilities for 4 ships and each ship leaves at the end of the same day that it arrives, what is the probability that the facilities will be filled?

Review Exercises

In the following exercises, the student will need to determine which distribution is applicable before solving.

5.36 In a manufacturing division of a company, 80 per cent of the employees are paid by the hour and 20 per cent receive a salary.
 (a) Four names are selected at random. What is the probability that exactly 3 are hourly?
 (b) Determine the mean and standard deviation for samples of 4 selected at random.

5.37 Mr. Jack is considering the purchase of an apple orchard. He is told that 84.13 per cent of the trees give more than 14 pecks and 2.28 per cent give less than 12 pecks of apples. If there are 1,200 trees in the orchard, what is the expected yield? Between what two limits do the middle 95 per cent of the tree yields fall?

5.38 A company fills cans with fruit juice and as long as the weight falls between 15.8 and 16.2 oz, the process is considered acceptable. The process has consistently produced with a $\sigma = .05$ oz. In a day of checking 100 samples, there are 3 below 15.8.

What do you conclude and recommend as to resetting the process mean?

5.39 An assembly line has 8 grinders. At least 6 of these must be in operation in order to maintain adequate material for the assembly line to produce on schedule. The probability that a grinder will be inoperative on any day because of mechanical failure is .15. There is also the probability of .10 that any operator will be absent. Any operator can use any machine. What is the probability that the production line will not meet its schedule on any particular day because of the grinder section?

5.40 A high-speed machine produces polyethylene sheeting. It produces an average of 5 flaws in each 100 yards of the material. If the inspector finds 11 or more flaws in a 100-yard piece, he will stop the machine.
  (a) What is the probability that the machine will be stopped upon inspection of a 100-yard piece? Assume a Poisson distribution.
  (b) What is the probability that a 50-yard piece will contain less than 4 flaws?
  (c) What is the probability of producing at least 7 but not as many as 13 flaws in a 200-yard piece?

5.41 Lots in a development have been laid out at random to have an area of 40,000 sq ft and a standard deviation of 2,000 sq ft.
  (a) A person selects a lot that he likes. What is the probability that the lot will have over 37,500 sq ft?
  (b) What is the probability that a lot selected at random will be over 1 acre (43,560 ft^2)?

5.42 A car dealer receives a shipment of 68 cars. In the past, 60 per cent have had automatic transmissions. What is the probability that 35 or more of the cars have automatic transmissions?

5.43 Trucks stop at a highway weigh station at an average of 64 per 8-hour day. Assume that arrivals are in a Poisson distribution.
  (a) If the operator goes to lunch for 1 hour, what is the probability that no trucks will arrive in that period?
  (b) If he takes a 1½-hour lunch period, what is the probability that 10 or more trucks will be missed?
  (c) The operator says that he will bet even money that not more than X trucks will go by in a 45-minute period. What should X be to give the operator better than an even chance of winning the bet?

Learning Objectives

Sampling

Advantages of Sampling
    Other Factors Related to Sampling

The Population

Random Versus Nonrandom Sampling

Simple Random Sample
    Replacement Versus Nonreplacement
    Table of Random Numbers

Other Random Sampling
    Systematic Sampling
    Stratified Sampling
    Cluster Sampling

Nonrandom Sampling

Single, Double, and Sequential Sampling

Sampling Distribution of the Mean
    Normal Population
    Distinguishing Between s and $\sigma_{\bar{x}}$ and Other Measures
    Population Nonnormal

Small Populations: Sampling Distributions

Sampling Distributions of Proportions

Sampling and Sampling Distributions

Learning Objectives

It is extremely important to recognize when sampling is appropriate, to determine how large a sample is needed, and to figure out how best to select the items that are to make up the sample. All of these factors are important in arriving at a valid conclusion. This chapter is intended to provide the tools needed to carry out basic sampling projects.

The student or business person must understand not only the advantages of sampling but also the limitations or risks associated with sampling. These must be presented as part of any sample evaluation along with the conclusions obtained. One must also learn how to interpret the sample data and to estimate the characteristics of the population from which the sample came. In addition, it is necessary to know methods of determining the reliability of the sample information for a particular intended purpose. Not only is it important to learn the sampling procedures, but our learning objectives must include means of handling economic tradeoffs related to sampling and to selecting economically feasible approaches.

The primary learning objectives for this chapter are:

1. To recognize when sampling is appropriate, including an understanding of the economic trade-offs related to sampling.
2. To understand how large a sample is needed and what procedure to use in selecting the sample items.
3. To learn how to interpret sample data and to ascertain limitations and risks in the use of sample data.
4. To understand the sampling distribution of the mean and its relationship to the population distribution.

Sampling

A sample is a subset of a total population and is usually taken for the purpose of learning more about the population. It is preferable that those persons responsible for analysis of the sample data also determine the manner in which the sample is selected. In some cases, however, we may need to analyze sample data that have been given to us. Sampling theory provides many practical and useful guides for establishing how large a sample should be, as well as procedures for selecting samples from a population. It is important to consider potential sample data usage, as well as the nature of the population itself when designing a sample procedure. Our objective in selecting a sample is to provide data having optimum value for analysis; however, in choosing a sample procedure and sample size we must also consider economic factors. Money and other resources to be expended in taking a sample and in the analysis of the resulting data are usually limited, and should be used efficiently in order to arrive at better information from which to make decisions.

The stated purpose of sampling is to learn more about the characteristics of a population. It is therefore very important that the sampling be properly done so as to provide adequate information to carry out valid analyses and aid in making good decisions. This requires that the size of sample be adequate and also that the sample data represent the population. Take an example where a manufacturer receives a shipment of 10 kegs of galvanized bolts and a sample is to be selected and tested to determine resistance of the bolts to corrosion. A sample of 30 taken from the top of one of the kegs may be adequate as to sample size but would not really represent the whole population contained in the 10 kegs. A portion of the kegs could have come from the supplier's older inventory produced at different times. Our goal, then, in sampling is to select a sample that is representative of the whole population. We recognize that there are risks in any sampling plan and we seldom obtain complete representation; however, there are advantages to sampling. Sampling can provide accurate data in addition to the economic advantages. It becomes important, therefore, that we recognize the risks associated with sampling along with the merits of sampling. Later, however, we shall see cases where we actually have no choice except to sample.

Advantages of Sampling

In many business situations, it is either impractical or impossible to examine the entire population. Consider some examples where it is mandatory to sample or to use only a portion of the population if we are to make an evaluation.

The polling of all households in a city of 100,000 population would be costly and present substantial obstacles. If one person could poll 10 homes per hour (equal to 80 per 8-hour day), it would take 10 people a total of 125 working days to do the job. At $3 per hour, this would cost $125 \times 10 \times 8 \times 3$ or $30,000. This does not even consider the problem of identification of all households, difficulties in finding people at home, or even changes in the population over the 125-day period resulting from people moving in or out. As we recall, if we did take a survey, or enumeration, of the entire population, it would be called a census. Usually, however, sampling would be adequate or better for most purposes.

As a second example where measurement of the total population is impossible, consider a production process for making TV sets. At some point in time, some sets in the population may have been produced and shipped, whereas another unknown quantity has not yet been produced. In this case, any group of one or more sets selected for evaluation would be a sample. As another manufacturing example, consider the case where the examination of steel bolts may require a test that destroys or damages the part. Strength tests, shear tests, and chemical analysis tests are typical examples of this type of test. Again, sampling is our only possibility.

The analysis of a sample can also be much more extensive and detailed than would be practical if the entire population were examined. The part could be damaged in the test or, in the case of a life test, the part would be completely used up or worn out. One further advantage of sampling should also be recognized. In addition to prohibitive costs and other limitations, the examination of large numbers of items is more readily subject to human error. A person can take greater care when inspecting a small quantity of data or parts. The element of human fatigue or human error comes into the picture when extensive, tedious inspections or repetitive tasks are involved. It is well known that 100 per cent inspection does not find all defects. In addition to these considerations and limitations, the possible computational errors and inaccuracies in processing a large quantity of data also provide a disadvantage for the census and an advantage for sampling. The human element again causes us to go to sampling for more accurate results.

Other Factors Related to Sampling

When decisions are being made, an adequate degree of precision is needed in the data to result in a satisfactory conclusion. The estimation of a population mean or variance from sample data must be adequate for the purpose of the decision. This does not mean, however, that we should seek the greatest possible accuracy. In estimating the average monthly income of households in a town, for example, we would certainly not need data to the nearest cent. Perhaps for our purposes a possible error of $30 per month would be acceptable. If we have sufficient accuracy for making our decision, any further accuracy would have little or no economic value. Later we shall evaluate the need for more precise data than that available as a trade-off against the cost of obtaining these more precise data.

Decisions also require timely data. If information for a decision is needed in 3 days, there may be no time for extensive data collection or compilation. Perfect information has little value if it is received too late and the decision has been made or the action has already been taken. Again, sampling may allow us to overcome obstacles such as these; however, the process of sampling must be carried out carefully.

The Population

Knowledge of the population (or universe) is important to the reliability of any sampling process. Therefore, it is important to highlight factors which must be considered when sampling is to be carried out. These can be considered part of the planning process for establishing the sampling procedures.

First of all, what is the population in which we are interested? If a poll is being taken, are we interested in all persons or only in those of a certain age group or other specific groupings? Once this is determined, we must select the sample to represent this particular population. A classical example of misuse is the *Literary Digest* poll conducted in 1936 to predict the results of the forthcoming presidential election between Roosevelt and Landon. The *Digest* used telephone directories and automobile registration lists to obtain names to be included in the sample. Based upon the sample results, the *Literary Digest* projected Landon as the winner in the coming election. In the actual election, Roosevelt won by a landslide. The *Literary Digest* was hurt considerably by the incorrect prediction and ceased publication soon after. In looking for the reason for the *Digest*'s error, we see that the sample taken reflected views of middle- and upperclass people and not the views of the total voting population. During this period in the middle of the Great Depression, a large proportion of voters did not own automobiles nor did they have telephone service. The sample, therefore, did not represent all persons eligible to vote, which was really the population that the *Digest* should have been testing. This illustrates the importance of clearly establishing the population of interest.

As a second important factor, the sampling unit must also be clearly identified. As an example, consider a restaurant owner evaluating her business. She would be concerned with headcount if she were evaluating receipts to determine the average meal check. If she were planning parking facilities, however, the number of customer groups arriving by automobile would be the more important unit. If she were making decisions on table sizes and chair arrangements, both person count and group count would be important. This illustrates the need to identify the sampling unit. If our sample size is 50, for example, do we mean 50 persons or 50 groups? If we do not determine this unit correctly, our sample may indicate the incorrect answer. Having discussed these two important factors, we next consider different ways of taking the sample.

Random Versus Nonrandom Sampling

A sampling process can be either random or nonrandom. These two categories of sampling can be further subdivided; but before we consider this aspect, it is important to recall the basic differences between random sampling and nonrandom sampling.

In random sampling (also called probability sampling), each element that makes up the population has a known chance of being selected as part of the sample. Furthermore, this chance or probability for any item to be selected can be determined. The chance of items in the population need not be equal, which was a requirement for the simple random sample.

In nonrandom sampling (or *nonprobability sampling*) the units to be included in the sample are *not* determined by chance, but rather their selection is based on a judgment decision of the investigator. Those units are selected which the investigator feels would serve as a good representative sample. As an example, viewer ratings of TV programs are made from carefully selected samples. They are not taken at random because the pollsters believe that a sample selected by judgment will provide a more rational result. Further characteristics of random and nonrandom sampling are discussed in greater detail in the following sections. In the sections to follow, each of these two groupings is further divided into categories.

Simple Random Sample

Some of the most basic elements of sampling can best be illustrated by examples involving simple random samples. A *simple* random sample (or *unrestricted* random sample) must meet two criteria:

1. Each item in the population has an equal chance of being selected, and the probability of being selected can be determined.
2. Each sample combination has an equal probability of occurrence.

Consider a city council of 11 members. An acting chairman and an advisory committee of 3 persons is to be selected at random. If the 11 names were written on cards and 1 card drawn at random from a hat, each person would have a 1/11 chance of being selected as acting chairman. Assume, then, that after the chairman is selected, the advisory committee is to be selected at random from the remaining 10 persons. The possible combinations of 3 persons out of the 10 is given as

$$_{10}C_3 = \frac{10!}{3!(10-3)!} = 120$$

Therefore, there are 120 possible combinations of 3 persons from the population of 10. If A, B, C, D, E, F, G, H, I, and J, represented the 10 members, each member, such as A, would have a 36/120 chance of being on the committee. We can illustrate how this figure is arrived at by enumeration of the possibilities. Disregarding order, the following 36 groups of 3 would contain A:

ABC	ABI	ACH	ADH	AEI	AGH
ABD	ABJ	ACI	ADI	AEJ	AGI
ABE	ACD	ACJ	ADJ	AFG	AGJ
ABF	ACE	ADE	AEF	AFH	AHI
ABG	ACF	ADF	AEG	AFI	AHJ
ABH	ACG	ADG	AEH	AFJ	AIJ

There are no other combinations of 3 persons which would contain A. Therefore, the probability of any one of these sample combinations is 1/120 and the probability of A being on the committee is 36/120 or .3. The probability of any other individual council member being on the committee could be established by the same procedure and would also be .3. In this example the two requirements for a random sample are met. Each person on the council has an equal chance (.3) of being selected to serve on the committee. Second, each sample combination (or group of 3) has an equal probability (1/120) of occurrence.

Replacement Versus Nonreplacement

When carrying out sampling, the concept of replacement versus nonreplacement must also be considered. If in the situation above we desired to obtain the probability of selecting A, B, and C in that exact order for the committee, we would obtain

$$P = 1/10 \times 1/9 \times 1/8 = 1/720$$

Once A has been selected, the chance of drawing B is one out of the remaining 9 names. After B is selected, C is one of the 8 then left. In this case we are concerned about order of selection, and we do not have replacement. Once a name has been selected, it is not available for later selection. If we had replacement, the same name could be drawn again.

Since in our actual selection of 3 committee members we are not concerned about order of selection, we can determine the probability of selecting A, B, and C (in any order) based on the following concept:

$P =$ 3/10	×	2/9	×	1/8	=	1/120
probability of either A, B, or C on first drawing		probability of either of 2 remaining being selected, given 1 of the 3 was selected on the first draw		probability of the 1 remaining person being selected, given that 2 of 3 were selected on the first draw		probability of A, B, and C being selected in any order

Assume a different type of situation in which there are 10 persons and 3 prizes to be given out by drawing. The probability of any person, such as A, winning the first draw is 1/10. If we have replacement, the name of the first winner is replaced; and again on the second draw, each person (including A) has a probability of 1/10. If we used the constraint that any person could not win more than once, we would have nonreplacement. On the second draw, the probability of a win by A would be 1/9, and on the third draw it would be 1/8. The probability of A winning one prize would then be:

$P =$	1/10	+	9/10	×	(1/9)	+ (9/10 × 8/9)×	(1/8)	=	3/10
$P =$	probability of winning on first draw	+	probability of not winning on first draw	×	probability of winning on second draw	+ probability of not winning on either of first 2 draws	× probability of winning on third draw	=	probability of winning a prize

Table of Random Numbers

Drawing from a hat or bowl was a satisfactory procedure when only a few units made up the population. When larger populations are involved, there are greater difficulties associated with obtaining a random sample.

In cases of larger populations, a table of random numbers helps overcome this difficulty in selecting a random sample. Such a table is Appendix J.

Suppose that we desire to obtain a random sample of 50 from a student body of 10,000. The first step would be to assign numbers from 0 to 9,999 to each student by name. A table of random numbers will then be utilized to select the random sample of 50. Our largest number, in this case 9,999, has 4 digits. We should then break the table into columns of 4 digits. We can start anywhere in the table as long as we decide on a systematic approach. A suitable approach here would be to start (Appendix J) going down the third column using the last 4 digits. The first number would be 4,520 (the 7 is discarded). We proceed until we have enough numbers to give us the desired size sample. In this case we would start out with:

4,520	first
4,894	second
9,645	third
9,376	fourth
.	.
.	.
.	.
1,249	49th
7,637	50th

The students whose names were assigned these numbers would make up the sample of 50. If the sample were greater than 90, we would then go to the fourth column and again use the last 4 digits (starting at 3,586).

What if the student body numbered 1,000? We can still use the same table, but we would use only the last 3 digits. (We could have chosen the first 3 — it makes no difference as long as we are consistent.) The table of random numbers can be used for any size of population up to the limits of the table. As an example of a more complicated situation, consider a student population of 6,843. In this case we would still proceed as we did for the 10,000 students except that each time we encountered a number in the table that was 6,843 or higher, it would be passed by and ignored. We would proceed as before (but skipping these numbers) until we had 50 numbers of 6,842 or less. Using the same column, our first 10 numbers would be 4520, 4894, 157, 4072, 5571, 2051, 5325, 3529, 4905, and 6288. The fact that we discarded some numbers does not affect the equality of opportunity for each of the 6,843 numbers to be selected.

This process of random sampling from a table of random numbers can be used where population units can easily be identified and numbered. The population must also be homogeneous, however, and not be excessively large. If we were working with the population of a city of 200,000, it would be difficult to identify and number each unit (person) in the population. Furthermore, if we used street addresses to work with, the higher or lower income units would probably be grouped together and a homogeneous population would not exist. In addition, if a person were the unit, there would be more than one person at each street address. We can then see that there are limitations on the use of random-number tables. In many situations, however, a table of random numbers is of considerable help in selecting a simple random sample.

Other Random Sampling

The procedures just discussed for random-sample selection are not very practical for large populations or cases where it is difficult to identify units. There are other approaches to random sampling that help to solve some of these difficulties in identifying units. These other approaches help in obtaining a random sample from a larger nonhomogeneous population. These methods are often referred to as *restricted random sampling* and they include systematic sampling, stratified sampling, and cluster sampling.

Systematic Sampling

Take an example where a university has a total student population of 10,000 students and we need a random sample of 500 students. Instead of selecting these as before, we can set up what is called a *systematic sampling procedure*. An example of systematic sampling is to take each 20th unit, since (10,000/500 = 20). *Systematic sampling* consists of taking every nth unit, once the units have been somehow arranged in a sequence or list. In order to give each element in the population a chance at being selected, the starting item from those numbered 0 to 19 must be selected at random initially. This can be done by drawing from a bowl of chips numbered 0 to 19. If 8 were the first number selected, we would proceed, by systematic sampling, to sample students numbered 8, 28, 48, 68, etc. Systematic sampling is very convenient when an unnumbered list already exists, such as with a telephone directory. We could just go down the list and take each nth name. Polling every 15th person entering a store, or inspecting every 50th item off a production line would also constitute systematic sampling. If we took every nth address from a city directory, this would help us in our problem of obtaining a representative sample of residence units from the nonhomogeneous city population. All areas of the city would be represented and each address would have an equal chance at being selected.

Systematic sampling should not be used if there is a likelihood of cyclical tendencies. Sampling traffic at 5 P.M. each day will count people coming home from work, whereas a sample of 2 P.M. would catch people in the process of their day-to-day work or activities. In this case the traffic pattern takes on the form of a cycle, where each cycle is similar from one day to the next. Therefore, we would not carry out systematic sampling by taking, for example, one automobile every 15 minutes throughout the day.

Stratified Sampling

The term *stratified sampling* is derived from the process of separating the population into groups, or *strata*. Units are placed into groups such that items within a group are homogeneous. Thus each stratum will contain like items, while the differences between strata will be as great as possible; however, a stratum may be homogeneous for only one or more specific characteristics. In our case of taking a sample of families from a city population, the strata grouping could be by income level, where we group all high-income families together in one group or stratum. Other strata might be middle-income and low-income families. The intent, then, is to take subsamples from each stratum or group and combine these subsamples to make up our overall sample. This procedure would assure that each income level was represented.

If we determined that 5 per cent of the homes in the area being sampled were in the expensive range, 25 per cent in the upper range, 40 per cent in the middle, and 30 per cent in the lower range, we would have different proportions in each strata. In this case we may use proportionate sampling by drawing from each group or stratum a quantity of items such that the total overall sample would also contain this actual proportional representation.

We had *proportional stratified sampling* when we took items from each stratum in proportion to the percentage that stratum took in the total population. Where 30 per cent of the families were in the low-income range, the final total sample would consist of 30 per cent from the low-income stratum. This is a good procedure when each stratum is homogeneous, in that it comprises similar items.

Suppose that we are collecting data about students in a university where there are both day and evening classes. Suppose also that we separate the students into 2 strata, day students (80 per cent) and evening students (20 per cent). The day students may be more alike than the night students in that the evening-student stratum would be made up from a wide variety of ages and backgrounds. Assume that our total population is 10,000 (with 2,000 night students) and our sample size is 200. We may decide that the resulting sample of 40 from the evening stratum is too small to reflect the wide variation within the night strata. In this case, we could use *disproportionate sampling*. We may select 100 students from each stratum. If we did this, we would then weight the 100 day students by a factor of 4 to reflect the fact that there are 4 times as many day students as evening students.

Cluster Sampling

Cluster sampling uses a completely different approach. An attempt is made to form groupings such that all clusters are as similar to each other as possible. Thus each cluster would reflect the diversity of the whole population. Assume that we are concerned with the opinions of people in midwestern towns in the population range of 15,000 to 25,000, and we are going to use sampling to poll these opinions. In using cluster sampling we would first attempt to obtain a random selection of towns in this size range. Assume that we are able to select 6 representative towns. The second step would be to obtain random samples from each of those 6 towns. These random samples would then be combined to make up our total overall sample. The purpose in following this cluster-sampling procedure was to be able to obtain a random sample without the necessity of working with all persons from all towns at one time.

As another example, suppose that we desire to take a sample from all service stations in a large city. We first divide the city into clusters by geographical area. Then a random sample would be taken from selected clusters, and these would be combined to form a total sample. A comparison between cluster sampling and stratified sampling will show the differences. In stratified sampling the objective was to make each stratum homogeneous within itself; that is, all items in a stratum should be somewhat similar. Cluster sampling is different in that our objective is to make each cluster contain all the typical nonhomogeneous elements, and at the same time to have each cluster similar to all other clusters. Having established each cluster, we then select clusters at random to obtain groups which each contain all the varieties of elements. Where we designated towns as clusters, each town would contain families with a wide variety of income levels. Cluster sampling might be thought of as a two-stage sampling process. The first stage is to select one or more sample clusters; the second stage is to take a sample from each of the selected clusters.

A reduction in costs of sampling is the principal advantage of cluster sampling. A second advantage of cluster sampling is that it permits us to sample without having a complete list of elements. In the case of the towns, we did not need to identify all families in all towns. Only those families in the selected town needed identification or listing so that the second stage could be completed. In the case of the service stations we never needed to identify all service stations in the whole city. We only needed to tabulate those in the specific area clusters selected. Sampling from a population that comprises a large geographical area is one of the important applications of cluster sampling.

Nonrandom Sampling

In nonrandom (or nonprobability) sampling we cannot determine the probability of selecting specific units within the population. Hence we do not have the statistical methods available to measure the risks associated with the sampling results. There are different types of nonrandom sampling and some varieties are dealt with briefly in the following paragraphs.

Quota sampling is the first type of nonrandom sampling. For example, interviewers may be told to interview a specified number of people but be allowed to determine how their samples are to be selected. One interviewer might then stand on a street corner and arbitrarily pick 10 people. This procedure is allowable in quota sampling. In another situation, company representatives may be asked to obtain opinions from 10 local businesspeople. Each representative could then obtain the sample in a way most suitable and convenient.

Quota sampling is used frequently in market surveys because it is less costly per unit to conduct than is random sampling. If it is carefully planned, it carries some of the advantages of stratified sampling. Errors in quota sampling often exist, of course, because each interviewer may work to his own convenience and may select persons who are not typical of the population.

Convenience sampling is used in cases where certain elements of the population can be reached more easily. A manufacturer may include a refund coupon in some packages of his product. The customer would fill in certain information and return the coupon to receive the refund. The manufacturer may be looking for customer preference information related to his product. He may ask buyers of his breakfast cereal whether they prefer large boxes or individual packages. The customers reached in this procedure would be a convenience sample. The coupon provides a convenient way to get customers to respond to the query. Many companies ask purchasers of home appliances to fill out special data on warranty cards when returned.

When an expert uses his judgment to select a sample, we have another type of sampling, called *judgment sampling*. This type of sampling may be less costly than random sampling and is often superior when used in pilot studies or small-scale surveys. If a university placement officer wants to estimate the starting salaries currently being paid to graduates of university business programs, he may call 5 to 10 businesses that he feels are representative. He will probably make a good selection in taking the sample; however, there is no way to measure statistically the reliability of the results. He relies only on his judgment to select the sample, hence "judgment sampling."

Judgment sampling is useful in pilot studies designed to pretest a questionnaire to be used in a planned, larger survey. The objective here is to evaluate possible unforeseen difficulties with the questionnaire. Sample validity is not very important in order to accomplish this objective. Judgment sampling is also used in compiling information for index numbers. For example, the U.S. Bureau of Labor Statistics uses a selected sample of goods and services for the Consumer Price Index.

Single, Double, and Sequential Sampling

The discussion so far has dealt with *single sampling*. In some cases it becomes more economical to sample in steps.

Double sampling takes place when we take an initial sample but then have the option of taking another sample. If a manufacturer receives a lot of 5,000 items from a supplier, he may take an initial sample of 50 items to inspect for quality. Based on this inspection, he may then decide to accept the lot, to reject it, or he may decide that he needs more information and so proceed to take a second sample.

Sequential sampling (or *multiple sampling*) is similar to double sampling, except that we can continue to take more samples. After each sample is inspected, a decision is made to accept, reject, or take another sample. The person doing the sampling will have certain rules as to when to discontinue the sequence of samples.

Exercises

6.1 Comment on the differences or similarities between:
(a) random and nonrandom sampling
(b) probability and random sampling
(c) quota sampling and stratified sampling
(d) proportional and disproportionate sampling in stratified sampling

6.2 You wish to obtain information on prescription prices in drugstores in a city of 600,000 population. You have selected 10 particular prescriptions to price. How would you proceed to carry out each of the following, and what are the advantages and disadvantages of each?
(a) cluster sampling
(b) random sampling
(c) stratified sampling
(d) judgment sampling
(e) census

6.3 In problem 6.2 assume you were told that "the errors in the survey are inversely proportional to the sample size." How would you respond to this statement?

6.4 *The YSU News* conducted a survey of some of its alumni to determine opinions of their college work. Questions were sent to 840 alumni who graduated between 1965 and 1976 and whose names began with C and D. What type of sampling would this be called? What are its advantages and disadvantages?

6.5 YSU has a total of 10,000 students. It is desired to take a random sample of 20 by use of the random-number tables in Appendix J. Where should you start in the tables? For purposes of consistency in this exercise, each student should start at the first number in the appendix, using 0097 for the first number in the random sample of 20. Select the complete sample of 20.

6.6 In problem 6.5, select the sample of 20 if the YSU student body were each of the following:
(a) 1,000 (first number is 97)
(b) 7,740 (first number is 97)
(c) 18,692 (first number is 10097)
(d) 845 (first number is 97)

6.7 A survey was conducted in a city to study parking needs. Based on maps and aerial photographs, it was estimated that there were 10,000 residential structures in the city and 600 city blocks.

Some of the residences housed more than one family. The following results were obtained by selecting a random sample of 12 blocks:

Block Number	Number of Automobiles in Block	Estimated Number of Residences
1	42	23
2	24	17
3	19	8
4	30	14
5	33	20
6	16	8
7	55	28
8	72	40
9	43	11
10	40	18
11	54	24
12	36	26

(a) Estimate the total number of automobiles in the city by first calculating the average number of autos per residence.
(b) How would this change if each block were considered a cluster?
(c) What other approach might be used to determine the total number of automobiles?

Sampling Distribution of the Mean

When we work with samples taken from populations, it is important to use the sample data properly. The mean (or average) calculated from a sample is often used to estimate the population mean, which is unknown. For example, suppose that we are testing the braking capability of a new tire-tread design by testing a sample of 50 tires. We collect the data from the results of the test of the 50 tires and would then make a judgment about all tires of that tread design. In this case the total population would include tires not yet produced. The sample is being used to estimate parameters of the population, and perhaps to decide whether or not to produce the new tread design.

Samples are also used to draw other inferences about the population from which the sample was selected. In other situations we may desire to make a judgment as to whether or not a sample could likely have come from a particular known population. As an example, a trucking company may periodically run mileage checks on a batch of tires received from the manufacturer to check if the tires are as good as those tested in the past. In this case the present sample is compared to tires evaluated in the past. Manufacturers frequently test samples from production to determine if batches of products are equivalent in quality to previous batches.

When we select different samples from a population, each time we select a new sample it obviously consists of different items than did past samples. If the items making up any particular sample are averaged, a mean value for that sample will result; however, any sample will likely have a mean value different from that of any other sample. It seems logical, however, that these sample means should group closely together since they have all been taken from the same population. Any extreme values that occur in a sample tend to be offset by other values in the same sample, especially if the sample is large. Our objective in this chapter will be to understand the relationship among the sample means and the original population from which the sample was selected so that we can better use the samples to make judgments about the population. These relationships are based on simple random sampling. We shall label the sample means $\bar{X}_1, \bar{X}_2, \bar{X}_3, \ldots$ and study the distribution of these values obtained from samples taken from a population. This distribution formed from all possible sample means ($\bar{X}$'s) will be called the *sampling distribution of the mean.* Suppose that in the case of the tires we took 10 samples of 5 tires each. For each sample we average the braking distance of the 5 tires to obtain the $\bar{X}$. These $\bar{X}$'s then form a sampling distribution of the mean. In establishing rules to study the sampling distribution of the mean, it will be necessary to systematically consider a number of different types of situations. The following factors determine the approach used in each situation as well as the formulas that apply:

1. Is the population normal or some other shape?
2. Is the population infinite or finite in size?
3. Do we replace items as the sample is selected?
4. What is the effect of the sample size itself? In accordance with statistical theory, we shall assume that a sample is small if it numbers less than 30 and large if it numbers 30 or more. (This consideration will be discussed in Chapter 7.)
5. Is the standard deviation of the population known? (In most cases it is not.)

In addition to these considerations, we shall develop special formulas to handle those cases in which percentages or proportions are involved.

Normal Population

The normal curve, as we have learned, is a good representation of the frequency distribution for the population data in many day-to-day business problems. It has been found that many of these data from business or industry are quite well represented by a normal curve.

Consider an illustration in the food-canning industry. A company has found from past evaluations that weights of one size of canned tomatoes tend to be normally distributed as in Figure 6-1, with a mean of 30 oz and a standard deviation of .5 oz. On a particular day an inspector takes a sample of 5 cans and weighs each can to obtain the following:

```
    sample A    30.64 oz
                29.28
                30.45
                30.77
                29.26
        Total  150.40
```

$$\bar{X}_A = \frac{150.40}{5} = 30.08$$

Later he takes two more samples, as follows:

```
    sample B   29.25 oz     sample C   30.08 oz
               30.20                   30.62
               31.21                   30.18
               30.02                   29.37
               29.12                   29.90
        Total 149.80 oz              150.15 oz

        X̄_B = 29.96 oz           X̄_C = 30.03 oz
```

If we took more samples and calculated the mean of each sample, we would find that the means of the samples would continue to vary somewhat from the mean of the whole population, which was 30.0 oz. These sample means, however, would tend to group more closely about the

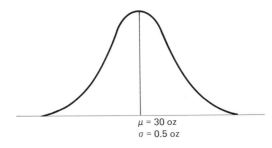

FIGURE 6-1. Normal Distribution of Weights of Tomato Cans

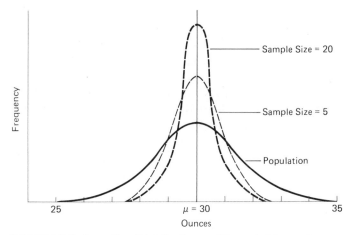

FIGURE 6-2. Sampling Distributions of Sample Means Compared to the Normal Population

population mean of 30 oz than would the individual can weights from the population. This is shown in Figure 6-2. The normal distribution of individual can weights is wider (greater dispersion or variance). It can further be shown that these sample means would also form a normal distribution of their own, and that the sampling distribution mean $\mu_{\bar{x}}$ would be the same as the mean of the total population μ (i.e., $\mu_{\bar{x}} = \mu$). The distribution of the sample means would, however, be narrower than the original population. If larger samples were taken, such as 20 (instead of 5), we would find that the means of these larger samples would cluster even closer to the population mean. This is logical since, with a larger sample, more items are being averaged and any extreme items would tend to average out. Three distributions are shown together in Figure 6-2 as an example of how the variance decreases as the sample size increases. Each of these is simply a normal distribution. We shall now show that if we take several samples of size n and determine the distribution of the sample means, this information can be used to estimate the population mean and standard deviation. We concluded that the sample means are normally distributed and have the same overall mean $\mu_{\bar{x}}$ as the original population μ. We now introduce a relationship between dispersions of these distributions given by the following formula:

$$\sigma_{\bar{x}} = \frac{\sigma}{\sqrt{n}}$$

where

σ = standard deviation of the population
n = sample size
$\sigma_{\bar{X}}$ = standard deviation of the sampling distribution of the mean

The $\sigma_{\bar{X}}$ is usually called the *standard error of the mean*. The standard errors of the mean can now be calculated for the various sample sizes in the tomato-canning example.

Sample of size 5: $\sigma_{\bar{X}} = \dfrac{\sigma}{\sqrt{n}} = \dfrac{.5}{\sqrt{5}} = .224$

Sample of size 20: $\sigma_{\bar{X}} = \dfrac{.5}{\sqrt{20}} = .112$

Given a constant population standard deviation σ, as the sample size increases, the smaller will be the standard error of the mean $\sigma_{\bar{X}}$, which indicates a narrower distribution and closer clustering of the sample means about the population mean. This is illustrated by the relationship in Figure 6-2. By the same logic, we can use this relationship to estimate the dispersion of an unknown population based upon samples. Where the sample size becomes larger, we can expect the sample mean to be a better estimate of the population mean because of this lesser dispersion.

As another example, consider a tire manufacturing company that has extensively tested and kept records of a certain type of tire. The mean mileage μ based on this recorded data is 44,600 with a standard deviation of 6,240 miles. If a sample of 10 tires were taken and the mileage measured, we would expect the mean to be part of the distribution where $\mu_{\bar{X}}$ is 44,600 and

$$\sigma_{\bar{X}} = \dfrac{\sigma}{\sqrt{n}} = \dfrac{6{,}240}{\sqrt{10}} = 1{,}975 \text{ miles}$$

If a sample of 100 tires were tested and the mean mileage measured, then we would expect the mean to be a part of the distribution where $\mu_{\bar{X}}$ is 44,600 and

$$\sigma_{\bar{X}} = \dfrac{\sigma}{\sqrt{n}} = \dfrac{6{,}240}{\sqrt{100}} = 624 \text{ miles}$$

Again, the larger sample has produced a narrower distribution and a resulting smaller standard error of the mean.

The examples discussed so far have assumed that the population standard deviation σ was known. Actually in most real situations σ is unknown; we can, however, use information from the sample (or samples) to estimate σ. The best procedure is to use the sample standard deviation s as an approximation to the population standard deviation σ. Then we would use s, calculated as

$$s_{\bar{X}} = \frac{s}{\sqrt{n}}$$

as an estimator of $\sigma_{\bar{X}}$. This $s_{\bar{X}}$ would then be utilized in the same way that we have used $\sigma_{\bar{X}}$.

Distinguishing Between s and $\sigma_{\bar{X}}$ and Other Measures

It is easy to confuse s with $\sigma_{\bar{X}}$, since both deal with samples. It is important, however, to recognize the difference between them.

We remember that s is a statistic and represents the standard deviation of a single particular sample. It is calculated from the actual values measured in the sample, as we studied in Chapter 4. The parameter $\sigma_{\bar{X}}$ is entirely different. If we take several samples, the means for these samples will form a distribution with a standard deviation of $\sigma_{\bar{X}}$. As noted previously, this $\sigma_{\bar{X}}$ is called the standard error of the sample mean, and is calculated from the population σ.

Another way of looking at $\sigma_{\bar{X}}$ is that it is a measure of sampling error. This error decreases as the sizes of samples taken increase. It can also be interpreted as a degree of precision with which the true population mean (μ) can be estimated from a sample mean $\bar{X}$.

We have dealt with a number of types of standard deviations up to now, and it seems worthwhile to summarize these at this point so as to avoid confusion in our later discussions.

σ: the population standard deviation. This is the standard deviation of the entire population.

$$\sigma = \sqrt{\frac{\Sigma (X - \mu)^2}{N}}$$

s: the standard deviation of a sample taken from a population. We used a correction factor ($n - 1$) in the computation of s as follows:

$$s = \sqrt{\frac{\Sigma (X - \bar{X})^2}{n - 1}}$$

$\sigma_{\bar{X}}$: the standard deviation of the sampling distribution of the mean — that is, the standard deviation of the sample means computed for all possible samples which could be selected from the population (also called the standard error of the mean). It varies with the sample size and is calculated by

$$\sigma_{\bar{X}} = \frac{\sigma}{\sqrt{n}}$$

$s_{\bar{X}}$: the estimator of $\sigma_{\bar{X}}$ obtained from samples:

$$s_{\bar{X}} = \frac{s}{\sqrt{n}}$$

These formulas have been based on a normal population, so we will now proceed to discuss populations that are not normal.

Population Nonnormal

In order to observe what takes place when we sample from a nonnormal population, assume the discrete rectangular distribution of Figure 6-3. In this example, each of the values 1, 2, 3, and 4 has an equal probability of being selected. This could be represented by 4 chips in a box numbered 1, 2, 3, and 4. Next assume that we shall take samples of 3 from the 4 items. There are 64 possible permutations of 3, or samples of size 3, assuming that we are sampling with replacement. If we selected samples and obtained the mean of each sample, the lowest possible mean we could obtain would be to draw the chip marked 1 three times in succession, giving:

$$\bar{X} = \frac{1 + 1 + 1}{3} = 1$$

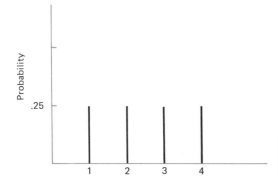

FIGURE 6-3. Rectangular Distribution Having Equal Probabilities for 1, 2, 3, or 4

FIGURE 6-4. Sampling Distribution of the Mean for Samples of Size 3 from Rectangular Population in Figure 6-3

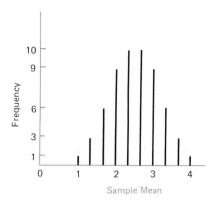

There is only one sample that has a mean of 1; this is plotted in Figure 6-4. There would be 3 cases in which the sum of the sample equals 4:

$$\begin{array}{l} 112 \\ 121 \\ 211 \end{array} \quad \text{where } \bar{X} = \frac{4}{3} = 1.33$$

There would be 6 cases in which the sum of the sample would be 5 and the mean would be 1.67, as follows:

$$\begin{array}{l} 122 \\ 212 \\ 221 \\ 113 \\ 131 \\ 311 \end{array} \quad \text{where } \bar{X} = \frac{5}{3} = 1.67$$

There would be 9 cases when the sum of the samples would be 6 and the mean would be 2, as follows:

$$\begin{array}{l} 123 \\ 132 \\ 213 \\ 231 \\ 312 \\ 321 \\ 114 \\ 141 \\ 411 \end{array} \quad \text{where } X = \frac{6}{3} = 2.0$$

If all the possible samples of 3 were compiled in this manner and the means tabulated, it would result in the distribution of Figure 6-4. It is apparent that although we started with a rectangular distribution in Figure

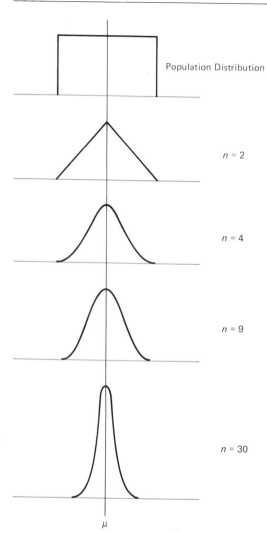

FIGURE 6-5. Sampling Distributions of the Mean for Various Size Samples from a Rectangular Distribution

6-3, the distribution of sample means took a form tending toward the normal curve shape. Actually, in taking samples from a rectangular distribution, the distribution of the sample means tends more and more toward a normal curve as the sample size increases, as shown in Figure 6-5.

We have established that when samples are taken from a normal population, the distribution of the sample means is also normal but with a narrower dispersion than the basic population. It is also true that when the original population is not normal, the sample means will still form a normal distribution as long as the sample size is adequately large (i.e.,

30 or more). This relationship is known as the *central limit theorem*, and is one of the important basic concepts of statistics. We can then state this as a theorem.

Central Limit Theorem: *Regardless of the shape of the population, as the sample size increases, the means of random samples taken from the population tend toward a normal distribution with mean and standard deviation given by*

$$\mu_{\bar{X}} = \mu$$

and

$$\sigma_{\bar{X}} = \frac{\sigma}{\sqrt{n}}$$

The central limit theorem is valid for all population distributions, including square, triangular, skewed, *U*-shaped distributions, and other forms. This theorem is very important because it permits us to make judgments about populations from samples without knowledge of the shape of the population distribution. It is important because in most real situations we do not know the shape of the population from which we are sampling.

EXAMPLE
A compilation was made of the medical expenses of all families in a medium-sized city. The census showed a skewed curve, as shown at the top of Figure 6-6, with a mean of $420 and a standard deviation of $100. What is the probability that when a sample of 64 families is taken, that the mean of the sample will differ from $420 by more than $20?

SOLUTION
The central limit theorem states that even though the basic population is skewed, we can assume that the sampling distribution of the mean for samples of size 64 drawn from the population will form a normal distribution. This distribution of sample means will have a mean of $420 also, and a standard error of the mean

$$\sigma_{\bar{X}} = \frac{\sigma}{\sqrt{n}} = \frac{100}{\sqrt{64}} = \$12.50$$

From the tables of areas of a normal curve, we could conclude that 68.26 per cent of the means of samples drawn from this population will lie between $420 ± $12.50. In the particular example, we ask for the

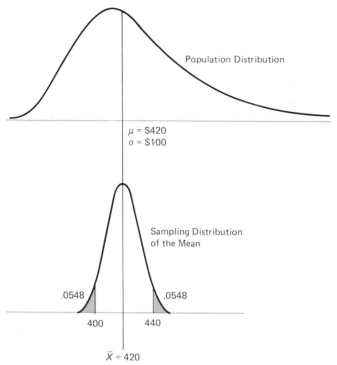

FIGURE 6-6. Distributions Corresponding to Example

probability that the mean of the sample will differ by $20 or more from the population mean, then

$$z = \frac{\bar{X} - \mu}{\sigma_{\bar{X}}} = \frac{20}{12.50} = 1.6$$

The required answer is then 2(.5000 − .4452), or 2(.0548), or .1096. This area is shown in Figure 6-6.

Exercises

> **6.8** The average purchase in a furniture store over the past 24 months has been $118 with a standard deviation of $24. If a sample of 36 sales were selected at random, what is the probability that the average will be $110 or less for the sample?

6.9 The average number of orders per day placed at a dry-cleaning establishment is 484 with a standard deviation of 32. During a sample week of 6 days, what is the probability that the average number of orders per day will exceed 500? Assume that the population is normal.

6.10 All students entering the graduate program at YSU are required to take an aptitude test. Over the past year the mean scores have been normally distributed with a mean score of 120 and a standard deviation of 16.
  (a) What is the probability that any one score will exceed 160?
  (b) A sample of 64 students take the test in one week. What is the probability that the mean will be less than 100?
  (c) What is the probability that the average of a sample of 100 students will fall between 119 and 121?

6.11 A census was taken of the ages of students in elementary schools in a large city. It was determined that the ages formed a rectangular distribution with a mean of 10 years. What can you say about the likely mean and shape of the sampling distribution of the mean for samples of size 100 drawn from the total population of elementary students? What if the sample size were 4 instead of 100?

6.12 Insulation on a type of wire has been shown to withstand an average of 2,800 volts with a standard deviation of 300 volts. A sample of 9 pieces are to be selected at random and tested. What is the probability that the mean of the sample of 9 falls below 2,530 volts?

6.13 A sample of size 81 is taken from a population having a mean $\mu = 160$ and a standard deviation $\sigma = 14$. Determine the probability that the average of the 81 items in the sample will fall between 157 and 164.

6.14 A distributor has been packing grass seed into packages, and has found that they average 2.04 lb with a standard deviation of .08 lb. A sample of 25 bags is taken. What is the probability that the average of the sample of 25 will fall below 2.00 lb?

6.15 A type of 100-watt light bulb has been found to have a mean life of 2,200 hours and is skewed to the right. Over the past year the manufacturer has taken 200 samples of 36 and found the standard error of the mean to be 18 hours. What is the probability that a particular sample of 81 will have an average life of fewer than 2,180 hours.

Small Populations: Sampling Distributions

In our problems so far it has been assumed that the total population is large in comparison to the sample size. When sampling from a large population, the withdrawal of items to make up the sample does not have a significant effect on the makeup of the population. The situation is different, however, if the total population is *finite* and we sample *without replacement*. In this case the items withdrawn as part of the sample make a significant impact on the population. If we sample *with replacement* from a finite population, the equations for $\sigma_{\bar{x}}$ are the same as for the infinite population.

To assist us in understanding sampling distributions in this special case, we shall assume a very small (or finite) population and that samples are to be taken from this small population. Consider 5 partners in a law firm whose ages are 26, 34, 44, 58, and 62. If a sample of 2 were selected from the total group, the outcomes in Table 6-1 are possible. Note that although the original population spread was from 26 to 62, the means of samples of 2 can vary only from 30 to 60.

Now consider Table 6-2, where samples of 3 and 4 are taken. The means of the samples of 3 range from 34.67 to 54.67, which is a narrower

TABLE 6-1. Means of All Possible Samples of 2 from a Population of 5 Law Partners

X_1	X_2	Sample Mean
26	34	30
26	44	35
34	44	39
26	58	42
26	62	44
34	58	46
34	62	48
44	58	51
44	62	53
58	62	60

TABLE 6-2. Means of Samples of 3 and 4 from Population of 5 Law Partners

Sample of Three				Sample of Four				
X_1	X_2	X_3	Sample Mean	X_1	X_2	X_3	X_4	Sample Mean
26	34	44	34.67	26	34	44	58	40.5
26	34	58	39.33	26	34	44	62	41.5
26	44	58	42.67	26	44	58	62	47.5
26	34	62	40.67	34	44	58	62	49.5
26	44	62	44.00					
34	44	58	45.33					
26	58	62	48.67					
34	44	62	46.67					
34	58	62	51.33					
44	58	62	54.67					

band than for the samples of two. When samples of 4 are taken, the range of the sample means is further narrowed to the range of 40.5 to 49.5. This finding is consistent with our earlier conclusions when we studied means of samples from normal populations. In that case we also found that the band of values for the sample mean decreased with increasing sample sizes. It turns out that in practical problems we can use the same approaches as we use for larger populations; however, where the population is small, $\sigma/\sqrt{n}$ is multiplied by a finite population correction factor. The standard error of the mean for a small population when sampling without replacement is expressed by

$$\sigma_{\bar{X}} = \frac{\sigma}{\sqrt{n}} \sqrt{\frac{N-n}{N-1}}$$

As before, if the population standard deviation is unknown, then the sample standard deviation s is used as an estimator of σ in the above equation.

Consider an example to show the impact of this correction factor upon $\sigma_{\bar{X}}$. Assume that town A has a population of 8,000 and that the mean income of the families is $14,000 with a standard deviation of $1,800. Town B has a population of 2,400 and the average income is also $14,000 with a standard deviation of $1,800. A sample of 400 is taken from each town.

$$\text{For town } A: \sigma_{\bar{X}} = \frac{\$1,800}{\sqrt{400}} \sqrt{\frac{8,000 - 400}{8,000 - 1}} = 90 \sqrt{\frac{7,400}{7,999}} = 86.56$$

For town B: $\sigma_{\bar{x}} = \dfrac{\$1{,}800}{\sqrt{400}} \sqrt{\dfrac{2{,}400 - 400}{2{,}400 - 1}} = 90\sqrt{\dfrac{2{,}000}{2{,}399}} = 82.18$

Although the population of town A is 3 times greater than the population of town B, the standard error of the mean is only about 8 per cent greater when calculated by the formula using the correction factor. When n/N becomes small, the correction factor approaches 1 and can be neglected. As a general rule, the correction factor can be omitted if n/N is less than or equal to .05.

Sampling Distributions of Proportions

Quite often it is necessary to make estimates of populations based on proportions. Examples might include a company's need to determine the proportion of employees living outside the city limits, the proportion of customers who are single, or the proportion of defective items produced by a process. The objective in each instance is similar to that of our previous problems—that is, to estimate the population characteristics from a sample. As before, we shall examine the sampling distributions of the proportions based on sample size and also in relation to the population from which the samples were taken.

The italicized p will be used to indicate the sample proportion and π the population proportion. In using proportions there are usually two characteristics, such as good or bad items in a lot or yes or no voters in a poll. The symbol p is the proportion in the sample having the desired characteristic, which can be either of the two as long as we are consistent. As another example, π could represent the proportion of acceptable items in a lot (population) of manufactured material. If there are more than two characteristics, π represents the proportion with the desired characteristic and $1 - \pi$, the proportion of all other characteristics. For example, assume that there are red, green, and yellow labels in a box. We are interested in the proportion of red labels, so this will be represented by π. The remaining colors are then represented by $1 - \pi$. The value of π is calculated by taking the number of items in the population with the desired characteristics and dividing by N, where N is the number in the population.

Just as in our prior discussions, the sampling distribution of the proportion has the same mean μ_p as the population mean π ($\mu_p = \pi$). If the population is infinite, the standard deviation of the sampling distribution is

$$\sigma_p = \sqrt{\frac{\pi(1-\pi)}{n}}$$

If the population π is not known, then we calculate:

$$s_p = \sqrt{\frac{p(1-p)}{n}}$$

and use s_p as the estimator of σ_p. If the population is finite (small relative to the sample selected) and sampling is performed without replacement, the same corrective factor is used as in the previous section. This gives the equation for the standard error of the proportion as

$$\sigma_p = \sqrt{\frac{\pi(1-\pi)}{n}} \cdot \sqrt{\frac{N-n}{N-1}}$$

where

π = the population proportion
N = the population size
n = the sample size

When the sample sizes are adequately large—that is, n is 30 or greater—the central limit theorem applies and the sampling distribution of the proportion can be approximated by a normal distribution.

Consider an example where 10 per cent of the assemblies received by a manufacturer from a supplier require adjustment. The population proportion of good items π is .90. Samples of 100 are taken. The mean of the sampling distribution $\mu_p = \pi = .90$ and the standard deviation of the sampling distribution

$$\sigma_p = \sqrt{\frac{\pi(1-\pi)}{n}} = \sqrt{\frac{.9(1-.9)}{100}} = \sqrt{.0009} = .03$$

Since the sample size is 30 or over, the sampling distribution of the mean is considered to be normal. From the normal curve data in Appendix A the probability of a sample proportion of good items in a sample being between .90 and .93 is .3413 since .93 is one standard deviation above the mean.

Take an example where the population is infinite and π is unknown. A sample is taken giving 12 defectives and 48 nondefectives. The sample proportion is calculated:

$$p = \frac{48}{60} = .8$$

The μ_p is estimated by P so that

$$\mu_p \cong p = .8$$

$$s_p = \sqrt{\frac{p(1-p)}{n}} = \sqrt{\frac{.8(.2)}{60}} = \sqrt{\frac{.16}{60}} = .052$$

This value of s_p is used as the estimator of σ_p.

Take another example where the population consists of 300 items and a sample of 40 is taken giving 36 good items and 4 defectives.

$$\mu_p \cong p = \frac{36}{40} = .9$$

$$s_p = \sqrt{\frac{p(1-p)}{n}} \sqrt{\frac{N-n}{N-1}} = \sqrt{\frac{.9(.1)}{40}} \sqrt{\frac{300-40}{300-1}} = .0474 \sqrt{\frac{260}{299}} = .0442$$

Then s_p is used as an estimator of σ_p.

Exercises

6.16 A professional association took a census of all its 8,462 members and determined that the mean number of years of experience was 14.6 years with a standard deviation of 3.8 years. They also took a random sample of 10 members from the membership and obtained the following years of experience after they had been arranged in order: 2, 5, 7, 8, 10, 14, 17, 19, 23, 32. Define or determine μ, $\bar{X}$, σ, s, and $\sigma_{\bar{X}}$ (for samples of 10) for the situation.

6.17 Calculate $\sigma_{\bar{X}}$ for the following where $N = 1800$.
   (a) $\mu = 24.9$, $\sigma = 1.9$, $n = 100$
   (b) $\mu = 448$, $\sigma = 30$, $n = 300$
   (c) $\mu = 10.4$, $\sigma = 3.2$, $n = 50$

6.18 The daily receipts from a retail store have a mean of $12,200 and a standard deviation of $1,400. Assume the receipts to be normally distributed. A sample of total receipts for each of 30 days is taken. What are the expected mean and standard error of the mean for the sample of 30? What if the population were not normal?

6.19 The weights of Oakest cereal packages are normally distributed with a mean of 25 oz and a standard deviation of .4 oz. A random sample of 25 packages is taken and the sample mean is 26 oz. What is the probability that a sample of 25 from the original population will give a mean of 26 oz or larger?

6.20 Consider again the retail store of problem 6.18 whose mean sales are $12,200 with a standard deviation of $1,400. If a sample of 49 days were taken, between what limits would the mean of a sample of 49 fall 99.7 per cent of the time?

6.21 A salesman receives magazine subscriptions by mail in response to a coupon inserted in a newspaper advertisement. Respondents can either send in $2, $4, $6, or $8. The salesman has found these to be of equal probability (.25) and therefore of a rectangular distribution.
  (a) Plot the population distribution and determine the mean.
  (b) Samples of size 2 are taken at random without replacement from a population consisting of 4 items numbered 2, 4, 6, 8 (i.e. we cannot have two 2s in the sample). Determine all the possible permutations and the probability of each value of the sample mean.
  (c) Plot the distribution of the sample means for the sample size of 2.
  (d) What distribution is obtained if samples of size 2 are drawn from (a)? Compare this to the result in part (c).

6.22 A chinaware company produced bowls that had a mean weight of 5.0 lb and a standard deviation of .20 lb. The weights of the bowls were normally distributed. A random sample of 100 bowls was selected to fill a customer's order.
  (a) What is the $\mu_{\bar{x}}$ and $\sigma_{\bar{x}}$ for the sampling distribution of the mean for samples of size 100?
  (b) What is the expected weight of the 100 items?
  (c) Between what weight could the mean weight of the sample be expected to fall 99.7 per cent of the time? Between what weight would the total shipment fall 99.7 per cent of the time?
  (d) If a single item is selected from the population, what is the probability that it will fall between the limits determined in (c)?

6.23 In problem 6.22, assume that the inventory of bowls was at 300 when the order of 100 was to be filled. Use the finite correction factor to determine $\sigma_{\bar{x}}$. What now is the range between which the mean of the bowls in the shipment of 100 will be 99.7 per cent of the time?

6.24 A population with a continuous distribution has a mean of 117 and a variance of 200, but the shape of the population is unknown. A sample of 50 is taken.

(a) What is the probability that the sample mean will be less than 120.2?
(b) What is the probability that the sample mean will be between 115.4 and 116.8?

6.25 A company produces insulators with a mean breakdown voltage of 2,000 and a variance of 125,000. An order of 5,000 is shipped to a customer who will take a sample of 50 and accept the shipment if the sample mean is between 1,900 and 2,120. What is the probability that the shipment will be accepted?

6.26 A company produces tires with a lifetime mileage that is normally distributed with a mean of 52,000 miles and a standard deviation of 2,000 miles. A taxi company has ordered 64 tires for test purposes and is sent a random sample of 64. The taxi company says that if the average of the 64 is greater than 51,400 it will purchase these tires. What is the probability that the purchase will be made?

6.27 In problem 6.26, what would be the probability of purchase if only 16 tires were tested using the same limits?

6.28 In problem 6.26, how would the result be changed if the 64 were selected from a special run of 320 tires?

6.29 In a large university, 20 per cent of the students are in the graduate program. Determine the shape, mean, and standard deviation for the sampling distribution of the proportion for all possible random samples of 100 students.

6.30 In problem 6.29, what per cent of all possible samples will fall in the range from 15 to 25 per cent?

6.31 In a certain town, 64 per cent of the households subscribe to the *Daily News*. A random sample of 100 households is selected.
(a) What is the probability that the sample will give a proportion of .60 subscribers or less?
(b) How would the probability change if the random sample were 400?

6.32 A retail store manager has determined that 60 per cent of his customers are female. If a random sample of 100 is taken, what is the probability that the proportion of males will fall between .32 and .45?

6.33 A lot of 400 clocks was packaged individually. Forty per cent of the clocks had red cases. A sample of 64 clocks were selected at random to fill a customers order. What is the probability that the customer received between 30 and 35 red clocks?

Learning Objectives

Statistical Inference

Estimation

Point Estimates
  Point Estimates of the Mean
  Point Estimates of the Proportion

Interval Estimates

Frequently Used Procedures for Interval Estimation
  Large Sample: Population σ Known
  Estimating μ
  Determining n
  Large Sample: Population σ Unknown
  Finite Population
  Interval Estimate: Proportion
  Interval Estimates of μ from Small Samples

Properties of Good Estimators
  Unbiased Estimator
  Efficiency
  Consistency

Selecting the Correct Equation

Hypothesis Testing

Basic Need for Hypothesis Testing

Steps in Testing a Hypothesis
  Rationale for the Decision

One-Sample Hypothesis Tests
  Population Normal with Standard Devlation Known
  Two-Tailed Tests
  Population Normal with Standard Deviation Unknown
  Population Distribution and Standard Deviation Both Unknown
  One-Sample Tests for Proportions

Two-Sample Hypothesis Tests
  Difference Between Two Means (Sample Size 30 or Larger)
  Difference Between Two Means Based on Small Samples ($n < 30$)
  Difference Between Two Proportions

Selecting the Correct Hypothesis-Test Equation

Risks and Errors in Hypothesis Tests
  Risks (or Errors)
  Operating Characteristic Curve

Design of the Hypothesis Test
  Significance Level
  Balancing α and β
  Setting the Critical Value: An Example
  Summary of Sample-Testing Procedure
  Variation in Critical Value
  Sample-Size Variation

Limitations in the Use of Hypothesis Tests

Estimation and Hypothesis Testing

Learning Objectives

We have pointed out several times that a complete census is either impossible or economically prohibitive in most real-life problems. The procedures to be developed now will provide a means of estimating the actual population parameters from sample data. We shall also learn how to evaluate statements that at least appear to be valid. The procedures will provide techniques for establishing the risks or likely errors that may be encountered in these estimation and hypothesis-testing procedures.

In any statistical procedure it is important to be able to design a sampling plan that considers trade-offs between cost of collecting greater amounts of data and the degree of error acceptable when smaller samples are used. One of the important objectives in this chapter is to learn how to design a sampling plan so that all the costs can be related to the expected degree of precision. It is important to consider the costs and potential risks before the experiment or survey is started so that the procedure can be carefully planned. The costs can be traded off against the accuracy or risks before starting to collect the data.

The primary learning objectives for this chapter are:

1. To learn how to estimate population parameters from sample data.
2. To be able to determine the validity of these estimates.
3. To learn how to design simple sampling plans to achieve objectives.
4. To understand how to evaluate data to determine if the data could have reasonably come from a given population.
5. To understand how to evaluate two or more sets of data to determine which of two methods is better or worse.
6. To understand the risks related to decision making based on sample data and test results.

Statistical Inference

Many decisions must be made under conditions where complete information is not readily available. In some situations the complete information does not exist; in other cases, it may be too costly to obtain all the data. The process of drawing conclusions from available data when we lack complete information is known as *statistical inference*. In this chapter we deal with two means of making statistical inferences—estimation and hypothesis testing.

Estimation from sample data is an important phase of statistics.

Here we desire to estimate population values (such as the mean or standard deviation) based on sample data.

In *hypothesis testing,* we deal with statements that we have reason to believe are true. We then test these statements based on the limited evidence at hand. Both estimation and hypothesis testing deal with making *inferences* about an unknown population based on limited data. After making the inference, we then need to know how accurate the inference is. Therefore, our study will also include the evaluation of errors or risks associated with an inference.

Estimation

Estimation is the first area that we shall study under the general topic of statistical inference. Our discussion will center primarily on the use of sample information to estimate the mean and the dispersion of a population.

In many business or managerial situations we merely have a sample from the population. Seldom do we have the complete information desired. A sample is referred to as incomplete data because it is only a small part of the total population. We wish to use this information from the available sample to make estimates about the overall population from which the sample came. We are making an inference about the unknown population using sample data.

Estimation is important in business, government, and many scientific endeavors. Anyone can make an estimate; however, the accuracy of an estimate is of crucial importance. As an example, accurate estimates or forecasts of sales for a proposed new product are vital to the business. Estimates of the effectiveness of new medical products are vital to the company considering manufacture. The difference between good or poor estimates may mean the difference between profit or loss for a company, its survival or failure, and even between life and death for the product users.

The first portion of this chapter will discuss ways of making the most reliable estimates from sample data. In other words, with a limited amount of available data, we wish to make our estimate as accurate as possible. In other instances, we may be able to pay to obtain more data but still desire to get the best information from a set amount of data.

In attempting to understand the nature of a population distribution, one of the most important measures is the mean of the population (μ).

A population can be described quite well by its mean and standard deviation along with additional information on the shape of the distribution. We will not know the value of the true mean (μ) since complete information on the population is not available. Therefore, it becomes necessary to estimate the population mean μ based on the sample data that are available or that we can obtain.

We previously determined that the sample mean was an estimate of the population mean, where the symbol $\cong$ means approximately equal to,

$$\mu \cong \bar{X}$$

We also determined previously that the population variance could be estimated by the sample variance

$$\sigma^2 \cong s^2$$

where

$$s^2 = \frac{\Sigma(X - \bar{X})^2}{n - 1} \quad \text{for a set of sample data}$$

For our purposes, we will be satisfied with the estimate of σ from s; however, we shall investigate further the preciseness with which $\bar{X}$ approximates μ. This estimate of the population mean μ can be expressed in either of two ways: (1) as a point estimate, or (2) as an interval estimate.

Point Estimates

The simplest form of an estimate of a population mean is a *point estimate*. As an example, assume that we are interested in a particular new model of automobile and the miles per gallon (mpg) that it achieves for highway driving. Suppose that we took a sample of 40 cars of that model and obtained an average of 21.6 mpg. This would be a point estimate. The results of our tests may have varied from a low of 18.8 on one of the 40 cars in the sample up to a high of 27.6 on another of the cars. If we picked one more car at random and ran a test, we would not expect to obtain our exact mean value of 21.6. If we wished to use a single point as our best guess, however, the 21.6 mpg would be this best estimate. It is the best single value that we have available to use as our estimate.

In this example, the sample mean $\bar{X}$ has been used as an estimator of the population mean μ. The 21.6 mpg was our point estimate of the population mean.

Point Estimates of the Mean

When we make a statement such as "our best estimate of the population mean is 2.6," we are making a point estimate of the population mean. We could from the above data make the statement that our car gives an average of 21.6 mpg. From previous discussions of the sampling and sampling errors, it was shown that a larger sample will normally provide a better estimate. We also learned that the expected mean ($\bar{X}$) of a random sample equals the population mean (μ), although individual sample means ($\bar{X}$'s) will vary from this expected value. This statement was shown to be true regardless of the shape of the population. Therefore, we can say that the sample mean ($\bar{X}$) is an *unbiased estimator* of the population mean (μ). The median and the mode of the sample are certainly valuable in estimating the shape of the population distribution; however, we could not normally expect them to coincide with the population mean, especially if the population were not normal. Another important factor to consider is that we have available a formula $\bar{X} = (\Sigma X)/n$, which we can use to calculate the sample mean $\bar{X}$ and use it as an estimate of μ. In contrast, we do not have any procedure for estimating μ from the sample median or mode.

We conclude, therefore, that $\bar{X}$ is the best estimator of the population mean μ; at the same time, however, it is necessary to recognize that error will exist when we approximate μ from an individual sample $\bar{X}$.

Point Estimates of the Proportion

The procedure for estimating a population proportion (π) from a sample is similar to the method for estimating a population mean. Suppose that we desire to estimate the proportion of students at YSU who drive to school. A sample of 80 is taken and the poll shows that 16 of these drive to school, so that the sample $p = .20$. Our best estimate is then that 20 per cent of the total student population drive to school.

Interval Estimates

Any point estimate of the mean will almost certainly be different from the true population mean. This will be true even though we take careful steps to reduce possible errors and make the estimate as accurate as possible. This then brings us to the use of an *interval,* rather than a single point, to estimate the population mean μ. The use of an interval will permit us to recognize and evaluate the uncertainty and lack of precision associated with any estimate of the population mean.

The statement "the true population mean lies between 3.62 and 3.68" makes an estimate of the population mean by stating an interval in which this mean is expected to lie. This interval could also be stated in the form

$$3.62 < \mu < 3.68$$

This notation indicates that μ lies between the two specified values. As another way of expressing this interval, we could have made use of the point estimate 3.65 and stated that "the true population mean lies in the interval $3.65 \pm .03$." This statement also introduces the interval as an estimate to take into account the possible error and uncertainty present.

The use of an interval has several advantages over the use of a single point. First, the interval emphasizes that there is some uncertainty as to the exact position of the true population mean. Second, it provides a specific range over which we expect the true population μ to fall. Third, the interval will permit us to define a confidence in our statement, expressed as a probability that the true mean actually lies in the specified interval. If we use a wider interval, naturally we can have greater confidence in our statement, but the estimate is less precise. A point estimate could be thought of as an extremely narrow interval, and we would have a low confidence that the true population mean was exactly at that point.

The estimate of an interval is illustrated in Figure 7-1 using a normal

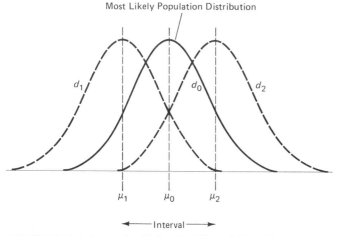

FIGURE 7-1. Interval for Estimated Population Mean

distribution. If a point estimate of the mean were used, μ_0 would be the best estimate and the associated population distribution is portrayed as d_0. The use of the interval estimate indicates that the population mean could be as low as μ_1 or as high as μ_2, and the associated population distribution would lie somewhere between d_1 and d_2 as illustrated.

We have seen that if we use a point estimate for making a decision as to a population mean, it is very unlikely that we will be *exactly* correct. The larger the sample we take, however, the more precise we are likely to be. If we could tabulate data and actually calculate the mean of the population, we would expect it to be within the range of the value estimated by a sample. The method for determining this range will be covered in this chapter. In the case of the automobile mileage example, we might be quite confident that the true mean would lie in the interval 21.6 ± 4.0, or the range of 17.6 to 25.6. We would be less confident, however, that the true mean would lie in a shorter interval of 21.6 ± 2.0. Thus, for a given set of data, we can place less confidence in our statement as we narrow down the interval on either side of our estimate. In other words, as the interval becomes wider, the probability increases that the true mean will fall within the interval. Therefore, the use of the interval estimate permits us to draw a conclusion that a true population mean lies in a defined interval, and to further state a probability or confidence level in our statement. This type of statement would be more useful as a basis for a management decision than would a mere point estimate. It clearly points out to the user of the information that there is a set degree of uncertainty or risk connected with the use of the estimated value. This range of values in which the actual mean might lie is the *confidence interval*. With each interval we then associate a *confidence coefficient*. In the problem above we might have stated that "with 95 per cent confidence, the population mean lies in the range 17.6 to 25.6." Typical confidence coefficients used in statistics for decision making are 90, 95, or 99 per cent. The greater the confidence desired, for a specific analysis, the wider the interval. Our objective, of course, is to try to obtain a narrow interval, as well as a high degree of confidence. How well we can achieve this objective will now be examined.

Frequently Used Procedures for Interval Estimation

When trying to determine information about an unknown population, we might desire to estimate the population mean and standard deviation, or sometimes the proportion of items in a population having particular characteristics. The correct formula to use for making an estimate of a population parameter will vary depending on a number of factors.

Some of these factors are summarized as follows:

1. Is the population standard deviation σ known or unknown?
2. Is the shape of the population distribution known? If known, is it normal or some other shape?
3. Is the sample large (30 or greater) or small (less than 30)?
4. Is the population large (assumed infinite) or finite?

The procedures to be introduced will apply to different situations, depending on whether or not certain information is known, or what the actual known information indicates. Since the formulas are different for each of these cases, we shall systematically consider each situation so that we will have a procedure for covering the various combinations. The formulas will be used to estimate an interval for the population mean in each of these various situations. Exhibit 7-1, which follows our discussion of estimation, summarizes the formulas in a decision tree for use in evaluating problems and selecting the correct formula. As each alternative is studied, its position in Exhibit 7-1 should be reviewed.

Large Sample: Population σ Known

The first procedure to be developed will cover the situation in which a large sample (30 or more) is to be used to estimate a population mean and where the standard deviation of the population is already known. We can recall that when samples of size n are drawn from a normal population having a mean μ and a standard deviation σ, the means of these samples form a sampling distribution that is normal with

$$\mu_{\bar{X}} = \mu$$

and

$$\sigma_{\bar{X}} = \frac{\sigma}{\sqrt{n}}$$

We expect these sample means to fall in the distribution shown in Figure 7-2c. The original population distribution and the distribution of a typical sample are also illustrated in Figure 7-2. If we took many samples and calculated the mean of each, these means would form the distribution of sample means. Since the distribution of sample means was earlier shown to be normal, we can state further that 68.26 per cent of the time the

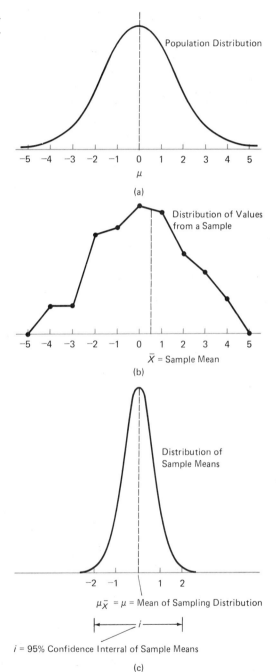

FIGURE 7-2. Three Distributions Involved in Estimating the Mean

population mean will fall in the interval of $\bar{X} \pm \sigma_{\bar{X}}$. We further learned that for large samples the sample means formed a normal distribution regardless of the shape of the parent population. Therefore, the method developed here will not depend on the shape of the original population.

Estimating μ

As a practical example, consider a machine used to fill cans in a tomato canning factory. From past experience it is known that weights of cans filled by the machine are normally distributed with a standard deviation of 0.4 oz. During maintenance, the mean setting position of the machine has required adjustment and the quality engineer desires to reset it. He takes a sample of 64 cans and finds

$$\bar{X} = 24.20 \text{ oz}$$

What can he now say about the cans being filled? He can use the known population standard deviation to estimate $\sigma_{\bar{X}}$. Since

$$\sigma_{\bar{X}} = \frac{\sigma}{\sqrt{n}} = \frac{.4}{\sqrt{64}} = .05$$

he can state that there is a .6826 probability that the population mean falls at a point between $24.20 \pm .05$ oz, which can also be stated as "from 24.15 to 24.25 oz." The formula is then

$$\text{interval}(\mu) = \bar{X} \pm z \frac{\sigma}{\sqrt{n}} = \bar{X} \pm z\sigma_{\bar{X}} \qquad (7.1)$$

If he wanted to have 95 per cent confidence instead of the 68.26 per cent, a wider interval would be formed. From the normal curve table, 95 per cent represents ± 1.96 standard deviations. He could then state with 95 per cent confidence that the population mean lies in the interval

$$\text{interval}(\mu) = 24.20 \pm (1.96 \times .05)$$
$$= 24.20 \pm .098$$

or, stated in another way,

$$24.102 \leq \mu \leq 24.298$$

This procedure has allowed us to establish an interval and to establish an associated confidence level for our statement. The confidence level represents a probability that our statement is true. Since a number of confidence levels could have been chosen, the question arises as to what confidence level should be used. Our choice is somewhat arbitrary; however, our primary objective is to present our results so that they are understandable. Although it is standard practice to use 99, 95, or 90 per cent, other values can be used. The procedure developed so far applies to any shape distribution if the population standard deviation is known and the sample is large (30 or more). If the population is normal and σ is known, the procedure applies to any size sample.

Determining n

As another example, suppose that we are attempting to determine the sample size needed to provide a certain precision in the result. We would expect a larger sample to give us greater precision, but the larger the sample, the greater the cost of collecting the data. Assume that we desire to estimate the average annual household earnings in St. Louis and that we are willing to have an error (e) of $50 at the 95 per cent confidence level. We know from prior investigations that $\sigma = \$600$. What size sample do we need to take?

$$z\sigma_{\bar{x}} = e$$

and

$$\sigma_{\bar{x}} = \frac{\sigma}{\sqrt{n}}$$

Substituting the second equation into the first, we have

$$z \cdot \frac{\sigma}{\sqrt{n}} = e$$
$$\sqrt{n} = z \cdot \frac{\sigma}{e}$$

(7.2)

At the 95 per cent confidence level $z = 1.96$. Therefore,

$$\sqrt{n} = 1.96 \cdot \frac{600}{50} = 23.52$$
$$n = 553.19$$

Thus, we would need to take a sample of 554 in order to obtain the estimate of the mean annual earnings and to allow a possible error of $50 at the 95 per cent confidence level.

Large Sample: Population σ Unknown

The previous procedure assumed knowledge of the population standard deviation. In most situations, however, the population standard deviation σ is usually not known. When this case is encountered, we must use the sample standard deviation s, as calculated from our sample, as an approximation for σ. The standard error of $\bar{X}$ is estimated by

$$s_{\bar{X}} = \frac{s}{\sqrt{n}}$$

The value $s_{\bar{X}}$ is used in place of $\sigma_{\bar{X}}$.

The formula developed in the last section now takes the following forms:

$$\text{interval } (\mu) = \bar{X} \pm z \frac{s}{\sqrt{n}} \qquad (7.3)$$

or

$$\bar{X} - z \frac{s}{\sqrt{n}} \leq \mu \leq \bar{X} + z \frac{s}{\sqrt{n}}$$

As an example where this formula can be applied, suppose that a radio station wished to estimate the average weekday listening time by families in that city. A sample of 81 families was taken at random. From the sample results, the mean $\bar{X}$ was 2.40 hours and the standard deviation s was calculated as .7 hours. Using the above formula, and assuming that a confidence level of .90 was desired,

$$\begin{aligned}
\text{interval } (\mu) &= \bar{X} \pm z \frac{s}{\sqrt{n}} \\
&= 2.40 \pm 1.645 \frac{.7}{\sqrt{81}} \\
&= 2.40 \pm .128
\end{aligned}$$

or, expressed in a different form,

$$2.272 \leq \mu \leq 2.528$$

This procedure allows us to calculate an interval estimate of the population mean with an associated confidence level in cases when the population standard deviation is unknown and the sample is large (30 or over).

Finite Population

The procedures developed so far have assumed a large (essentially infinite) population. If the population is finite and the sample size n is 5 per cent or more of the population N, a finite population correction factor must be used such that our formula is now

$$\text{interval } (\mu) = \bar{X} \pm z \frac{s}{\sqrt{n}} \sqrt{\frac{N-n}{N-1}} \qquad (7.4)$$

where

N = the size of the finite population
n = the sample size
s = the sample standard deviation (if σ were known, it would be used instead of s)

In our previous example, assume that the poll was being taken in a small town of 900 families. Then the confidence interval is

$$\text{interval } (\mu) = \bar{X} \pm z \frac{s}{\sqrt{n}} \sqrt{\frac{N-n}{N-1}} = 2.40 \pm 1.645 \frac{.7}{\sqrt{81}} \sqrt{\frac{900-81}{900-1}}$$
$$= 2.40 \pm .128(.954)$$
$$= 2.40 \pm .122$$

This can also be expressed as an interval estimate of μ by

$$2.278 \leq \mu \leq 2.522$$

It is observed that the finite population correction always reduces the estimate of μ to an interval more narrow than if the correction factor were not used.

Interval Estimate: Proportion

Often it is desired to make an estimate of a population proportion based on a sample, just as we have seen the need to estimate a population mean from a sample. For example, we may desire to estimate the proportion of usable items in a lot of material, or we might want to determine the proportion of employees in a company who live in the city limits. In either of these situations, an interval estimate of the proportion is to be made based on data collected in a sample.

We learned earlier that if a population of size N has a proportion π not close to either zero or one [i.e., $N\pi$ and $N(1-\pi)$ both must exceed 5], then the normal approximation can be used to calculate μ and σ as follows:

$$\mu = N\pi$$
$$\sigma = \sqrt{N\pi(1-\pi)}$$

If $N\pi$ or $N(1-\pi)$ are less than 5, the binomial table must be used.

If we take a sample of size n and compute the proportion p in the sample, then this p is the best estimate of the population proportion and also the distribution of the sample means. The estimates of the mean and standard deviation (standard error) for the distribution of sample proportions are

$$\mu_p = p$$

$$\sigma_p = \sqrt{\frac{p(1-p)}{n}} = \sqrt{\frac{pq}{n}}$$

The central limit theorem holds for the means of samples from the binomial distribution; therefore, the distribution of the proportion for samples can be considered normal.

As an example of an application of this concept, suppose that a newspaper is attempting to determine reader preference. The paper polls 1,600 of its readers. In the sample, 960 respondents prefer that the Saturday edition of the paper be delivered in the morning and 640 prefer the evening. The paper wants to set up a 95 per cent confidence interval to estimate the true population proportion.

The sample proportion can be easily calculated as $p = 960/1{,}600 = .60$. The sampling distribution for p can be approximated with a normal distribution, since p is not close to either 0 to 1.0.

$$\text{interval } (\mu_p) = p \pm zs_p \quad \text{or} \quad p \pm z\sqrt{\frac{p(1-p)}{n}} \quad (7.5)$$

$$= .60 \pm 1.96 \sqrt{\frac{.60 \times .40}{1{,}600}}$$
$$= .60 \pm 1.96 \sqrt{.00015}$$
$$= .60 \pm 1.96(.01225)$$
$$= .60 \pm .024$$

or, expressed in a different way,

$$.576 \leq \mu_p \leq .624$$

or

$$.576 \leq \pi \leq .624$$

We can then conclude that with a confidence coefficient of .95, the population proportion will lie in the interval .576 to .624.

If the population is finite and the sample is selected without replacement, we apply the same correction factor as we used for the continuous variable in Eq. 7.4. This would result in

$$\text{interval } (\pi) = p \pm z \sqrt{\frac{p(1-p)}{n}} \sqrt{\frac{N-n}{N-1}} \qquad (7.6)$$

We are thus able to estimate the interval in which a population proportion lies based on the sample.

Interval Estimates of μ from Small Samples

Frequently we need to estimate an interval for a population mean when the sample is small (less than 30). In some cases only a few data are available; in other cases the cost of taking a bigger sample may be excessive; in still other instances time may not permit collecting the larger sample. Destructive tests to evaluate manufacturing processes, or life tests on expensive equipment are typical instances in which dollar costs are the limiting constraint. In other cases it might be impossible to obtain a bigger sample. If we were evaluating the ability of a drug to control an unusual illness, there might be very few people who have the disease. For these reasons it is often necessary to draw conclusions with the limited data available.

When the population standard deviation σ is unknown and the sample is small (less than 30), inherent inaccuracies do not permit us to use the normal curve to represent the sampling statistic. For small samples, a new distribution is used instead of the normal curve. It is called the *t distribution* or *Student's t distribution*. This distribution was first defined by W. S. Gosset, a statistician who published his work under the name "Student." In Figure 7-3 it can be seen that the *t* distribution with 5 degrees of freedom is not extremely different from the normal distribution. The shape of the *t* distribution varies with the number of degrees of freedom (df) which in turn depends on the sample size n. We define degrees of freedom in this case as equal to one less than the sample size, or

$$df = n - 1$$

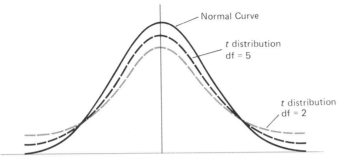

FIGURE 7.3. Relative Shape of t Distribution to Normal Distribution for df = 2 and df = 5

Essentially, the t for the t distribution is used in the same way as the z for the normal distribution. The values are given in Appendix B. Our formula for the confidence interval of the population mean μ then becomes

$$\text{interval } (\mu) = \bar{X} \pm t \frac{s}{\sqrt{n}} \qquad (7.8)$$

or, expressed in a different form,

$$\bar{X} - t \frac{s}{\sqrt{n}} \leq \mu \leq \bar{X} + t \frac{s}{\sqrt{n}}$$

The value for t is selected from Appendix B based on the number of degrees of freedom (df $= n - 1$) and the confidence level that is selected.

Consider an example where a salesperson kept records of expenses for meals and lodging for 9 days as follows:

$66	$63	$68
71	42	49
59	55	59

From these data, we can determine the sample mean $\bar{X}$ and standard deviation s:

$$\bar{X} = \$59.11 \qquad s = \$9.33$$

It is desired to estimate the population mean cost from the sample data. We construct a 95 per cent confidence interval for the average daily expenses. For 8 df (df $= n - 1 = 9 - 1 = 8$), the value of $t_{.025}$ is 2.31.

$$\text{interval } (\mu) = \bar{X} \pm t \frac{s}{\sqrt{n}}$$

$$= 59.11 \pm 2.31 \frac{9.33}{\sqrt{9}}$$

$$= 59.11 \pm 7.18$$

or

$$51.93 \leq \mu \leq 66.29$$

This procedure provides us a method for estimating the population mean μ from a small sample when the population is normal and the population standard deviation is unknown.

Properties of Good Estimators

There are three properties which a good *estimator* should possess: (1) unbiasedness, (2) efficiency, and (3) consistency. These factors will be discussed here only briefly, since a more involved exploration is reserved for a more advanced statistics text.

Unbiased Estimator

We learned previously that as we took a sample and calculated $\bar{X}$, then the $\bar{X}$ is a best estimate of the population mean. We can also say that it is an *unbiased estimator* since the mean of all possible sample means computed from random samples is equal to the population mean μ. The term *bias* as used here is not the same as the term "bias" used when we talked about bias in selection of samples. An unbiased estimator is a statistic having an expected value equal to the population value being estimated.

In contrast, the formula

$$s^2 = \frac{\Sigma (X_i - \bar{X})^2}{n}$$

is *not* an unbiased estimator of σ^2, the population variance. We learned earlier that in order to obtain an unbiased estimator of s^2, we use $n - 1$ rather than n in the denominator.

Efficiency

One estimator is said to be more efficient than another if the variance (or standard error) is less for the same sample size. As an example, let us compare the sample mean to the sample median in terms of the efficiency of each as an estimator of the population mean. We have shown previously that the population σ is related to the standard error $\sigma_{\bar{x}}$ as follows:

$$\sigma_x = \frac{\sigma}{\sqrt{n}}$$

Although we shall not deal with the standard error of the sample median σ_m, it can be shown for one case that

$$\sigma_m = 1.253 \frac{\sigma}{\sqrt{n}}$$

Thus the standard error of the sample median is greater by a factor of 1.253. We can conclude then that the sample mean is a more efficient estimator of the population mean μ than is the sample median. Stating this conclusion in another way, we can say that when the sample mean is used as the estimator, we shall likely obtain an estimate closer to the true population mean. Therefore, we can say that the sample mean is more *efficient* than the sample median as an estimator of the population mean.

Consistency

We define a *consistent estimator* as one that, as the sample size increases, comes closer to the population parameter being estimated. From the previous analysis,

$$\sigma_{\bar{X}}^2 = \frac{\sigma^2}{n}$$

Therefore, as the sample size n increases, we can expect $\sigma_{\bar{x}}$ to become smaller. This means that the variability in $\bar{X}$ becomes smaller as the sample size increases, so that we can define $\bar{X}$ as a consistent estimator of μ.

Selecting the Correct Equation

Several equations have been presented in this chapter for estimating a confidence interval. Various factors must be considered before selecting the equation for a particular problem. Exhibit 7-1 shows a tree diagram

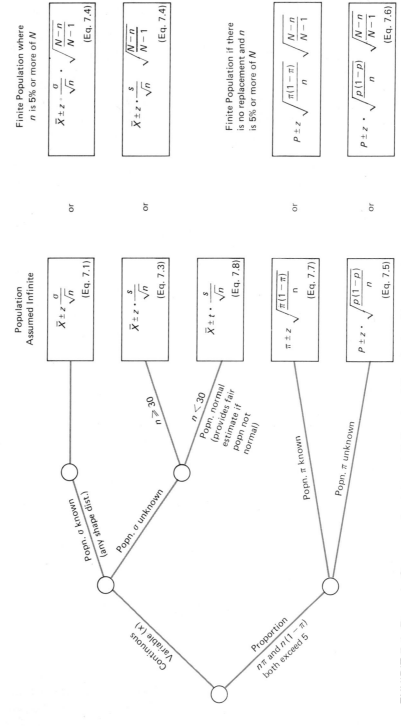

EXHIBIT 7-1. Formula Summary for Interval Estimate of Population Mean or Proportion

that can be used to make the correct equation selection. By starting at the left, and selecting each appropriate branch, the student will be able to select the right equation to fit each problem.

Exercises

7.1 A manufacturer applies a plastic coating to a metal surface. An inspector takes 36 separate readings of the thickness and the mean is 20.22 mm. He knows from past experience that the process applies a normally distributed coating with a standard deviation of 5.4 mm. Determine the .95 confidence interval for the true thickness in the process.

7.2 Account balances at a department store are known from past records to have a slightly positive skewness. The standard deviation has stayed at about $8.00 rather consistently over the past few years. A random sample of 64 accounts has resulted in a current mean balance of $64.50.
 (a) What is the 99 per cent confidence interval estimate of the mean value for all accounts?
 (b) What if the confidence were 95 per cent?
 (c) What sample size is needed at the 95 per cent confidence level to obtain an estimate within $\pm$ 1.00?

7.3 A newspaper owner desires to know the average family income of readers. He collects a random sample of 100 readers, which shows an average of $14,200 and a sample standard deviation of $2,650.
 (a) What can the owner say with a confidence of 90 per cent?
 (b) What if the confidence were 95 per cent?

7.4 In problem 7.3, if the newspaper owner desired to estimate the mean income within $150 with a 95 per cent confidence level, how large of a sample must he take? What assumptions is he making?

7.5 A sample of 400 items from a production process is inspected and it contains 24 rejects.
 (a) If lots of 400 are selected, what is the best estimate of the number of rejects per lot?
 (b) What would be the mean and standard error for the distribution of the proportion of rejects in samples of 400?

7.6 In problem 7.5, make a 95 per cent confidence interval estimate of the number of rejects in lots of 400.

7.7 A sample of 400 light bulbs are tested giving a sample mean life of 380 hours and a standard deviation of 60 hours.
  (a) What is the 95 per cent confidence interval for the population mean?
  (b) If it was desired to determine the sample mean within an interval of ± 3 hours at 95 per cent confidence, what sample size is needed, assuming that the 60 hours represented the population standard deviation?

7.8 A company has manufactured a total of 300 of a new type of signal lamp. A sample of 36 is tested to give a mean life of 3,100 hours with a sample standard deviation of 200 hours.
  (a) Determine a 90 per cent confidence interval for the mean life of the population of 300.
  (b) What would be the interval if the population were infinite?
  (c) What would be the interval if the population were 400?

7.9 The weights of a sample of 6 bags of sand filled by a machine were checked giving 103, 104, 106, 107, 106, and 110 lb. Determine the 99 per cent confidence interval for the mean weight filled by the machine.

7.10 A process standard deviation is known to be $\sigma = 100$. It is desired to estimate the population mean from a sample with an error less than 5.0 and a risk of .01 (.99 confidence level).
  (a) Determine the necessary sample size.
  (b) If the sample had to be limited to 400 due to costs, what confidence level would be provided for the 5.0 allowable error?

7.11 A sample is to be used to estimate the actual proportion in a large population. It is felt that π is about .2 and an error of .05 is acceptable with a risk of .05. What size of sample is needed?

7.12 A manufacturing company desired to estimate the mean weight of copper used in a certain design circuit breaker to a 99.5 per cent confidence level. To obtain this low risk (.005) a sample of 10,000 was taken and an $\bar{X}$ of 148.32 oz was obtained with a sample standard deviation of 4.5. Calculate the interval for the estimate of the population mean.

7.13 An engineer had a sample of 15 bolts made from a new type of alloy. Strength tests gave a mean sample pull strength of 3.45 tons with s equal to 2.1. Determine a 90 per cent confidence interval for the population mean. If the company needed to be 99 per cent confident that the strength was at least 3.0 tons, what sample size is needed?

7.14 It is known from past usage that the standard deviation of the life of a signal lamp is 100 hours. A new type of lamp is being developed and there is reason to believe that this standard deviation will be the same. A sample of size 100 is taken and the sample mean is 2,024 hours. Estimate the mean life for confidence coefficients of 95 and 99 per cent.

7.15 A farm contains 240 acres. A sample of 30 acres shows an average yield of 96 bushels with a variance of 4,000. Use a 95 per cent confidence interval to estimate the total bushels yield on the firm.

Hypothesis Testing

In estimation, we inferred or estimated the population mean or variance based on data from a sample. In hypothesis testing, we usually make a comparison in order to arrive at a decision. For example, a company that manufactures tires wishes to compare the performance of a modified tread design with the performance of regular tires. Our approach is to collect data on regular tires and compare them to data taken from the modified tires. A hypothesis test is then used to evaluate the data and determine if there is really a difference between the new and the old designs.

In this testing process the performance data of the regular tires formed a population. We have information on the regular tires so that the mean μ and the standard deviation σ of the original population are known. We have shown that a statistic (such as $\bar{X}$) for the sample from a population will be equal to the equivalent population parameter (such as μ). Our hypothesis test, then, is a check to see if there is a resemblance between the sample of new tread-design tires and the regular tire population. If they do not resemble each other, we shall conclude that the new tread design is different (better or worse) than the original. We are making a test to determine if the sample could have reasonably come from the original population.

As another example, a school may want to compare a new teaching method with the standard method. We would collect test results from a class in which the new method was used and compare data on the new method with data compiled on the standard method. A hypothesis test is used to make the comparison. If the new method provides data which could likely have come from the standard population, we will conclude that the new method is neither better or worse.

A hypothesis can be defined as a statement of belief. Usually, however, there will be some basis for making the statement. In the first example above, we might state the hypothesis. "There is no difference between the modified product and the original." In the teaching example, we might state, "There is no difference between the new teaching method and the standard." The data are then evaluated in order to test whether or not the statement (or hypothesis) is valid.

Basic Need for Hypothesis Testing

Hypothesis testing has many useful applications in the management of a business or other enterprise where it is necessary to decide between possible alternative courses of action. The following represent some other typical cases where the need for a decision is encountered in management, and where the hypothesis-test procedure can be of use.

1. A new advertising approach is proposed for a firm and some trial data are collected. Is the new approach really better than the prior method? There may be a lot at stake in this decision. How much reliance can be placed on the data? The advertising manager will base his future strategy on the test results, and the success of the product (and possibly the company) may depend on his decision.

2. A company that services computer equipment is considering a new training program for its personnel. The effectiveness of the new program is to be compared with data taken previously under the existing program. The program used in the future by the company will depend on the outcome of the analysis, and the effectiveness of the service is important to future company sales and growth. It is important that a valid decision be rendered.

3. A modification to a product is proposed. Data are to be collected on the performance of a sample of items that contain the modification. The results are to be compared with data on the old design to determine if the average performance of the new design is better. A decision is to be made as to whether or not the new product will be produced. The decision is critical and will affect the future of the company. It is important that the executive understand the risks and make a proper decision.

4. A new drug is being tested. Results from use of the drug are to be

compared to another drug used in the past. Based on the test, the new drug will either be manufactured and marketed or abandoned.

5. Acceptance sampling is used by many companies to determine if the quality of shipments received from suppliers are up to the standard. Based on taking a sample from the shipment and statistical analysis of the sample data, the shipment is accepted or rejected. Rejections are costly to the supplier; however, the acceptance of defectives is costly to the using company. It is important that the decision to accept or reject be valid.

We can see from these situations that the analysis of information can affect decisions involving large amounts of money. These decisions may affect the future profitability or survival of a firm. We shall now see how to use hypothesis testing to help make better decisions.

Steps in Testing a Hypothesis

The process of hypothesis testing, as in any good statistical analysis procedure, should result in a good decision while at the same time allowing evaluation of the risks associated with the decision. The risk here is that the sample data will result in an incorrect decision. This possibility exists because, owing to sampling error, the sample might not represent the population. Therefore, along with the procedures for testing a hypothesis, we shall also study ways to recognize and decrease the risks so as to guard against making a bad decision.

An example will illustrate the steps in the hypothesis testing procedure. The Eating Place Restaurant has been recording a mean daily sales of $800 for the past 6 months with a standard deviation of $64. Daily sales are normally distributed. A new advertising program was started 10 days ago and the mean daily sales for the past 9 days has been $830. The owner has reason to believe that there is really no actual effect of the new advertising program. In other words, he believes that the average of $830 for 9 days was due to chance alone. The new advertising program costs more than the former and a decision is to be made whether or not to continue with it. Our objective, then, is to evaluate whether the results of the new program are significant, or if the higher sales could indeed be due just to chance. The following series of steps will be used to analyze the data and arrive at the decision:

Step 1: Establish the hypothesis.
Step 2: Set the significance level and/or critical value in standard deviation units (SDU: z or t).
Step 3: Calculate the appropriate standard error.
Step 4: Calculate the standard deviation units for the sample mean.
Step 5: Compare the result in step 4 to the value in step 2.
Step 6: Accept or reject the hypothesis.

Each step is discussed in greater detail in the following paragraphs.

STEP 1. ESTABLISH THE HYPOTHESIS

To evaluate the advertising program in the example above, we start out with a statement of the hypothesis that "There is no difference between sales with the new advertising program and prior sales." This is called a *null hypothesis* (H_0), indicating an assumption of no difference between the new and old methods. Our procedure will be to test the validity of H_0. If we cannot show a significant difference, we shall assume no difference and accept H_0. It is common practice to state H_0 in such a way that acceptance maintains the status quo. The *alternative hypothesis* (H_a) states that a difference does exist, and H_a will be accepted if the test does not show H_0 to be valid.

In some situations we shall ask "Is there a difference between a new method and the standard?" whereas in other cases we shall ask "Is the new method better than the standard?" In the second case we could have said "Is the new method worse?" but in the second case we are only checking one or the other. When we say "difference," we are checking both better and worse at the same time. This consideration is important since it will affect our procedure. Often we shall want to set up the test procedure before the data are collected, so we would select one or the other of these two objectives. In the case of the restaurant, let us assume that we are attempting to determine if the new method is better. This is called a *one-tailed test* since we will be working with one end of the distribution. (The two-tailed test will be considered later.)

In this particular example, if we accept the alternative hypothesis, we would conclude that the new advertising program actually does produce greater sales. Where μ_1 is defined as the mean of the new population using the new advertising program, we can restate these hypotheses as follows:

$$H_0: \quad \mu_1 = 800$$
$$H_a: \quad \mu_1 > 800$$

When accepting the alternative hypothesis H_a we are concluding that the mean for the new program is greater than the mean of the original population. In our test, if we cannot show that μ_1 is likely to be 800, μ_1 could possibly have other values such as 805, 815, etc.; however, at this time the 830 is the sample mean of the available data and will be our best estimate if we do conclude that the new program actually provides a higher level of sales than the original.

STEP 2. SET THE SIGNIFICANCE LEVEL AND/OR CRITICAL VALUE

In our testing process, we are attempting to test whether the null hypothesis (H_0) is true (i.e., that there is no difference). In other words, we shall attempt to find if there is a difference in the effectiveness of the new advertising program as compared to the old. There is a risk, however, that our conclusion will be wrong. To recognize this risk, we need to set a probability of rejecting H_0 if it is actually true. This is called the *significance level*. Figure 7-4 shows the normal distribution of the original sales around the mean of \$800 with $\sigma = \$64$. We know from Chapter 6 that if we take samples of size 9 from this population, we will obtain the sampling distribution of the mean shown in Figure 7-5, where the mean is \$800, the same as the population mean, and the standard error is calculated as follows, considering the 6-month period data as the population:

$$\sigma_{\bar{X}} = \frac{\sigma}{\sqrt{n}}$$

Our objective at this step in the hypothesis-testing procedure is to set up a critical value such that if our sample mean ($\bar{X}$) falls above this value and the null hypothesis is true, there will be a 5 per cent (or less) probability that the sample came from the original distribution in Figure 7-5. In other words, we are selecting a risk α of 5 per cent. The critical value of 1.645 SDU is shown in Figure 7-5, where the total area above the critical value α is equivalent to the selected probability of .05 (or 5 per cent). This value

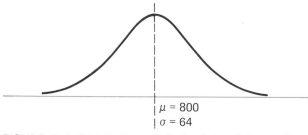

$\mu = 800$
$\sigma = 64$

FIGURE 7-4. Distribution of Daily Sales Prior to New Advertising Program

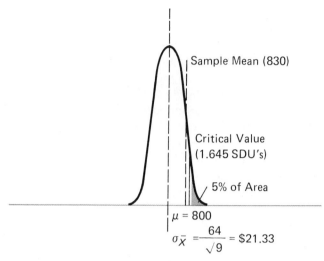

FIGURE 7-5. Distribution of Means of Samples (Size 9) from Population in Fig. 7-4

is taken from Appendix A showing areas under the normal curve. Other levels of significance can also be used.

STEP 3. CALCULATE THE STANDARD ERROR

The next step is to calculate the standard error $\sigma_{\bar{X}}$ for the distribution of sample means. In our problem, the distribution of sample means is shown in Figure 7-5 and $\sigma_{\bar{X}}$ is calculated as follows:

$$\sigma_{\bar{X}} = \frac{\sigma}{\sqrt{n}} = \frac{64}{\sqrt{9}} = \$21.33$$

This value of $\sigma_{\bar{X}}$ will now be used to determine how far out the mean of our particular sample ($\bar{X}$) falls from the mean μ.

STEP 4. CALCULATE THE STANDARD DEVIATION UNITS FOR SAMPLE MEAN

The next step is to determine if the mean of our particular sample fell above the critical value. Since the critical value is specified in terms of standard deviation units (SDU), it is necessary to also establish the distance of the sample mean from μ in SDU. This is calculated as

$$\text{SDU} = \frac{\bar{X} - \mu}{\sigma_{\bar{X}}}$$

For our example,

$$\text{SDU} = \frac{\bar{X} - \mu}{\sigma_{\bar{X}}} = \frac{830 - 800}{21.33} = \frac{30}{21.33} = 1.41$$

The relative position of the sample mean is shown in Figure 7-5.

STEP 5. COMPARE THE RESULT IN STEP 4 TO THE VALUE IN STEP 2

In step 2 we determined the critical value. In our problem we obtained 1.645 SDU for the critical value. In step 4 it was determined that the mean of our sample fell 1.41 SDU out from μ, which is closer to μ than is the critical value.

STEP 6. ACCEPT OR REJECT THE HYPOTHESIS

Our original null hypothesis (H_0) was that there is no significant difference between the sample mean and μ. Since the sample mean fell below the critical value, we accept H_0 (i.e., $1.41 < 1.645$). If our sample mean had exceeded 1.645, we would have rejected H_0 (and accepted H_a).

Rationale for the Decision

Our decision rule in this example was that if the sample mean falls above the critical value, we shall assume that it is unlikely (.05 or less probability) that the sample could have come from the original population if H_0 is true. Thus, if it falls above the critical value, we shall reject the null hypothesis (H_0) and assume H_a, that the sample did not come from the original distribution. If the sample mean is equal to or less than critical value, we shall not reject the null hypothesis. In the latter case, we are saying that there is insufficient evidence to convince us there is actually a difference between the sample and the original population. In other words, we are reserving judgment. We have not shown that the sample *did* come from the original population; we have merely said that we have not been able to adequately prove that it did not. Since we have not proven otherwise, we shall at this point continue to assume that the sample could likely have come from the original distribution. This assumption was our H_0, so we now accept H_0.

In our problem of evaluating the advertising program, the critical value could have been computed in terms of dollars. For a probability of .05, $z = 1.645$ and

$$1.645\ \sigma_{\bar{X}} = 1.645\ \frac{64}{\sqrt{9}} = 35.1$$

The boundary is then $800 + $35.1, or $835.10. Since our sample mean was $830, it is inside the boundary limit. Again we have not been able to show from our data that the new program is significantly better than the old at the .05 significance level selected.

One-Sample Hypothesis Tests

This discussion of *one-sample tests* will cover several types of cases where we have one sample and desire to compare the sample mean to a known population mean. These cases can differ, depending on whether the known population may or may not have a known variance. The population may also be normal, nonnormal, or of unknown shape. These considerations provide a number of possible combinations that must be dealt with, and the procedures vary for the different situations. In addition, the procedures will also vary depending on whether the sample is large (30 or more) or small (less than 30). A number of possible combinations will therefore be dealt with individually. These alternatives are shown in Exhibit 7-2 following our discussion of two-sample hypothesis tests. This tree can be used to select the proper formula for a particular problem. As each alternative is studied, the formula should be located on the tree.

Population Normal with Standard Deviation Known

In this case we know that the population is normal and we know the standard deviation of the population σ. The illustration of the restaurant and the new advertising program fits this category. Consider another example where a meat market manager is evaluating a supplier's claim that he has changed his procedures and trims his steaks so that there is less fat. The manager had determined that, over the past 6 months, for each 100 lb of beef purchased, the manager had needed to remove 16.4 lb of fat. The results were normally distributed with a standard deviation of 2.4 lb. To test this claim, the manager now obtains a sample of 36 (100-lb groups), giving a mean of 15.64 lb of fat. Was this significantly lower (at the .05 significance level) than the original 16.4 lb? In other words, could the new sample have logically come from the original population? We establish the hypothesis H_0 that there is no difference:

$$H_0: \mu_1 = 16.4$$
$$H_a: \mu_1 < 16.4$$

At the .05 significance level (from Appendix A), the critical value for z is calculated in terms of SDU:

$$z = 1.645$$

The standard error is calculated as

$$\sigma_{\bar{X}} = \frac{\sigma}{\sqrt{n}} = \frac{2.4}{\sqrt{36}} = .40 \text{ lb}$$

The position of the sample mean in SDU is calculated next.

$$\text{SDU} = z = \frac{\bar{X} - \mu}{\sigma_{\bar{X}}} = \frac{15.64 - 16.4}{.40} = -\frac{.76}{.40} = -1.9$$

As illustrated in Figure 7-6, the sample mean of 15.64 is at 1.9 SDU and is farther from the mean value (16.4) than the critical value of 1.645. Therefore, we reject the null hypothesis (H_0) and conclude that the new supplier does supply meat with less fat.

What if we had wanted a level of significance with less risk? We could have used a .01 significance level with the critical value of $z = -2.33$. In any case, the formula is summarized as

$$\text{SDU} = z = \frac{\bar{X} - \mu}{\sigma/\sqrt{n}} \qquad (7.9)$$

The mean of the sample (15.64) is at -1.9 SDU and is closer to the population mean than is this .01 critical value of -2.33. Therefore, at the 1 per cent level, we would not have rejected H_0. We would not have concluded

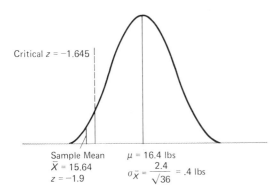

FIGURE 7-6. Meat Market Example Showing Critical Value Not Exceeded

the new source was better. When the .01 significance level is selected, a greater difference is required before the decision is made to accept the sample as being different from the original population.

As another example, assume that a tire manufacturer has been producing a passenger-car radial tire showing an average life μ of 50,000 miles with a normal distribution and a standard deviation σ of 4,000 miles. He has been experimenting with a modified synthetic additive, and a sample of 16 tires of the modified design gave a mean life $\bar{X}$ of 52,000 miles. Is it likely that the sample taken could have come from the original population? Based on the results he will decide if the new modification should be put into production.

We start with the hypothesis H_0 that there is no significant change in mileage.

$$H_0: \mu_1 = 50,000$$
$$H_a: \mu_1 > 50,000$$

We evaluate the risk we are willing to take and set a significance level (α) of .025, giving a critical value of 1.96. The standard error is then calculated as

$$\sigma_{\bar{X}} = \frac{\sigma}{\sqrt{n}} = \frac{4,000}{\sqrt{16}} = 1,000$$

The sample mean is located at a distance from μ as:

$$\text{SDU} = \frac{\bar{X} - \mu}{\sigma_{\bar{X}}} = \frac{52,000 - 50,000}{1,000} = 2$$

Our sample is 2 SDU from the mean, and this exceeds the critical value (1.96) as shown in Figure 7-7. We therefore reject H_0 and conclude that the new tire is better.

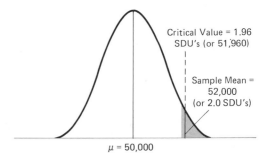

FIGURE 7-7. Illustration of Tire Example

Two-Tailed Tests

Initially, we discussed situations where the problem was to determine if a sample could likely have come from a given population. A test was conducted to determine if the sample mean was significantly larger (or smaller) than that which could be expected if the sample had come from the original population. In those cases only one tail of the distribution was tested. If the null hypothesis (H_0) was not accepted, the alternative hypothesis in a typical case was that the sample had come from another population with a larger (or smaller) mean.

In two-tailed tests we hypothesize that the unknown population mean is a certain value. Then a decision rule will be formulated such that when this hypothesis (H_0) is rejected, it will mean that the sample mean lies significantly outside (in *either* direction) of the limits established. In other words, our alternative hypothesis is that the sample mean is significantly *different*. It can be either larger or smaller.

Consider a fertilizer distributor who has been making shipments of fertilizer in 100-lb bags over a period of several months. He has determined that the population is normal and has a mean weight of 100.4 lb with a standard deviation of .60 lb. When a carload is ready to ship, he weighs a sample of 4 bags to determine if the weights are consistent with past items. He is interested in cases either of overweight or underweight. Limits are to be set up to test our hypothesis H_0 that there is no difference between the current sample and the original population. If the sample mean falls within these limits, the distributor will assume no significant change in the filling process. The null hypothesis (H_0) and alternative hypothesis (H_a) would then be stated as

$$H_0: \mu_1 = 100.4$$
$$H_a: \mu_1 \neq 100.4$$

Figure 7-8 shows the distribution of samples of size 4 taken from the original population. The standard error of the sample mean $\sigma_{\bar{x}}$ is calculated as follows:

$$\sigma_{\bar{x}} = \frac{\sigma}{\sqrt{n}} = \frac{.60}{\sqrt{4}} = .30$$

It is necessary to establish the significance level α, which we shall choose here as .05. This means that 5 times out of 100, a mean of a sample of 36 from our original population could exceed the limits in either direction by chance alone if the population mean was in fact equal to 100.4 lb. This is also the probability that we reject H_0, when actually it is true.

FIGURE 7-8. Distribution of Sample Means ($n = 4$) for Fertilizer Example

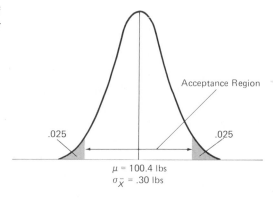

In this case we are letting $\alpha/2$ or (.025) of the area be in each of the two tails so that the total area of the normal curve outside the limits gives the α of .05. From Appendix A, the area of .025 in one tail gives a z of 1.96. Therefore, our limits will be at

$$\mu \pm 1.96 \text{ SDU}$$

The distributor can convert this to weight limits:

$$\begin{aligned} \text{limits} &= \mu \pm z\sigma_{\bar{x}} \\ &= 100.4 \pm 1.96(.30) = 100.4 \pm .588 \text{ (in pounds)} \end{aligned}$$

Then the acceptance region becomes 99.812 → 100.988. The test procedure for future use then is to take a sample of 4 and calculate the mean of the sample. If the mean falls outside either limit (99.812 or 100.988), H_0 will be rejected and we will conclude that this sample came from a different population. If it falls within the acceptance region, we will accept H_0 and assume that the shipment is not different from the original distribution.

Again, the possibility of an error exists in our decision. In this example, we can state that the probability of a sample being rejected when it should actually be accepted is α, or .05. In other words, there is a probability of $1 - \alpha$ (or .95 in our case) of accepting the null hypothesis H_0 when it is actually true.

The discussion will now proceed to develop other one-sample procedures that can be applied to either a one-tailed or a two-tailed test situation.

Population Normal With Standard Deviation Unknown

It is to our advantage to know the population standard deviation so that we can calculate $\sigma_{\bar{X}}$, but in most real-life problems we will not. If the population standard deviation σ is unknown, however, we can still make use of the sample data to estimate $\sigma_{\bar{X}}$. There are two procedures, depending on the size of the sample. Either procedure can be applied to either the one-tailed or two-tailed test, as long as we are careful in selecting α or $\alpha/2$, respectively.

Sample Size 30 or Larger. When the population is normally distributed, we know that the distribution of the sample mean is also normally distributed. Furthermore, when the sample size is 30 or larger, the sample standard error $s_{\bar{X}}$ will closely approximate $\sigma_{\bar{X}}$. The following can then be used as the acceptance region where we are using a two-tailed test and where μ_0 denotes the hypothesized mean:

$$\mu_0 \pm z_{\alpha/2} \times s_{\bar{X}}$$

Take an example where a new process is expected to deposit a normally distributed coating of 40 mils (or .040 inches). This is our hypothesized mean value μ_0. A sample of 36 is to be taken. What should be the acceptance region? The first sample gives $\bar{X}$ of 37.8 mils and $s = 7.2$ mils. This value of s is used as an estimator of σ. A significance level of .10 is desired for a two-tailed test, giving a z of 1.645 as the critical value.

$$H_0: \mu_1 = 40$$
$$H_a: \mu_1 \neq 40$$
$$s_{\bar{X}} = \frac{s}{\sqrt{n}} = \frac{7.2}{\sqrt{36}} = 1.2$$

We then calculate the number of SDU for our sample as follows:

$$\text{SDU} = z = \frac{\bar{X} - \mu}{s_{\bar{X}}} = \frac{\bar{X} - \mu}{s/\sqrt{n}} \qquad (7.10)$$
$$= \frac{37.8 - 40}{1.2} = -\frac{2.2}{1.2}$$
$$= -1.83$$

The z of 1.83 which we obtained exceeds the critical value of 1.645 and, therefore, we reject the hypothesis (H_0) that the sample came from a population with a mean of 40 mils. We would then accept the alternative hypothesis (H_a) that the sample is from a population with a different mean.

Sample Size Less Than 30. For a normally distributed population, unknown σ, and where the sample size is less than 30, it is necessary to utilize the t distribution, which was introduced previously. For example, suppose that a filling machine at a canning factory is intended to place 30 oz of fruit juice in each can. A sample of 9 is taken and the sample mean is 30.04 and the sample standard deviation is .08 oz. Could the sample logically have come from a population with a mean of 30.0? A significance level of .05 is to be used, giving 1.86 as the critical value of t for a one-tailed test with $n - 1$, or 8 degrees of freedom. Using Appendix B, we have

$$H_0: \mu_1 = 30$$
$$H_a: \mu_1 > 30$$

$$s_{\bar{x}} = \frac{s}{\sqrt{n}} = \frac{.08}{\sqrt{9}} = .027$$

Then the sample SDU are determined as follows:

$$\text{SDU} = t = \frac{\bar{X} - \mu}{s/\sqrt{n}}$$
$$= \frac{30.04 - 30.00}{.08/\sqrt{9}} = \frac{.04}{.027} \qquad (7.11)$$
$$= 1.48$$

The mean of the sample is at 1.48 SDU, which is less than the critical t value of 1.86. Therefore, we accept H_0 and conclude that the sample could have come from the population having a mean of 30 oz.

Population Distribution and Standard Deviation Both Unknown

If the population distribution is not known to be normal and σ is also unknown, then we must have a sample size of at least 50 in order to be able to set up a valid acceptance region. If the sample is less than 50, we cannot depend on both the central limit theorem and the sample estimate and we should try to obtain a larger sample.

One-Sample Tests for Proportions

Many problem situations necessitate the comparison of a sample proportion to a population proportion. Take an example where an automatic process produces 96 per cent good parts and 4 per cent defectives. The manufacturer is satisfied with this since new machinery would be very expensive. He does, however, want to be able to detect if something changes in the process which would increase this percentage of defects. One day he takes a sample of 400 items and finds 22 defectives, or 5.5 per cent defective. At the 5 per cent significance level, should we recommend that the process be stopped because of a significant difference? Using a one-tailed test, the critical value of z is 1.645.

$$\pi = \text{population proportion} = .04$$
$$p = \text{sample proportion} = .055$$
$$H_0: \ p = \pi$$
$$H_a: \ p > \pi$$

The value of z is calculated as

$$\text{SDU} = z = \frac{p - \pi}{\sigma_p} = \frac{p - \pi}{\sqrt{\frac{\pi(1-\pi)}{n}} \cdot \sqrt{\frac{N-n}{N-1}}} \tag{7.12}$$

When sampling from an infinite population or whenever the sample size is less than 5 per cent of the population, we can neglect the correction factor and use the approximation

$$z = \frac{p - \pi}{\sqrt{\frac{\pi(1-\pi)}{n}}} \tag{7.13}$$

From our problem,

$$\text{SDU} = z = \frac{.055 - .040}{\sqrt{\frac{(.04)(.96)}{400}}} = \frac{.015}{.0098} = 1.53$$

At the 5 per cent significance level, H_0 is rejected if the z is greater than 1.645. Since our 1.53 for this problem is less than 1.645, we accept the null hypothesis, H_0, and assume that the sample is from the original process. If the value of z were greater than the critical value of 1.645, we would probably stop the process and investigate for any possible causes.

We can summarize our procedure for testing a proportion for this problem as follows:

1. Establish the null hypothesis H_0 that the machine is now turning out parts with a mean proportion defective of .04.
2. The alternative hypothesis H_a is that $p \neq .04$.
3. Decision rule: Reject H_0 if $z > 1.645$ (at $\alpha = .05$).

Exercises

7.16 The mean of a normal population is $\mu = 120$ with $\sigma = 15$. State and test a hypothesis that a sample of 49 with $\bar{X} = 124$ came from the original population. Use a one-tailed test at the .05 significance level.

7.17 The mean of a normal population is 120 but σ is not known. A sample of size 49 has a mean of 124 and $s = 15$. State the hypothesis, rule, and decision in this case at the .05 level using a one-tailed test.

7.18 An insurance salesman is trying to decide whether or not to make calls in an area of a city. His decision will be yes if the average income level in the area is $15,000 or over. The σ is known to be $900. A random sample of 100 homes in the area is made and the $\bar{X} = \$14{,}620$. State the null hypothesis and set up the critical value and decision rule using a .025 significance level and one-tailed test. Should he make the calls?

7.19 An inspector is checking the foreign-matter level of the city water supply. He wants to make sure that the level does not exceed 10 grains per cubic foot. The weights have a σ of 0.2 grains and are normally distributed. Set up the decision rule for samples of 25 at the .01 level. A sample of 25 shows a $\bar{X}$ of 9.84 grains. Is the level being exceeded?

7.20 A population is known to have a mean of 59 and $\sigma = 7$. After a period of time has elapsed, a sample of 98 items is taken, giving a sample mean of 63. Perform a two-tailed test to test if there has been a shift in the distribution, using a .01 significance level.

7.21 In designing its parking lot, a company has determined that, formerly, 20 per cent of the employees drove to work alone. A current random sample of 400 cars shows 70 driving alone. Set up the null hypothesis at the 10 per cent level (two tails). Should H_0 be rejected?

7.22 The Bart Department Store claims that in the past the company's accounts receivable have averaged $190.00. An auditor takes a sample of 50 accounts and obtains a sample mean of

$180.00 and a sample standard deviation of $35.00. Test the claim at the 5 per cent level using a one-tailed test.

7.23 A company makes a car battery with a normally distributed mean life of 3.4 years. The company has tested a sample of 9 of a competitive battery which gave a mean life of 3.1 years and a sample standard deviation of .45 years. Can the company really claim a longer life, using a .05 significance level?

7.24 Records of a bank showed a mean checking account balance of $284 for 1977. This and prior measurements have shown σ to remain about constant at $44 over the years.
  (a) Prepare a test to judge whether or not there is a significant change from the 1977 base at a .02 level of significance.
  (b) This year the bank eliminated a service charge for lack of a minimum balance of $100. A sample of 49 accounts were selected and the mean balance was $216. What is your conclusion?

7.25 A machine has been producing rods (normal distribution) cut off at 8.0 inches. A random sample of 10 items shows a mean of 8.14 inches with a standard deviation of .35 inches. Set up a test to measure if the machine setting changed at the .10 significance level and use the test to evaluate the sample taken.

7.26 In a large city, a census had shown that 60 per cent of the people owned their own home. A publisher of the local newspaper took a random sample of 900 subscribers and determined that 62 per cent owned their own home. Is the proportion of subscribers who own their homes significantly greater than the proportion in the city (.01 level)? What would be the minimum sample size for which Eq. (7.13) could be used?

7.27 A machine shop has been operating over the years with 5 per cent of the output considered defective. This month a sample of 100 items was taken and 7.2 per cent were defective. What can you conclude? Use a one-tailed test at the .04 significance level.

7.28 A manufacturing process is designed to drill a hole 2.84 inches from the edge of a plate with a σ of .008. A sample of 4 pieces yields 2.88, 2.94, 2.83, and 2.90. Set up a test to measure if the process is out of control (.05 level). Do you conclude that the distances in the sample indicate a setting that is incorrect?

7.29 A manufacturing plant produces thread with a breaking strength of 18.4 lb and a standard deviation of 2.8 lb. Set up a test and rule to evaluate if there is a significant change using samples of

4 at the .01 level. A test is run on 4 pieces to give breaking strengths of 19.3, 17.2, 16.8, and 18.3. Evaluate the results.

7.30 A new thread process was developed and was supposed to provide a strength of 26 lb. A test was run and provided strengths of 25.4, 26.1, 24.3, and 25.6. Is the requirement being met? Use a .01 significance level one-tailed test.

7.31 A breakfast food company had held 28 per cent of the market for the past 3 years. This year a sample of 49 showed 25.4 per cent of the sales going to this company. Is this result significantly lower at the .01 level?

Two-Sample Hypothesis Tests

The applications of hypothesis testing which have been discussed have dealt with situations where a sample was being compared to a known or hypothesized population. In that case we had knowledge of the population mean and possibly also of the standard deviation and shape of the population distribution. We are now going to learn how to use hypothesis testing to make a comparison between two samples to determine if they could have likely come from the same population. These are called two-sample hypothesis tests. The null hypothesis (H_0) is that the samples came from populations with the same mean, whereas the alternative (H_a) is that the means are not equal:

$$H_0: \mu_1 = \mu_2$$
$$H_a: \mu_1 \neq \mu_2$$

Difference Between Two Means (Sample Size 30 or Larger)

Consider a trucking company that desires to compare the mileage it obtains from two brands of tires. Mileage records have been compiled on samples of both brand A and brand B. Since the trucking company uses a large quantity of tires, any significant cost saving due to better mileage would be important.

The trucking company had maintained records on the mileage obtained from 100 tires of brand A and also 100 tires of brand B. The average mileage of the brand A tires was 51,000 miles and the standard deviation

(s) computed from the sample was 2,000 miles. The standard error for the distribution of sample means ($n = 100$) is estimated as

$$s_{\bar{X}_1} = \frac{s_1}{\sqrt{n_1}} = \frac{2{,}000}{\sqrt{100}} = 200 \text{ miles}$$

Brand B tires averaged 50,100 miles with a standard deviation of 3,000 miles. Then,

$$s_{\bar{X}_2} = \frac{s_2}{\sqrt{n_2}} = \frac{3{,}000}{\sqrt{100}} = 300 \text{ miles}$$

Based on this information, it is claimed that brand A tires are better. We propose to use a hypothesis test to check this claim at the 1 per cent significance level. A one-tailed test is to be used because we are checking the hypothesis that brand A gives *greater* mileage, rather than merely that the two brands are *different*. We shall start with the null hypothesis assumption that there is no real difference in the brands. If we are able to *reject* the null hypothesis on the basis of our test results, then we shall *accept* the alternative hypothesis that brand A is actually better than brand B. In this case our hypothesis is that two means are equal; however, we do not specify to what value they are equal. This is different from our previous one-sample tests where we formed a hypothesis that a mean was equal to a specific value.

Proceeding with our analysis of the tire-mileage data, the observed difference in the means is

$$\text{difference} = \bar{X}_1 - \bar{X}_2 = 51{,}000 - 50{,}100 = 900$$

Again, we are dealing with sample data and different samples would likely yield other sample means and differences. Statistical theory has shown that if a number of pairs of samples (size n_1 and n_2) were taken and the difference between the means computed for each pair, then the differences would form a normal distribution with a mean of zero and a standard deviation or standard error σ_d equal to

$$\sigma_d = \sqrt{\frac{\sigma_1^2}{n_1} + \frac{\sigma_2^2}{n_2}}$$

Where we use sample data, and the population standard deviations are unknown;

$$\text{est. } \sigma_d = s_d = \sqrt{\frac{s_1^2}{n_1} + \frac{s_2^2}{n_2}} = \sqrt{s_{\bar{X}_1}^2 + s_{\bar{X}_2}^2}$$

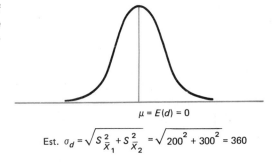

FIGURE 7-9. Distribution of Differences Between Sample Means in Tire Mileage Example

In our case, then,

$$s_d = \sqrt{200^2 + 300^2} = \sqrt{40{,}000 + 90{,}000}$$
$$= 360$$

This is illustrated by the normal curve in Figure 7-9. If there were actually no difference in the mean mileage of the two tire brands, then we could conclude that, for example, 68.2 per cent of the differences in pairs of sample means would fall in the range of ± 360 miles from the mean. In our case we have a difference of 900 miles, so

$$z = \frac{900}{360} = 2.5$$

The area on one tail beyond 2.5 standard deviations is found from Appendix A to be

$$\alpha = .500 - .4938 = .0062.$$

Therefore, we can conclude that the chance of obtaining such a large difference due to chance alone would be about 6 in 1,000, or less than our critical value (α) of .01. Based on this, we reject the original H_0 (that the brands are equal) and conclude that brand A is significantly better than brand B. The procedure can then be formalized as follows:

1. H_0: $\mu_1 = \mu_2$
2. critical value $z = 2.33$
3. standard error $= \sqrt{\dfrac{s_1^2}{n_1} + \dfrac{s_2^2}{n_2}}$

4. $$\text{SDU} = z = \frac{\bar{X}_1 - \bar{X}_2}{\sqrt{\frac{s_1^2}{n_1} + \frac{s_2^2}{n_2}}} \qquad (7.14)$$

$$= \frac{51{,}000 - 50{,}100}{\sqrt{\frac{(2{,}000)^2}{100} + \frac{(3{,}000)^2}{100}}}$$

$$= 2.5$$

5. $2.50 > 2.33$

6. Reject H_0 and conclude that A has a greater mean mileage than B.

The procedure just demonstrated applied to situations where the samples were 30 or larger. We shall next determine how to handle cases where we are limited to smaller samples, either because of costs or limited available data.

Difference Between Two Means Based on Small Samples ($n < 30$)

Our earlier discussions have pointed out that when the sample size is small, the t distribution is a more accurate representative of the sampling distribution than is the normal distribution. This is also true for the two-sample hypothesis test using two small samples.

For large samples we used

$$z = \frac{d}{\sigma_d} = \frac{\bar{X}_1 - \bar{X}_2}{\sqrt{\frac{s_1^2}{n_1} + \frac{s_2^2}{n_2}}}$$

For small samples we shall utilize

$$t = \frac{d}{\sigma_d} = \frac{\bar{X}_1 - \bar{X}_2}{\sqrt{\frac{(n_1 - 1)s_1^2 + (n_2 - 1)s_2^2}{n_1 + n_2 - 2}} \sqrt{\frac{1}{n_1} + \frac{1}{n_2}}} \qquad (7.15)$$

This assumes that the two samples are independent random samples and that their parent populations are distributed normally with equal variances.

In this process we are using $(n_1 + n_2 - 2)$ degrees of freedom in making use of the t distribution table. If in this case we are asking if there is a difference between the means of the two samples, it is a two-tailed test and is considered significant at the 5 per cent level when the calculated value of t exceeds $t_{.025}$ in either direction. The method is illustrated by the following example.

EXAMPLE

A class of 28 students is divided into two groups. The students were preparing to take a state real estate licensing examination. The first group of 16 took a 30-hour course, whereas the second group of 12 took a 24-hour course. After the courses were completed, a test was given with the following results:

30-Hour Course	24-Hour Course
$n_1 = 16$	$n_2 = 12$
$\overline{X}_1 = 76$	$\overline{X}_2 = 72$
$s_1 = 7$	$s_2 = 9$

We desire the answer to our question: Is the difference between the mean test scores significant at the .05 level? The initial hypothesis H_0 is that there is no difference. $H_0: \mu_1 = \mu_2$ and $H_a: \mu_1 \neq \mu_2$. Using the formula above, we have

$$t = \frac{76 - 72}{\sqrt{\frac{(16-1)7^2 + (12-1)9^2}{(16+12-2)}} \cdot \sqrt{\frac{1}{16} + \frac{1}{12}}}$$

$$= \frac{4}{\sqrt{\frac{735 + 891}{26}} \cdot \sqrt{\frac{28}{192}}} = \frac{4}{\sqrt{62.54} \cdot \sqrt{.146}}$$

$$= \frac{4}{7.91 \times .382} = 1.324$$

With 26 (16 + 12 − 2) degrees of freedom, the critical value of $t_{.025}$ is 2.056 (from Appendix B). Since our t value of 1.324 is lower, we conclude that the difference between the means of the two groups is not significant at the .05 level. In other words, we accept H_0, our null hypothesis which stated there was no difference. The evidence has not been sufficient to convince us that a significant difference exists between the two courses.

Difference Between Two Proportions

The previous method made a comparison test between two sample means. Frequently, our data are in a form such that we must work with proportions or percentages. A similar concept can be used if we desire to compare two proportions. Consider an example. A claim has been made that there are a greater proportion of female automobile owners in Columbus than there are in Dayton. Proceeding to test this claim, we assume the hypothesis H_0 that there is no difference in the proportion of female owners in the two cities. The hypothesis is to be tested at the 10 per cent significance level. We have available a random sample of 225 owners from Columbus and 144 owners from Dayton. Our data show that 81 of the 225 automobile owners in Columbus are female, or 36 per cent, giving $p_1 = .360$. We also find that 64 of the 144 owners in Dayton are female, so $p_2 = .444$.

With these facts, we set out to determine if the difference of 8.4 per cent is significant. We learned previously that the variance of a proportion based on a sample size of n was given by

$$s^2 = \frac{p(1-p)}{n}$$

In our case, we are attempting to test a hypothesis stating that there is no significant difference in the proportions. With this as an assumption, the best estimate of the proportion for the combined sample would be the weighted average of the given samples, or

$$\hat{p} = \frac{n_1 p_1 + n_2 p_2}{n_1 + n_2} = \frac{81 + 64}{225 + 144} = \frac{145}{369} = .393$$

Then the variances of the two proportions π_1 and π_2 are estimated as

$$s_1^2 = \frac{p(1-p)}{n_1} = \frac{.393 \times .607}{225}$$

$$s_2^2 = \frac{p(1-p)}{n_2} = \frac{.393 \times .607}{144}$$

$$s_d = \sqrt{s_1^2 + s_2^2}$$

$$= \sqrt{\frac{.393 \times .607}{225} + \frac{.393 \times .607}{144}} = \sqrt{.00272} = .052$$

Our actual difference was $.444 - .360$ or $.084$. How many standard deviations is this out from the hypothesized difference of zero?

$$z = \frac{p_1 - p_2}{s_d} = \frac{.084}{.052} = 1.62 \text{ standard deviations}$$

The probability associated with this is .0526. This probability is smaller than the .10 significance level, so we conclude that p_2 is significantly higher. (We could also have observed this by noting that the z of 1.62 was greater than the 1.28 necessary for a 10 per cent significance level.)

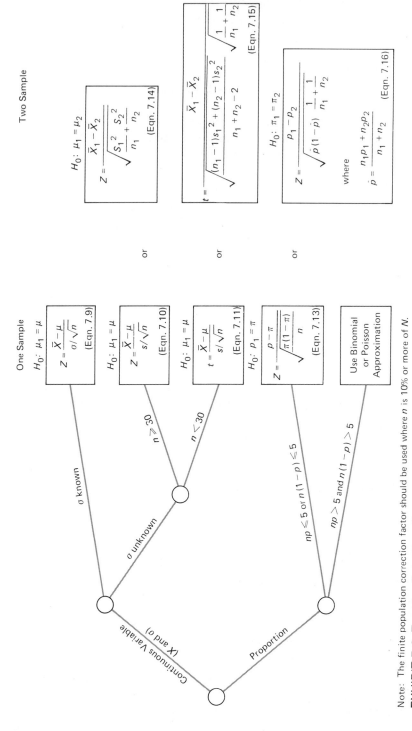

Note: The finite population correction factor should be used where n is 10% or more of N.

EXHIBIT 7-2. Tree for Selecting Equations for Hypothesis Tests

At this point, let us summarize the procedure used when testing the difference between two proportions.

$$H_0: \pi_1 = \pi_2$$

critical value of $z = 1.28$

$$\text{SDU} = z = \frac{p_1 - p_2}{\sqrt{\hat{p}(1-\hat{p})(1/n_1 + 1/n_2)}} \qquad 7.16$$

where

$$\hat{p} = \frac{n_1 p_1 + n_2 p_2}{n_1 + n_2}$$

In our example

$$z = \frac{.084}{\sqrt{(.393)(.607)(1/225 + 1/144)}} = 1.62$$

Since the 1.62 is greater than the 1.28, we reject H_0 and conclude that the proportion of women drivers in Dayton is higher.

Selecting the Correct Hypothesis-Test Equation

The equations used in hypothesis testing are summarized in Exhibit 7.2. This tree diagram is utilized in much the same way as Exhibit 7.1 was used in selecting the estimation equation.

Exercises

7.32 An argument was presented that women high school graduates did better on an aptitude test than men graduates. A random sample of 30 men scored 254 with s_1 of 93. A similar random sample of 30 women scored 278 with $s_2 = 90$. Set up a hypothesis that there is no difference between men and women. Test your hypothesis at the .05 significance level.

7.33 The manager of marketing is evaluating two sales approaches for selling insurance. A group of 30 salespeople using approach A for a month sold an average of 13.32 policies per person with $s_A^2 = 25.6$. When the group used approach B they sold an average of 15.30 policies per person with $s_B^2 = 32.8$. Set up a hypothesis test to measure if there is any significant difference (two-tailed test) at the .05 level. Evaluate the results.

7.34 Solve problem 7.33 if the sample size were 9 salespeople.

7.35 Mr. Jack is considering the purchase of two dairy farms of about equal price and size. He keeps records of the milk output on random days as follows:

Farm A	Farm B
$n_A = 11$ days	$n_B = 15$ days
$\bar{X}_A = 157.3$ gal	$\bar{X}_B = 151.5$ gal
$s_A = 4.14$ gal	$s_B = 6.92$ gal

Design a test to evaluate if A is better than B at the .05 level.

7.36 Our company claims that its new cheese mixture has a better taste than the competitor's. Fifty customers are given samples and 40 select our brand. Set up a hypothesis to test at the .01 level.
   (a) Is the result significant?
   (b) What if only 6 customers were sampled and all 6 selected our brand?

7.37 A feed supply wholesaler wishes to compare two seed mixtures from two suppliers. A sample of 600 seeds is taken from supplier A and 10 per cent do not germinate. A sample of 400 is taken from supplier B and 14 per cent do not germinate. Set up a hypothesis test to measure if there is any difference at the .05 significance level. Use the sample results to reach a conclusion.

7.38 Two bars are cut off by two different automatic saws. It is important that the bars be 1.0 inch different in length. The standard deviation of saw 1 is .01 inch and that of saw 2 is .02 inches. Data from two samples are as follows:

	Bar 1	Bar 2
Standard deviation	$\sigma_1 = .01$	$\sigma_2 = .02$
Sample size	$n_1 = 25$	$n_2 = 20$
Length $\bar{X}$	$\bar{X}_1 = 1.01$	$\bar{X}_2 = 1.99$

Form a hypothesis test at the .01 level which assumes that an error in either direction is important. What is your conclusion on the two samples?

7.39 A car rental agency is comparing two types of compact cars in terms of gasoline mileage. The following results were obtained.

	Brand A	Brand B
Sample size	$n_A = 31$	$n_B = 44$
Miles per gallon	$\bar{X}_A = 26.4$	$\bar{X}_B = 28.5$
Standard deviation	$s_A = 4.4$	$s_B = 6.2$

Is brand B significantly better than brand A at the .02 significance level?

Risks and Errors in Hypothesis Tests

Risks (or Errors)

In the example evaluating the proposed new advertising program, we were to decide either to adopt the new program or not to adopt it. In that example the restaurant had daily sales of \$800 (with $\sigma = 64$) for a 6-month period. When a new advertising program was started, the daily sales for 9 days amounted to \$830. In our example we accepted H_0 and concluded that the new advertising program was not better than the old. In this decision process, as is true in any statistical process, there is the possibility of making a wrong decision because of the probabilities involved. It is important to be aware of these risks and to understand them. The possible correct and incorrect decisions are tabulated in Table 7-1. The right-hand margin of the table shows our two possible decisions, either accept H_0 or reject H_0. The top of the table shows the two possible true situations: Either H_0 is true or it is not true. Two types of errors are shown in the body of the table. A type I error is made when H_0 is rejected, but it should have been accepted. A type II error is made when H_0 is accepted, but should have been rejected. The Greek letter α represents the maximum probability of a type I error; this is the same α we used for the significance level of the test. In our problem, we set α equal to .05. It was the probability of getting a sample mean greater than our critical value of 1.645 SDU (or \$835.08, as shown in Figure 7-3) if the true population mean is \$800 and the true standard deviation is \$64. The probability of a type II error is called β. We have not used β so far, so we will see how to determine this risk.

To calculate β, we must first assume that H_0 is false ($\mu \neq 800$). Sup-

TABLE 7-1. Analysis of Possible Decisions in Hypothesis Testing

Decision Based on Sample Test	Actual	
	H_0 True (New program is not really better)	H_0 False (New program is actually better)
Accept H_0 (Tests indicate no significant difference)	Correct decision: probability $= 1 - \alpha$	Type II error: probability $= \beta$ Incorrect decision rejects a new program that is really better.
Reject H_0 (Test indicates a significant difference)	Type I error: probability $= \alpha$ Incorrect decision accepts a new program that is not really better.	Correct decision: probability $= 1 - \beta$

pose, then, that the new advertising program is actually better and the actual mean sales for the new advertising program is $870 per day with $\sigma_{\bar{X}}$ equal to 60. Then $\sigma_{\bar{X}}$ equals $\frac{60}{\sqrt{9}}$ or 20. The probability of making a type II error is the probability of not detecting this new mean value (i.e., accepting H_0). The distribution of means of samples size 9 when sales have a mean of $870 is shown in Figure 7-10.

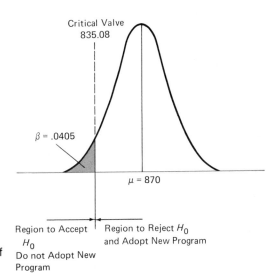

FIGURE 7-10. Illustration of Type II Error and β

The next step is to determine how far out the boundary value of 835.08 is from the mean of 870 in terms of z. In this case,

$$z = \frac{835.08 - 870}{\sigma_{\bar{X}}} = \frac{-34.92}{20} = -1.746$$

From Appendix A for areas of the normal curve, a z of 1.746 is equivalent to $\beta = .0405$, representing the area in Figure 7-10. This is interpreted as a .0405 probability of accepting H_0 when it should have been rejected. This is the type II error.

In setting up a test, it is desirable to make *both* α and β as small as practicable. In order to make both these errors smaller simultaneously, however, it becomes necessary to increase the sample size and thus incur the increased cost of collecting more data. If we cannot reasonably endure this added cost, we are faced with a trade-off between cost and probability of error. The steps in setting up tests and selecting α and β are dealt with in greater depth later in the chapter. At that time other factors important in the design of the test will be dealt with also.

Operating Characteristic Curve

So far a situation has been evaluated in terms of the probability α of a type I error and the probability β of a type II error. In setting up a decision-making rule we have worked with one or two critical values and the associated acceptance probabilities. Plans for acceptance or rejection of lots shipped from suppliers often use these concepts. Manufacturers also use them to efficiently control the quality level of purchased material.

Once a plan has been designed, it is often desirable to be able to answer the question: What will our sampling procedure do (accept or reject) if lots of a certain quality are submitted? The operating characteristic curve will now be explained, showing how it can be used to answer this question. The following example will be developed to illustrate the operating characteristic (OC) curve.

EXAMPLE

A manufacturer purchases 50-lb bags of a dry chemical for use in manufacturing. He uses a large quantity and receives shipments of 500 bags or more at a time. A sampling plan has been devised where a sample of 36 bags from a shipment is taken and weighed. The supplier and the user designed their plan based on past data showing a mean of 50.7 lb for the bags and a standard deviation of .9 lb. The distribution of individual bags is shown in Figure 7-11. He desires that bags be a minimum of 50 lb. If the mean of the distribution is 50.7, then

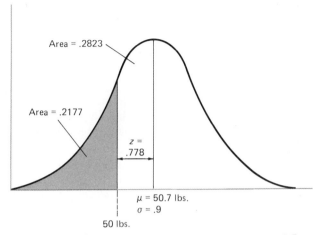

FIGURE 7-11. Distribution of Individual Bags of Dry Chemical

$$z = \frac{50 - 50.7}{.9} = -\frac{.7}{.9} = -.78$$

From Appendix A, the probability that an individual bag falls below a z of .78 is .5000 − .2823, or .2177. Assume that the manufacturer considers this risk to be acceptable.

The supplier and the manufacturer agree upon a plan whereby a sample of 36 is taken and the lot is accepted if the sample mean is 50.35 or more. This is halfway between the two values. Our question is: What happens when lots of various means weights are submitted to this plan?

Let us assume that a lot having a population mean of 50.50 lb is submitted to the sampling plan. The distribution of means of samples size 36 taken from this population is shown in Figure 7-12. The probability that the sample mean falls above 50.35 is calculated by first obtaining z.

$$z = \frac{\bar{X} - \mu}{\sigma/\sqrt{n}}$$

$$z = \frac{50.35 - 50.50}{.9/\sqrt{36}} = \frac{-.15}{.15} = -1.0$$

$$P(\bar{X} > 50.35) = .5000 + .3413 = .8413$$

Thus, the probability of accepting a lot having a μ of 50.50 is .8413 and this value appears in Table 7-2. We need to prepare the rest of the table now.

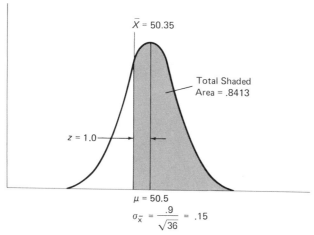

FIGURE 7-12. Distribution of Sample Means for Population Mean of 50.5 lbs.

Consider as the next point what would happen if a lot were submitted from a population having $\mu = 50.00$. The probability that the sample mean will fall at our boundary value of 50.35 or higher is calculated

$$z = \frac{50.35 - 50.00}{.15} = \frac{.35}{.15} = 2.33$$

$$P(\bar{X} > 50.35) = .5000 - .4901 = .0099 \qquad \text{(or about .01)}$$

TABLE 7-2. Probabilities of Accepting Lots from Various Populations Where Acceptance Value = 50.35

Population Mean μ from which Sample Is Taken	Probability of Acceptance Based on Sample
50.00	.0099
50.05	.0228
50.20	.1587
50.30	.3707
50.35	.5000
50.40	.6293
50.50	.8413
50.65	.9772
50.75	.9987

FIGURE 7-13. Distribution of Sample Means for a Population with $\mu = 50.0$

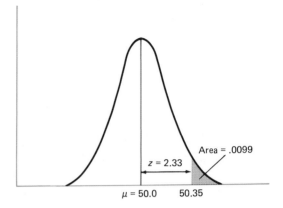

Figure 7-13 illustrates that the probability of accepting a lot where $\mu = 50.00$ and the boundary value is 50.35. The resulting .0099 appears in Table 7-2 also.

The other points on Table 7-2 are calculated in a similar manner. Logically, as the population mean increases, the probability of acceptance gets larger and approaches 1.0. These values from Table 7-2 are plotted in Figure 7-14. This curve is known as the operating characteristic curve. For any μ (average submitted weight), the probability of acceptance of the lot can be read from the curve.

FIGURE 7-14. Operating Characteristic Curve for Dry Chemical Example ($n = 36$)

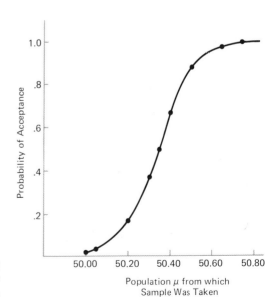

Design of the Hypothesis Test

We have learned how to select a test procedure for several different hypothesis-testing situations. Exhibit 7.2 summarized the selection process in the form of a decision tree. We have also examined types of errors which can occur in the hypothesis-test process. Some other important factors are yet to be considered. The design of a hypothesis test must consider all of these factors together. The form of the hypothesis, the setting of the acceptance boundary value and significance level, and the amount of data to collect (sample size) all interact with one another in the design of the test.

Significance Level

In the process of establishing a decision rule, a value for α (the significance level), must be set. We know that in any statistical process there is a chance of error. In this case we have established that the significance level α represents the probability of rejecting a true hypothesis H_0. In other words, it occurs when by chance the mean of a sample from the population falls outside the critical value(s). We shall discuss the impact of α upon β later in this discussion. Typical values used for α are .10, .05, .02, and .01, but sometimes even smaller values are used. In choosing one of these values, we must ask ourselves the question: What are the consequences of incorrectly rejecting our hypothesis? If our problem is to decide whether or not to place a modified product design into production, a substantial amount may be at stake and an incorrect decision could be costly. In this case the approval of an ineffective new product would be bad, just as the rejection of an effective product would be bad. Therefore, it becomes imperative to reduce the likelihood of a wrong test result. An α of .001 would provide one chance in a thousand that a good new product would be rejected. This may seem like a logical answer; however, the resulting failure to introduce the good new product may result in a considerable opportunity loss, especially if a competitor comes out with a similar product. We must, therefore, trade off the risks of an incorrect decision against the costs of obtaining enough data to reduce the possibility of error. If the costs of the product and of performing the test are low, it becomes much easier to design a plan for a higher level of significance within the cost constraints. We would just produce more items and test them. In other cases, the items may be expensive to build or to test, or they may just not be available.

Balancing α and β

Earlier in this chapter we learned how to determine a critical value and the corresponding decision rule. These concepts will now be further expanded. The final relation of α and β often becomes mostly subjective. It is difficult to make a quantitative comparison that considers all factors. Type I errors and type II errors are both undesirable, but both can occur, although not simultaneously in the same test. The seriousness of each type of error must be evaluated so that if one of the possible errors is more serious, the chance of that error must be reduced. If a drug company were evaluating a new drug, the introduction of an ineffective drug would be important to avoid; however, the introduction of a drug with possible harmful effects would be extremely serious and such an error would be costly if liability suits resulted. Again we find the quantitative tests using statistics able to provide ways to perform the evaluation; however, the final decisions must also consider many subjective aspects. As President Lincoln once said at his cabinet meeting, "The vote is 6 yea and 1 nay. I cast the no vote and the nays have it." The decision of an executive is often not a consensus of votes of others, nor it is merely conforming to the results of a formula. The subjective aspects must also be considered before the decision is made.

Setting the Critical Value: An Example

Let us consider an example where the critical value is to be set so that a shipment of material will be rejected if the mean of the sample falls below a set value. A manufacturer of camping lanterns includes a battery prior to shipping the lanterns to retailers. From past experience, he has set specifications that the batteries have a mean life of 6 hours and past data have shown the life distribution to be normal with $\sigma = .75$ hours. The manufacturer receives these batteries from a supplier in large lots. From each lot he takes a sample of 25. He desires to set up a test that will reject lots where the mean life is lower than 6 hours.

In considering the situation, it is apparent that the two types of errors can exist. These are shown in Table 7-3. In the type I error, a good lot is rejected. The type II error is the acceptance of a bad lot. Since the mean of a lot will rarely fall exactly on 6, it will be necessary to define a mean life value for a good lot and a bad lot. A good lot was therefore defined by the manufacturer (in coordination with the supplier) as having a mean of 6.3 hours, whereas a bad lot was set at 5.67 hours. This leaves a range of 5.67 to 6.3. The manufacturer is indifferent to life values in this range and is not too concerned whether the lot is accepted or rejected.

At this point we shall consider the probabilities of a type I error taking place (rejecting a good lot). Suppose that a good lot (from population where $\mu = 6.3$ and $\sigma = .75$) is received and subjected to our acceptance

Estimation and Hypothesis Testing

TABLE 7-3. Decision and Possible Errors in Battery Example

Decision Based on Sample	Actual Lot Quality (based on μ)	
	Good Lot ($\mu \geq 6.3$ hr)	Bad Lot ($\mu \leq 5.67$ hr)
Reject Lot	Type I error	Correct decision:
Accept Lot	Correct decision:	Type II error

plan. The distribution of sample means (size 25) is given in Figure 7-15. The probability that the mean of a sample will fall below our critical value of 6.0 is calculated

$$z = \frac{\bar{X} - \mu}{\sigma_{\bar{x}}} = \frac{6 - 6.3}{.75/\sqrt{25}} = \frac{-.3}{.15} = -2$$

From the normal curve tables in Appendix A, the probability of $\bar{X}$ being less than 6 is

$$\alpha = P(\bar{X} < 6.0) = .0228$$

Therefore, if we set our critical value at 6.0, the probability of rejecting a lot having an actual mean life of 6.3 is .0228. This is our type I error as defined in Table 7-3.

Consider next the probability of a type II error, which is the acceptance of a bad lot. In this case it would be the probability of accepting a lot with

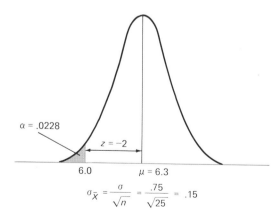

FIGURE 7-15. Distribution of Sample Means from a Good Lot ($\mu = 6.3$)

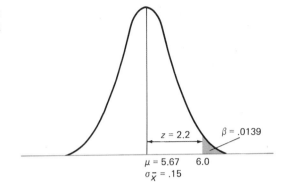

FIGURE 7-16. Distribution of Sample Means from a Bad Lot ($\mu = 5.67$)

a mean life of less than or equal to 5.67 hours. Consider an example where a lot is received having an actual mean life μ of 5.67 hours and a standard deviation σ of .75 hours. Although this information is not actually known by the company receiving the material, we shall attempt to see what the sampling plan will show when samples are taken from a lot of this quality. Although the lot is really bad, our sampling plan will accept the lot if the sample indicates an average life of 6.0 or less hours.

We are taking sample size 25 from each lot and, therefore, Figure 7-16 represents the distribution of sample means taken from the lot having an actual population mean of 5.67. The z value is calculated as

$$z = \frac{\bar{X} - \mu}{\sigma_{\bar{x}}} = \frac{6.00 - 5.67}{.15} = \frac{.33}{.15} = 2.2$$

The probability of accepting this lot will be equivalent to the area of the normal curve above 6.0. From Table I, with $z = 2.2$, the probability is .0139. Therefore, we conclude that our sampling procedure will accept a lot with an actual mean life of 5.67 hours with a probability of .0139. This is the type II error presented by β.

At this point we can restate our decision rules in Table 7-4, a revision of Table 7-3, showing actual numerical values. Our discussion can also now be restated in terms of hypothesis testing. The assumption that a lot received is a good lot would be the null hypothesis H_0. The alternative hypothesis H_a is then the assumption that the lot was bad.

$$H_0: \quad \mu \geq 6.3 \text{ hours}$$
$$H_a: \quad \mu \leq 5.67 \text{ hours}$$

If the procedure says to accept a lot, then H_0 is accepted and H_a is rejected. If, on the other hand, the sampling data and decision procedure

Estimation and Hypothesis Testing

TABLE 7-4. Summary of Decisions and Errors in Battery Example

Decision Based on Sample	Actual Lot Quality (based on μ)	
	Good Lot ($\mu \geq 6.3$ hr)	Bad Lot ($\mu \leq 5.67$ hr)
Reject Lot	Type I error: $\alpha = .0228$	Correct decision: $P = .9861$
Accept Lot	Correct decision: $P = .9772$	Type II error: $\beta = .0139$

say to reject the lot, then H_0 is rejected and H_a is accepted. With the same line of reasoning, the type I error exists when H_0 is rejected (a good lot is rejected). The type II error is made when H_0 is accepted, but the lot is actually bad.

Summary of Sample-Testing Procedure

For our problem, the complete set of decision rules is outlined below:

1. Select sample of size n (25 in our case) from a lot.
2. Compute the average life $\bar{X}$ of the sample of 25.
3. If $\bar{X} \geq 6.0$, accept the lot.
4. If $\bar{X} < 6.0$, reject the lot (where 6.0 is our boundary or critical value).

This set of rules is clear and easily followed by an inspector. He or she does not really need to understand the statistics underlying the sampling procedure. The quality engineer and supervisor should, of course, understand the risks and their effects.

Variation in Critical Value

In examining the decision rules and procedures, we can observe that the critical value and sample size are both critical elements in the sampling procedure. The risks (or errors) are dependent upon these two values. Therefore, in designing any sampling decision procedure, the selection of these two factors is of utmost importance. In order to help in this selection process, let us consider the effect of changes in these quantities upon the test results.

In our example, we had picked a critical value of 6.0 hours. What if this were adjusted to 6.1 hours? Assume again that a lot is received with a mean of 6.3 and $\sigma = .75$. This is a good lot but the sampling plan may reject it if $\bar{X}$ of the sample falls below our new critical value of 6.1

hours. The probability of this rejection (type I error in rejecting a good lot) is calculated below:

$$z = \frac{\bar{X} - \mu}{\sigma_{\bar{X}}} = \frac{6.1 - 6.3}{.15} = -\frac{.20}{.15} = -1.33$$

From the tables of a normal curve, the equivalent probability is $\alpha = .0918$. This .0918 probability is greater than the value .0228 which we obtained when the boundary was 6.0. This is logical since the critical value has been moved closer to the value for the good item of 6.3.

In an equivalent manner we can examine β, the probability of a type II error, under this new situation where the critical value is 6.1. Assume that the lot with the 5.67-hour average life were submitted. Then,

$$z = \frac{\bar{X} - \mu}{\sigma_{\bar{X}}} = \frac{6.1 - 5.67}{.15} = \frac{.43}{.15} = 2.87$$
$$\beta = .0021$$

The value of β, the type II error or probability of accepting a bad lot, has now been reduced considerably from the former value of .0139.

We observe, then, that when a critical value is adjusted, the probabilities of committing either of the types of error change. One probability increases whenever the other decreases. The ability to change the critical value then permits us to adjust the risk magnitudes; however, it does not permit us to reduce the two error probabilities, α and β, simultaneously.

Sample-Size Variation

We have seen that changing the critical value permits us to reduce one error probability only at the expense of increasing the other error probability. It will now be shown that if the size of the sample is increased, both types of error are reduced simultaneously. Continuing to use our battery-life example, let us change the original sample size of 25 to 49. At the same time, the original boundary value of 6.0 will be retained and the effects will be observed.

Consider first the probability of making a type I error:

$$z = \frac{\bar{X} - \mu}{\sigma_{\bar{X}}} = \frac{6.0 - 6.3}{.75/\sqrt{49}} = \frac{-.30}{.107} = -2.80$$
$$\alpha = .0026$$

This is much lower than the original α of .0228.

Next consider the probability of making a type II error:

$$z = \frac{\bar{X} - \mu}{\sigma_{\bar{X}}} = \frac{6.0 - 5.67}{.107} = \frac{.33}{.107} = 3.08$$

$$\beta = .0010$$

Thus, the value of β is also much lower than the original β of .0139 when the sample size was 25. This illustrates that increasing the sample size allows us to reduce both the α and β values, thus reducing the type I as well as the type II error simultaneously. By the same line of reasoning, a decrease in sample size would allow both risks to increase. The impact of the change in sample size can also be seen by examination of the operating characteristic (OC) curves. Figure 7-17 shows the OC curves for two sample sizes. The larger sample size can discriminate better between good and bad lots, as shown by the steeper OC curve.

Quite often in real situations the initial decision in selecting the size of sample to take is one of economics, since it costs more to take larger samples. The larger sample, however, produces results with lower error probabilities. The values are tabulated in Table 7-5. One of the primary applications of operating characteristic curves is the use of acceptance sampling by attributes. An example of a decision rule would be:

1. Take a sample of 100.
2. Accept if there are 3 or fewer defectives; otherwise, reject.

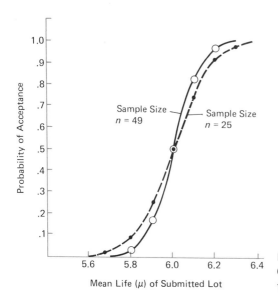

FIGURE 7-17. Operating Characteristic (OC) Curves for Battery Example

TABLE 7-5. Calculations for Figure 7-17

Populations Mean μ from which Submitted Lot is Taken	Probability of Acceptance ($n = 25$)	Probability of Acceptance ($n = 49$)
5.67	.0139	.0010
5.8	.0918	.0307
5.9	.2545	.1750
6.0	.5000	.5000
6.1	.7455	.8250
6.2	.9082	.9693
6.3	.9772	.9974

This is called sampling by attributes because we do not measure the values but merely count the defectives.

Limitations in the Use of Hypothesis Tests

The approaches developed in this chapter are applicable to a wide variety of decision-making situations. These can arise in management, whether it be the management of a business, a governmental organization, an educational institution, or a social institution such as a hospital. There are certain limitations, however, on the application of the previously discussed hypothesis-testing techniques.

One limitation is that they can be applied only where we are evaluating between two alternatives. In real-problem situations, there are often many alternatives that need to be considered. We shall discuss a technique in the next chapter by means of which an examination can be made of three or more alternatives. A second drawback to the hypothesis-testing procedure is its requirement of sample data. Any business manager knows that many decisions must be made without sample data. Frequently, it is costly to obtain samples; in other cases, sample data cannot be obtained at any cost. For example, a new drug may be an alternative treatment, but the new drug may have never been used on humans. In other decision-making situations, a decision may be needed immediately and there may not be time to take sample data.

A final limitation relates to cases where the results do not provide a clear decision one way or the other. For example, the results may be significant at the 15 per cent level. What should be done then? One way to resolve this is to take more data (a bigger sample). The size of the sample can be increased until a clear decision is apparent one way or another.

In spite of these drawbacks to hypothesis testing, the technique has many applications in management. Its use can be valuable; however, the limitations and risks must be recognized.

Exercises

7.40 For the dry-chemical example associated with Figures 7-11 and 7-12:
  (a) Calculate the probability of accepting a lot with population mean μ equal to 50.10.
  (b) Calculate the probability of accepting a lot with $\mu = 50.70$.
  (c) Compare the results of (a) and (b) to Figure 7-10.

7.41 For the dry-chemical example, calculate probabilities of acceptance (P_a) for a sample size 16. Plot the OC curve on the same sheet with the curve for $n = 36$. Plot both curves so that for $\mu = 50.35$, the $P_a = .50$.

7.42 In problem 7.41, plot the two curves on the same sheet if both curves have the same P_a at $\mu = 50.7$.

7.43 For the battery-life example plotted in Figure 7-16, plot OC curves for $n = 9, 16,$ and 100.

8

Learning Objectives

Introduction

Tabulation
   Simple Tabulation
   Cross Tabulations

Chi-Square Analysis
   Purposes
   Procedure
   Steps in Hypothesis Test
   Other Numerical Examples
   Cautions in Using Chi-Square Analysis

Analysis of Variance
   One-Way Analysis of Variance
   Two-Way Analysis of Variance
   Assumptions

Appendix

Tabulations, Chi-Square Analysis, and Analysis of Variance

Learning Objectives

As presented in a set of business examples and in discussion of the pertinent assumptions, the primary learning objectives for this chapter are:

1. To understand the basic format of tabulations and cross tabulations and why they are important.
2. To be able to work out a limited set of chi-square and analysis of variance problems by hand calculations.
3. To be able to identify situations where tabulation, cross tabulation, chi-square analysis, and analysis of variance are appropriate.
4. To understand the necessary assumptions and be able to identify when these assumptions are being violated to the point where the results will be negated.

Introduction

So far we have examined some basic concepts of statistics including frequency distributions, probability, standard deviation, variance, estimation, and hypothesis testing. This chapter serves as an extension of that material into several situations that differ from those we discussed earlier. Chi-square analysis and analysis of variance are new topics, but they will be shown to be similar to hypothesis testing. Tabulation is simply an extension of the previous work on frequency distributions. Applications of these two techniques are very common in business; however, the problems involved do not lend themselves readily to the techniques previously discussed.

Tabulation

Simple Tabulation

In this chapter we shall work through a business example for the tabulation and chi-square presentations and another example for the analysis of variance sections. The following paragraph establishes a real business example.

> Rental City, Inc. is a large-scale rental store in a city of 95,000 residents. The rental business is good in the city, but a major competitor seems to consistently get the largest share of the market. As in the bank example in Chapter 2, the management of Rental City, Inc. wanted to improve their market position. To do so it was felt that they first needed to have additional information about their current market, peoples' feelings about them, the impact of their advertising, and what types of people were "renters." After deliberation they decided on a telephone survey to collect the needed information. Some of the questions used are shown in Figure 8-1. The two tasks left after the data were collected were to make any necessary statistical calculations and then to put the data in a format that would be understandable and easy to read.

One of the first responsibilities is to report the frequency of response. Prior to the presentation of how to do and interpret tabulations, examine Figure 8-1 closely.

Tabulations of a large number of responses, such as in the current case, are generally made on a computer. In order to utilize the machine, responses to questions such as question A have to be keypunched or keytaped. The computer, in the main, deals with numbers and, as a result, individual responses have to be assigned a code number and subsequently coded. For example, each of the 501 households (response rate 85 per cent) had a response to question A. If the husband was the major rental decision maker for a household, a 1 was assigned to the response and a 1 was placed in a particular column on a code sheet and later keypunched in that same column of an IBM card. Each IBM card has 80 columns, and let us note that column 4 is reserved for question A. Therefore, if we have one IBM card for each of the 501 households, there will be a 1, 2, 3, or 4 in column 4 of each card. The computer then goes through the 501 cards and simultaneously counts how many 1, 2, 3, or 4 responses there are; it can then calculate percentages, cumulative percentages, and other necessary statistics.

Table 8-1 contains a number of interesting points:

1. As viewed individually, the husband is reported as being the major decision maker in approximately 37 per cent of the cases.

2. The wife is reported as the major decision maker in 19 per cent of the cases (rounded).

3. However, as we examine the cumulative frequency percentage column, we can see that the husband or wife individually is the major decision maker in only 56 per cent of the cases.

Rental City, Incorporated Questionnaire

A. Who in your household has the major influence on whether you rent products such as lawnmowers, hedge trimmers, party equipment, and carpet cleaners?
   _____ 1. Husband
   _____ 2. Wife
   _____ 3. Both
   _____ 4. Other (Please specify) _____

B. Would you tell me if you strongly disagree, disagree, neith agree nor disagree, agree, or strongly agree with the following statement:
   It is best to own equipment
   you occasionally need.   _____ _____ _____ _____ _____
                              SD   D   ND/NA   A   SA

C. Has anyone in your household *rented* equipment in town within the last two years?
   _____ 1. Yes
   _____ 2. No
   _____ 3. Don't know

D. Were you and/or other members of your household satisfied with renting?
   _____ 1. Yes
   _____ 2. No

E. When you or a member of your household rents equipment, who typically picks up and returns this equipment from the rental store?
   _____ 1. Husband
   _____ 2. Wife
   _____ 3. Son or Daughter
   _____ 4. Delivery
   _____ 5. Other (Please specify) _____

F. Where in town have you rented an item within the last two years?
   _____ 1. Carters
   _____ 2. Rental City, Inc.
   _____ 3. Klems
   _____ 4. Hardware Store (Name)_____
   _____ 5. Service Station (Name)_____
   _____ 6. Furniture Store (Name)_____
   _____ 7. Paint Store (Name)_____
   _____ 8. Other (Please specify)_____

G. What rental stores are you aware of in town?
   _____ 1. Rental City, Inc.
   _____ 2. Carters
   _____ 3. Other (Please specify)
   _____ 4. None

H. Which one of these letters best describes the category in which your age as of your last birthday falls?
   _____ 1. A (18-24)
   _____ 2. B (25-34)
   _____ 3. C (35-49)
   _____ 4. D (50-64)
   _____ 5. E (65 and over)

I. Which letter comes closest to your total yearly *family* income last year before taxes?
   _____ 1. A (Under $5,000)
   _____ 2. B ($5,000-$9,999)
   _____ 3. C ($10,000-$14,999)
   _____ 4. D ($15,000-$24,999)
   _____ 5. E ($25,000 and over)

J. What letter corresponds to the last level of education your household head obtained?
   _____ 1. A (Grammar School)
   _____ 2. B (1-2 years High School)
   _____ 3. C (High School Graduate)
   _____ 4. D (1-3 years College)
   _____ 5. E (College Graduate)
   _____ 6. F (Graduate Work)
   _____ 7. G (Trade/Professional School Graduate)
            (Please specify)_____

FIGURE 8-1. Rental City, Incorporated Questionnaire

TABLE 8-1. Tabulation 1: Question A—Who Has Major Rental Influence

Code	Absolute Frequency	Relative Frequency (%)	Cumulative Frequency (%)
1 (Husband)	186	37.1	37.1
2 (Wife)	94	18.8	55.9
3 (Both)	192	38.3	94.2
4 (Other)	29	5.8	100.0
	501	100.0	

4. An additional 38 per cent of the respondents indicate that there is a joint decision on renting, with 6 per cent of the responses revealing other influences such as children, friends, and parents.

Data presented in such a fashion can be very useful to the management of Rental City, Inc. First, they may have thought that decisions were predominantly made by the husband or the wife but not realized how much joint decision making took place. An understanding of this should have an impact on advertising content, in-store service, and employee training.

With respect to question B in Figure 8-1, we can see that it is an attitude question such as we discussed in Chapter 2. The responses to this question are shown in Table 8-2. The format of the tabulation shown in Table 8-2 is identical to that of Table 8-1. The question itself is considerably different, however. About 53 per cent of the respondents either agree or strongly agree that equipment which is needed only occasionally should still be owned. There are 21 per cent of the respondents who do

TABLE 8-2. Tabulation 2: Question B—Attitude Toward Owning Equipment

Code	Absolute Frequency	Relative Frequency (%)	Cumulative Frequency (%)
1 (Strongly Agree)	21	4.2	4.2
2 (Agree)	243	48.5	52.7
3 (Neither Disagree or Agree)	103	20.6	73.3
4 (Disagree)	130	25.9	99.2
5 (Strongly Disagree)	4	.8	100.0
	501	100.0	

TABLE 8-3. Tabulation 3: Question C—Rental Within the Last 2 Years

Code	Absolute Frequency	Relative Frequency (%)	Cumulative Frequency (%)
1 (Yes)	343	68.5	68.5
2 (No)	158	31.5	100.0
	501	100.0	

not have an attitude one way or the other, with about 27 per cent either disagreeing or strongly disagreeing with the statement. This can show management that there is a larger group of people who *feel* that they should own equipment than there is of people who feel that they do not have to. The implication in this case is that we need to draw people from the neutral group in order to improve attitudes toward renting.

Prior to moving on to the topic of cross tabulation, two more tables will be shown to demonstrate other ways that simple tabulations can be used. Question C examines the number of respondents who indicate that someone in their household has rented equipment within the last 2 years. In this case Table 8-3 indicates a simple frequency count of the yes and no answers. The group of yes answers can further be examined as shown in Table 8-4.

From Table 8-4 we can separate renters from nonrenters and note that 325 of the 343 renters were satisfied with their renting experience, while 18 were not. As compared to the 501, the relative frequencies are as shown in the table. We can go ahead and ask what percentage of the *renters* were satisfied or dissatisfied and find that $325/343 = 94.8$ per cent

TABLE 8-4. Tabulation 4: Question D—Renters Satisfied with Their Renting Experience

Code	Absolute Frequency	Relative Frequency (%)	Cumulative Frequency (%)
0 (Not Renters)	158	31.5	31.5
1 (Yes)	325	64.9	96.4
2 (No)	18	3.6	100.0
	501	100.0	

TABLE 8-5. Tabulation of Remaining Questions

Code	Absolute Frequency	Relative Frequency (%)	Cumulative Frequency (%)
Question E			
0 (Don't Rent)	158	31.5	31.5
1 (Husband)	202	40.3	71.8
2 (Wife)	75	15.0	86.8
3 (Son or Daughter)	9	1.8	88.6
4 (Delivery)	10	2.0	90.6
5 (Other)	47	9.4	100.0
	501	100.0	
Question F[a]			
1 (Carters)	197	42.1	42.1
2 (Rental City)	163	34.8	76.9
3 (Klems)	14	3.0	79.9
4 (Hardware Store)	9	1.9	81.8
5 (Service Station)	4	.9	82.6
6 (Furniture Store)	4	.9	83.4
7 (Paint Store)	8	1.7	85.0
8 (Other)	69	14.7	99.7
	468	100.0	
Question G[b]			
1 (Rental City)	373	41.7	41.7
2 (Carters)	443	49.5	91.2
3 (Other)	58	6.5	97.7
4 (None)	21	2.3	100.0
	895	100.0	

[a] Recall that those people who have rented can have multiple renting experiences over the past 2 years.
[b] People can be aware of more than one rental store.

were satisfied and $18/343 = 5.2$ per cent were not. The other tabulations appear in Table 8-5.

Cross Tabulations

The information given in the previous tabulations can be very useful to the management of Rental City, Inc., in their decision making. These decisions can pertain to alteration of strategies, reanalysis of the market, and changes in promotional materials. Although Rental City, Inc., can get a fairly clear picture of the descriptive profiles offered in the tabulations, additional questions need to be answered. Management should want to know more about these people who are their customers or potential customers. By this we do not mean names, but general cate-

TABLE 8-5. Continued

Code	Absolute Frequency	Relative Frequency (%)	Cumulative Frequency (%)
Question H			
1 (18–24)	91	18.2	18.2
2 (25–34)	127	25.3	43.5
3 (35–49)	117	23.4	66.9
4 (50–64)	99	19.7	86.6
5 (65 and over)	67	13.4	100.0
	501	100.0	
Question I			
1 (Under $5,000)	40	8.0	8.0
2 ($5,000–$9,999)	76	15.2	23.2
3 ($10,000–$14,999)	110	22.0	45.2
4 ($15,000–$24,999)	151	30.0	75.2
5 ($25,000 and over)	67	13.4	88.6
6 (Refusal)	57	11.4	89.6
	501	100.0	100.0
Question J			
1 (Grammar School)	19	3.8	3.8
2 (1–2 years High School)	31	6.2	10.0
3 (High School Graduate)	116	23.1	33.1
4 (1–3 years College)	114	22.8	55.9
5 (College Graduate)	95	19.0	74.9
6 (Graduate Work)	96	19.1	94.0
7 (Trade/Professional School Graduate)	30	6.0	100.0
	501	100.0	

gories such as age, income, education, and other variables. When they examine these data the types of people who have given particular responses can be more accurately pinpointed. This leads us to the very important management tool of cross classification.

For example, management may wish to know what proportion of those people who had rented within the last two years were aware of Rental City, Inc., and of their major competitor, Carters. This calls for what is called a *two-way cross classification*. Essentially, this means exactly what it says in that the responses to two questions are cross-classified against one another.

In Table 8-6 the firm is interested in finding out those people who are aware of Rental City as compared to those persons who rented over a 2-year period. Prior to drawing conclusions, let us examine the content

TABLE 8-6. Two-Way Cross Classification: Renters × Awareness of Rental City

Rented in Past 2 Years	Aware of Rental City		Totals
	Yes	No	
Yes	272 79.3% (row) 72.9% (column) 54.3% (total)	71 20.7% 55.5% 14.2%	343 68.5%
No	101 63.9% 27.1% 20.2%	57 36.1% 44.5% 11.4%	158 31.5%
Totals	373 74.5%	128 25.5%	501 100.0%

of the table. First, it is a 2 × 2 table, as there are 2 rows and 2 columns. In all it contains 4 "cells." In the upper left cell we see that 272 people who rented in the last 2 years (from anyone) were also aware of Rental City. The 272 respondents in this cell made up 79.3 per cent of the 343 total renters. Also shown is that the 272 total renters were 72.9 per cent of the 373 people aware of Rental City. In addition, the 272 respondents made up 54.3 per cent of the total respondents in the sample. Other cells can be interpreted in a similar manner.

Two major conclusions can be drawn from the results. First, of those people aware of Rental City, only 27.1 per cent had not rented (somewhere) in the last 2 years. However, of those people who had rented over the last 2 years, 20.7 per cent were not even aware of Rental City. This has serious implications to management in that a customer must be aware of a store's existence before making a decision to rent from that store.

However, as we examine the table for the major competitor, Carters, some different results surface. Table 8-7 indicates that 90.4 per cent of the people who rented were aware of Carters (as compared to 79.3 for Rental City). This leaves only 9.6 (100 − 90.4) per cent of the renters who were not aware of Carters, which is good awareness penetration. Of those people aware of Carters, 70.1 per cent had rented, which is roughly comparable to Rental City's results. This table is very revealing to Rental City as they examine the reasons for having a smaller market

TABLE 8-7. Two-Way Cross Classification: Renters × Awareness of Carters

Rented in Past 2 Years	Aware of Carters		Totals
	Yes	No	
Yes	310 90.4% (row) 70.1% (column) 61.9% (total)	33 9.6% 55.9% 6.6%	343 68.5%
No	132 83.5% 29.9% 26.3%	26 16.5% 44.1% 5.2%	158 31.5%
Totals	442 88.2%	59 11.8%	501 100.0%

share than Carters. One of the prime reasons appears to be awareness. We should note that management could then test whether the difference between 90.4 and 79.3 per cent was due to more than chance alone. Hypothesis-testing methods for doing this were presented in Chapter 7.

Prior to discussing a method of analysis useful with these types of tables (chi-square analysis), two more cross-classification tables yield interesting results.

Of the 128 people not aware of Rental City in Table 8-8 (row 1, sum of frequencies), it appears that the "weak points" are in the 18 to 24 and 25 to 34 age groups, where 23.4 and 24.2 per cent, respectively, are unaware of Rental City. The 35 to 49 age group is quite aware in that 99/117, or 84.6 per cent, of its members are aware of Rental City, Inc. as compared to roughly 74 per cent of the overall sample. Further analysis shows that 33 per cent of the 18 to 24 age group is not aware and 37.3 per cent of the 65 and over group is not aware. We should be careful to not simply look at the number who are aware in each cell, because the age-group sizes vary and we would expect more relative awareness in the 25 to 34 group than in the 65 and over group. It is most important to examine the proportion of the group who are aware as compared to the proportion for the overall sample who are aware. As differences in these proportions are noted, weak points of awareness can be spotted. Without further discussion, the results are useful in helping to pinpoint weaknesses in the market.

TABLE 8-8. Two-Way Cross Classification: Awareness of Rental City × Age

Aware of Rental City	Age					Totals
	18–24	25–34	35–49	50–64	65 and over	
No	30 23.4 33.0 6.0	31 24.2 24.4 6.2	18 14.1 15.4 3.6	24 18.8 24.2 4.8	25 19.5 37.3 5.0	128 25.5
Yes	61 16.4 67.0 12.2	96 25.7 75.6 19.2	99 26.5 84.6 19.8	75 20.1 75.8 15.0	42 11.3 62.7 8.4	373 74.5
Totals	91 18.2	127 25.3	117 23.3	99 19.8	67 13.4	501 100.0

TABLE 8-9. Two-Way Cross Classification: Awareness of Rental City × Occupation

Aware of Rental City	Occupation[a]								Totals
	1	2	3	4	5	6	7	8	
No	35 27.3% 28.9% 7.0%	9 7.0% 15.8% 1.8%	14 10.9% 12.5% 2.8%	6 4.7% 25.0% 1.2%	5 3.9% 14.7% 1.0%	7 5.5% 20.0% 1.4%	17 13.3% 51.5% 3.4%	35 27.3% 41.2% 7.0%	128 25.5%
Yes	86 23.0% 71.1% 17.2%	48 12.9% 84.2% 9.6%	98 26.3% 87.5% 19.6%	18 4.8% 75.0% 3.6%	29 7.8% 85.3% 5.8%	28 7.5% 80.0% 5.6%	16 4.3% 48.5% 3.2%	50 13.4% 58.8% 10.0%	373 74.5%
Totals	121 24.1%	57 11.4%	112 22.4%	24 4.8%	34 6.8%	35 7.0%	33 6.6%	85 17.0%	501 100.0%

[a] 1. Professional
2. Managers and owners
3. Skilled/unskilled labor
4. Clerical/secretarial
5. Sales
6. Service
7. Student
8. Retired/unemployed

One final table will be presented for limited discussion. Briefly, Table 8-9 can be used for the same purpose as previous tables. Although a number of statements can be made, the following are examples:

1. The clerical/secretarial and student groups show a low degree of awareness as compared to the other occupational classes. Twenty-five per cent of the former and 51.5 per cent of the latter classes are unaware of Rental City. If these groups represent a viable market, extensive new business may be available from this group.

2. Managers and owners as well as sales personnel seem to demonstrate somewhat greater awareness than other groups.

3. Although professional people may or may not be in the market for renting, they are a large group and 28.9 per cent were not aware of Rental City. The same conclusion applies to retired and unemployed persons.

In summary, it is hoped that the need for tabulation and cross-tabulation analysis is clear and that its usefulness in decision making is also clear. The topic sounds so simple that it is often skipped, although it is an excellent and pertinent area of statistics.

Exercises

8.1 Interpret the following tabulations from question A in Figure 8-1.

(a)

	Absolute Frequency	Relative Frequency (%)	Cumulative Frequency (%)
Husband	187	37.3	37.3
Wife	94	18.8	56.1
Both	192	38.3	94.4
Other	28	5.6	100.0
	501	100.0	

(b) To respondents who reported knowing about Carters and Rental City, the question was asked, "How did you find out about these stores?"

	Absolute Frequency	
	Carters	Rental City, Inc.
Newspaper	52	57
Radio	42	47
Friend	89	86
Saw when driving by	149	149
Don't remember	35	64
Other	71	34
	438	437

8.2 Interpret the following cross classifications.

(a)

Aware of Rental City	Marital Status				
	Single	Married	Separated/Divorced	Widowed	Totals
No	30	71	9	18	128
Yes	43	283	24	23	373
Totals	73	354	33	41	501

(b)

Aware of Rental City	Income		
	No	Yes	Totals
Under $5,000	17	23	40
$5,000–$9,999	25	51	76
$10,000–$14,999	32	78	110
$15,000–$24,999	30	121	151
$25,000 and Over	7	60	67
Refusal	17	40	57
Totals	128	373	501

Chi-Square Analysis

We have just completed examining a set of data in tabular form. A cross-classification table, such as age versus awareness or occupation versus awareness, is also known as a *contingency table*. This type of contingency table can be called a $r \times c$ table, where r represents the number of rows and c represents the number of columns. For example, our occupation by awareness data in Table 8-9 had 2 rows and 8 columns, making it a 2×8 contingency table with 16 cells.

Purposes

Although we have previously examined the cross-classification (contingency) tables, we have not really asked the basic questions: Are age and awareness independent of one another, or are occupation and awareness independent of one another? The reason we wish to examine this independence more closely is to determine if the differences across cells exist because occupation or age is related to awareness, *or* if the cell differences exist simply because of chance alone or sampling error. Once again, then, we are faced with accounting for possible variations due to sampling error.

As we have seen in previous examples, results obtained in a survey or another type of study do not always correspond to what we would expect. In other words, results do not seem to conform to rules of probability. To examine whether the results deviate extensively from what may be expected because of chance alone, we can perform a chi-square test.

Procedure

We perform a χ^2 (chi-square) test by comparing the observed set of data to another set of data operating under the null hypothesis that there are no relationships between the factors being studied. For example, we may set up a null hypothesis that there is no relationship between age and awareness. The χ^2 formula is as follows:

$$\chi^2 = \Sigma \left[\frac{(f_o - f_e)^2}{f_e} \right]$$

where f_o refers to the observed frequency in a cell of the table, and f_e is the expected cell frequency for a cell in the table. More will be explained about the χ^2 distribution and its properties later, but first let us examine exactly how the χ^2 formula works.

For our original contingency table shown in Table 8-6, we had:

Rented in Past 2 Years	Aware of Rental City		Totals
	Yes	No	
Yes	272	71	343
No	101	57	158
Totals	373	128	501

Now, what we are testing is whether awareness and rental behavior are independent. The null hypothesis states:

Awareness of Rental City and renting behavior in the last 2 years *are* independent of one another.

If we operate under this null hypothesis, it is the same as having the following table:

Rented in Past 2 Years	Aware of Rental City		Totals
	Yes	No	
Yes			343
No			158
Totals	373	128	501

In other words, if these are the only values we have, what would we calculate the cell frequencies to be? We know that 343/501 of the *total* respondents rented in the last 2 years and that 158/501 did not rent. There were 373/501 of the respondents aware of Rental City with 128/501 not aware. Using these figures, we can calculate the expected frequencies under the null hypothesis. If 343/501 of the *total* respondents rented in the last 2 years, we would also expect 343/501 of the respondents aware of Rental City to have rented in the last 2 years:

$$343/501 \times 373 = 255.3672$$

This represents the expected number of aware respondents to have rented in the last 2 years. Applying the same formulation to all cells in the analysis, Table 8-10 can be derived. (In the present case we are dealing with a 2×2 table. Applications to other than 2×2 tables will be treated later.)

Next, given the expected cell frequencies f_e, we can set up Table 8-11, which compares expected and actual frequencies. From this table we can calculate the χ^2 statistic:

TABLE 8-10. Expected Cell Frequencies: Renters × Awareness of Rental City

Rented in Past 2 Years	Aware of Rental City		Totals
	Yes	No	
Yes	$\frac{343}{501} \times 373 = 255.37$	$\frac{343}{501} \times 128 = 87.63$	343
No	$\frac{158}{501} \times 373 = 117.63$	$\frac{158}{501} \times 128 = 40.37$	158
Totals	373	128	501

TABLE 8-11. Expected and Observed Frequencies

f_{o_1} 272	f_{e_1} 255.37	f_{o_2} 71	f_{e_2} 87.63
$\frac{(f_{o_1} - f_{e_1})^2}{f_{e_1}} = 1.08$		$\frac{(f_{o_2} - f_{e_2})^2}{f_{e_2}} = 3.16$	
f_{o_3} 101	f_{e_3} 117.63	f_{o_4} 57	f_{e_4} 40.37
$\frac{(f_{o_3} - f_{e_3})^2}{f_{e_3}} = 2.35$		$\frac{(f_{o_4} - f_{e_4})^2}{f_{e_4}} = 6.85$	

$$\chi^2 = \Sigma \left[\frac{(f_{o_1} - f_{e_1})^2}{f_{e_1}} + \frac{(f_{o_2} - f_{e_2})^2}{f_{e_2}} + \frac{(f_{o_3} - f_{e_3})^2}{f_{e_3}} + \frac{(f_{o_4} - f_{e_4})^2}{f_{e_4}} \right]$$

$$= \Sigma \left[\frac{(272 - 255.37)^2}{255.37} + \frac{(71 - 87.63)^2}{87.63} + \frac{(101 - 117.63)^2}{117.63} \right.$$

$$\left. + \frac{(57 - 40.37)^2}{40.37} \right]$$

$$= 13.44$$

Once we have calculated the expected cell frequencies, we are ready to move on with the χ^2 test. First, we need to note a few points about these types of tests. If we have a 2×2 contingency table as shown below,

	B_1	B_2	
A_1	X	X	5
A_2	X	X	7
	4	8	

and we are given one of the cell values, say the upper left value,

	B_1	B_2	
A_1	1	(4)	5
A_2	(3)	(4)	7
	4	8	

then the circled values are all fixed, given that we know the marginal totals. If we know $A_1 B_1 = 1$, then each other value is automatically determined. Because of this, we say that only one cell is free to vary. The term for the number cells free to vary is called *degrees of freedom* and in χ^2 analysis, it refers to the total number of cells free to vary in your contingency table. More generally, the formula for calculating the total number of cells that can independently vary is

$$df = (r - 1)(c - 1)$$

where r and c stand for the number of rows and number of columns, respectively. In this case we have

$$df = (2 - 1)(2 - 1)$$

Because each independent cell is free to take on any value within the boundaries of the problem, the more independent cells, the more variation due to sampling error. In other terms, we expect there to be a larger computed χ^2 simply due to chance as the number of independent cells in the problem increases.

Therefore, when we examine Appendix G, we can note a few important points. As we did for the t distribution, we have to calculate df prior to using the table. We find the df values down the far left column. Across the top row we see our now familiar probability values. Within the body of the table are the tabled χ^2 values. Note that, unlike the t distribution, the tabled values *increase* as df increases. The reason for this is as discussed: The more independent cells, the more sampling error expected. For example, we have 1 df; for a 4 × 3 table, there would be 6 df. The χ^2 value at the .05 level for 1 df = 3.84 and for 6 df = 12.6. Note that the .05 refers to the area under the right-hand tail of the distribution. Therefore, using general terminology, the χ^2 values in the table represent *tolerance limits* of sampling error. If we obtain a computed χ^2 greater than the table χ^2 at a given significance level, we shall reject the null hypothesis and say that the factors are not independent of one another. Recall that α and β error possibilities are still operant and that we have an α error probability equal to our significance level. In terms of our example problem, we had a calculated χ^2 of 13.44. The tabled value of $\chi^2 = 3.84$. Because our computed value is greater than the tabled value, we reject the null hypothesis. What this means is that rental behavior and awareness are not independent of one another and therefore *are* related.*

Steps in the Hypothesis Test

Before returning to more numerical examples, we can set up the specific steps used in a χ^2 test. We can immediately see that these steps are very similar to those discussed in Chapter 7.

1. Establish the null hypothesis. (In our case, it usually says that two factors are not related or are independent of one another.)
2. Determine the level of significance (.10, .05, .02, and .01 are commonly used).
3. Compute the expected frequencies under the null hypothesis as shown.
4. Calculate the χ^2 value using $\Sigma[(f_o - f_e)^2/f_e]$, and determine the degrees of freedom.
5. Compare the computed χ^2 value to the tabled χ^2 value at the appropriate df and α level.

* For additional information on dealing with a 2 × 2 chi-square table, see the appendix at the end of this chapter.

6. If the computed χ^2 is greater than the tabled value, reject the null hypothesis. If it is equal to or less than the tabled value, accept the null hypothesis.

Other Numerical Examples

As we should recall from the cross-classification section, we also examined the relation between awareness and age as well as awareness and occupation. We never did examine whether these factors were independent of one another. A key question is: Could the results be due simply to sampling error? We can, and should, also run χ^2 tests on these contingency tables. The age table, giving the actual frequencies, was as follows:

Aware of Rental City	Age					Totals
	18–24	25–34	35–49	50–64	65 and over	
No	30	31	18	24	25	128
Yes	61	96	99	75	42	373
Totals	91	127	117	99	67	501

Table 8-12 shows the calculations of expected cell frequencies (f_e).

TABLE 8-12. Expected Frequencies: Awareness of Rental City × Age

Aware of Rental City	Age					Totals
	18–24	25–34	35–49	50–64	65 and over	
No	$\frac{128}{501} \times 91$ = 23.25	$\frac{128}{501} \times 127$ = 32.45	$\frac{128}{501} \times 117$ = 29.89	$\frac{128}{501} \times 99$ = 25.29	$\frac{128}{501} \times 67$ = 17.12	128
Yes	$\frac{373}{501} \times 91$ = 67.75	$\frac{373}{501} \times 127$ = 94.55	$\frac{373}{501} \times 117$ = 87.11	$\frac{373}{501} \times 99$ = 73.71	$\frac{373}{501} \times 67$ = 49.88	373
Totals	91	127	117	99	67	501

$$\chi^2 = \Sigma \left[\frac{(30 - 23.25)^2}{23.25} + \frac{(31 - 32.45)^2}{32.45} + \frac{(18 - 29.89)^2}{29.89} + \frac{(24 - 25.29)^2}{25.29} \right.$$

$$\left. + \frac{(25 - 17.12)^2}{17.12} + \frac{(61 - 67.75)^2}{67.75} + \frac{(96 - 94.55)^2}{94.55} + \frac{(99 - 87.11)^2}{87.11} \right.$$

$$+ \frac{(75 - 73.71)^2}{73.71} + \frac{(42 - 49.88)^2}{49.88} \Bigg]$$

$$= 1.96 + .065 + 4.73 + .066 + 3.63 + .67 + .02 + 1.62 + .02 + 1.24$$
$$= 14.02$$

Therefore, following good form, we can solve the problem.

1. H_0: Age and awareness are *not* related (are independent). H_a: Age and awareness are related (are dependent).
2. Let us test at .05.
3. The expected and observed frequencies are as shown previously.
4. The calculated $\chi^2 = 14.02$ with $(r-1)(c-1)$ df or $(2-1)(5-1) = 1 \times 4 = 4$ df.
5. The calculated χ^2 (14.02) is greater than the critical value found in the table (9.49).
6. We reject the null hypothesis and accept the alternative.

We can also examine the relation between awareness and occupation in a similar fashion.

1. H_0: Awareness and occupation are not related. H_a: Awareness and occupation are related.
2. α level = .05.
3.

	\multicolumn{8}{c}{Expected Frequencies}							
	1	2	3	4	5	6	7	8
0	$\frac{128}{501} \times 121$ $= 30.91$	$\frac{128}{501} \times 57$ $= 14.56$	$\frac{128}{501} \times 112$ $= 28.61$	$\frac{128}{501} \times 24$ $= 6.13$	$\frac{128}{501} \times 34$ $= 8.69$	$\frac{128}{501} \times 35$ $= 8.94$	$\frac{128}{501} \times 33$ $= 8.43$	$\frac{128}{501} \times 85$ $= 21.72$
1	$\frac{373}{501} \times 121$ $= 90.09$	$\frac{373}{501} \times 57$ $= 42.44$	$\frac{373}{501} \times 112$ $= 83.39$	$\frac{373}{501} \times 24$ $= 17.87$	$\frac{373}{501} \times 34$ $= 25.31$	$\frac{373}{501} \times 35$ $= 26.06$	$\frac{373}{501} \times 33$ $= 24.57$	$\frac{373}{501} \times 85$ $= 63.28$

$$\chi^2 = .54 + 2.12 + 7.46 + .003 + 1.57 + .42 + 8.71 + 8.12 + .19 + .73$$
$$+ 2.56 + 0 + .54 + .14 + 2.99 + 2.79$$

4. $\chi^2 = 38.88$, df = 7.
5. Critical $\chi^2_{.05, df=7} = 14.1$.
   Computed $\chi^2 >$ tabled χ^2.
6. Therefore, reject H_0.

Cautions in Using Chi-Square Analysis

For any statistical technique there are certain assumptions and/or conditions that have to be met in order to make a correct application. The technique of χ^2 analysis has four conditions necessary to make it a useful technique. Violation of these conditions has led to many misapplications in industry, government, and academic problem solving.

1. The sample observations must be collected independently of one another. Essentially, what this means is that an observation in one cell should have no influence over observations in another cell; for example, your answer on age should have no influence over someone else's answer.
2. The data should be shown as frequencies and not in percentages or ratios. Percentages can always be converted back to original units and ratios put back in their original format, but the important point to make is that in almost all cases this transformation back to frequencies must be made prior to the χ^2 computations.
3. As was the case for much of our prior work, the sample used must be a random sample. The reasoning for this has been presented previously.
4. The sample should have at least 50 observations with no fewer than 5 expected observations in any single cell. Generally, chi-square tests should not be performed on contingency tables with cells that contain zero observations. In this instance cells can be consolidated or collapsed together. For example, consolidate $0 to $4,999 income and $5,000 to $9,999 into a category of $0 to $9,999. Because small samples are subject to a great deal of sampling error, and because we perform many squaring operations in χ^2 analysis, the reasons for a large, well-distributed sample should be evident.

We have dealt with 2×2, 2×5, and 2×8 tables. Procedures for χ^2 calculations where there are more than 2 rows are exactly the same as shown, with the computation of df still being $(r-1)(c-1)$.

Exercises

8.3 For the marital-status cross-classification table shown in problem 8.2(b), perform a χ^2 test at the .05 level. Be sure to follow good form.

8.4 Based upon your conclusions in problem 8.3 and/or material shown in the chapter, what recommendations can you make to management?

8.5 *Golf Life* magazine is interested in whether the purchase of a new brand of registered golf ball is related to readership of their magazine. Through use of random sampling procedures, they have collected the following data:

	User	Nonuser	Totals
Reader	277	494	771
Nonreader	197	532	729
Totals	474	1,026	1,500

Test the appropriate hypotheses using good form and χ^2 analysis.

8.6 A large automobile manufacturer is interested in predicting purchase patterns for a new sporty automobile they are producing. They are currently producing the car in four colors and want to relate color preference to sex of purchaser. The following data have been collected:

	White	Metallic Green	Off-Red	Silver Mist	Totals
Male	261	237	176	421	1,095
Female	132	197	237	339	905
Totals	393	434	413	760	2,000

Test the appropriate hypothesis and interpret your conclusion.

8.7 The following data on product usage and age have been collected by a small beer distributor:

	18–24	25–34	35–49	50–64	65 and over	Totals
Use Product	17	16	3	12	7	55
Don't Use Product	12	7	21	0	8	48
Totals	29	23	24	12	15	103

(a) Can χ^2 analysis be applied to these data? Why or why not? If so, please perform the appropriate test.
(b) What recommendations do you have for management?

Analysis of Variance

Another important aid in decision making is analysis of variance. Analysis of variance also involves hypothesis testing as presented in Chapter 7 and the χ^2 analysis material presented in this chapter. *Analysis of Variance* (often called ANOVA) measures whether there are differences among three or more factors. For example, the material presented in Chapter 7 showed us how to examine the significance of differences between two sample means or between a sample mean and a parameter. In the present case, we want to test differences among three or more sample means.

Once again, as we saw in previous hypothesis testing, we are going to take a ratio of a number as compared to an estimated sampling error. The situation that we find ourselves in is as follows:

As a reasonably large residential air conditioning and heating company, we maintain a fairly large fleet of service and installation vans. We are soon to make a decision on the purchase of a new group of vans and need to have some accurate data on miles per gallon (mpg) for 4 makes of truck. The following data are available:

Truck 1 (mpg)	Truck 2 (mpg)	Truck 3 (mpg)	Truck 4 (mpg)
14	15	10	13
11	13	12	11
12	12	11	12
11	16	10	15
11	12	12	13
$\bar{X}_1 = 11.8$	$\bar{X}_2 = 13.6$	$\bar{X}_3 = 11.0$	$\bar{X}_4 = 12.8$

We have taken a sample from the 4 brands of trucks and recorded observations for 5 trucks of each of the 4 brands. After compiling the data, we found the sample means for each brand of truck. As we examine the sample means, we note that the rank order by magnitude (in mpg) is:

Truck 2: 13.6
Truck 4: 12.8
Truck 1: 11.8
Truck 3: 11.0

However, we also know from our past work that the differences among these sample means could simply be due to sampling error. As a result, we need to test a hypothesis regarding whether the difference could be due to sampling error.

Prior to discussing how we test this hypothesis, let us go back to the original data and examine them for a minute. As we look from column to column we see that there is a variation among the trucks. For example, in row 1 as we go across columns, the mpg go from 14 to 15 to 10 to 13. The same applies to an examination of the differences by column for rows 2, 3, and 4. Therefore, we can conclude that the differences *among* trucks appears to be substantial. This difference is the one we are interested in examining. Further, though, when we look down the columns of the data we can see that there are substantial differences for each individual truck. Truck make 1 had 14, 11, 12, 11, and 11, with truck 2 having 15, 13, 12, 16, and 12, and so forth. The differences we are discussing now are differences *within* trucks. The differences within trucks cannot be attributable to mpg differences *among* differing brands of truck. As a result, we can use these within-truck differences as an estimate of our sampling variability or sampling error. Therefore, we derive a ratio as follows:

$$F = \frac{\text{sampling error} + \text{variance due to differences in make of truck}}{\text{sampling error}}$$

This ratio is called the *F ratio* and would be expected to equal 1 if there were no differences among trucks. However, as differences among trucks become important, the right side of the numerator will increase and the overall *F* ratio will increase. As the *F* ratio increases, we become more and more convinced that there are significant differences among trucks. Once again, we can establish a critical value and when the *F* ratio exceeds the critical value, we shall reject our null hypothesis.

One-Way Analysis of Variance

The steps in conducting an analysis of variance are identical to those discussed previously for hypothesis testing. In the present case, we can say that μ_1, μ_2, μ_3, and μ_4 are the true population means for each brand of truck. The null hypothesis is that $\mu_1 = \mu_2 = \mu_3 = \mu_4$ and the alternative is that μ_1, μ_2, μ_3, and μ_4 are not all equal.

The theory of analysis of variance and long methods of computation are well covered elsewhere. As our emphasis is on interpretation, we shall utilize the "shortcut" method of ANOVA. This procedure is followed by using formulas that deal with calculations of sums of squares. Prior to examining the formulas, let us agree to use the following terms:

SST = sum of squares total
SSA = sum of squares among
SSW = sum of squares within
k = number of factors (trucks) being examined
n = number of observations per factor
X_i = individual observation within the data matrix
T = sum of all observations in the data matrix

Using these terms, we have

$$\text{SST} = \sum_{i=1}^{k} \sum_{j=1}^{n} X_{ij}^2 - \frac{1}{kn} \cdot T^2$$

$$\text{SSA} = \frac{1}{n} \cdot \sum_{i=1}^{k} T_i^2 - \frac{1}{kn} \cdot T^2$$

$$\text{SSW} = \text{SST} - \text{SSA}$$

Following a step-by-step procedure, let us examine how we test the one-way ANOVA hypothesis. It is called one-way because we are only examining one cause of variability: brand of truck.

STEP 1:
Square each individual value in the data matrix and sum these values:

$$\sum_{i=1}^{k} \sum_{j=1}^{n} X_{ij}^2 = \Sigma \, (14^2 + 11^2 + 12^2 + 11^2 + 11^2 \cdots + 12^2 + 15^2 + 13^2)$$

$$= 196 + 121 + 144 + 121 + 121 + 225 + 169 + 144$$
$$+ 256 + 144 + 100 + 144 + 121 + 100 + 144$$
$$+ 169 + 121 + 144 + 225 + 169 = 3{,}078$$

STEP 2:
Add all values (unsquared) in the data matrix:

$$T = (14 + 11 + 12 + 11 + 11 + 15 + 13 + 12 + 16 + 12 + 10 + 12$$
$$+ 11 + 10 + 12 + 13 + 11 + 12 + 15 + 13) = 246$$

STEP 3:
Square the results of step 2 and divide by the total number of observations in the data matrix; calculate SST:

$$\frac{1}{kn} \cdot T^2 = \frac{1}{4 \cdot 5} \cdot 246^2 = 3{,}025.8$$

$$\text{SST} = 3{,}078 - 3{,}025.8 = 52.2$$

STEP 4:
Sum the data for each factor (truck); in this case;

14	15	10	13
11	13	12	11
12	12	11	12
11	16	10	15
11	12	12	13
59	68	55	64

STEP 5:
Square each of these factor sums and sum them, and then divide the total by the number of observations per factor:

$$\frac{1}{n} \sum_{i=1}^{k} T_i^2 = \frac{1}{5}(59^2 + 68^2 + 55^2 + 64^2)$$
$$= \frac{1}{5}(3{,}481 + 4{,}624 + 3{,}025 + 4{,}096)$$
$$= \frac{1}{5}(15{,}226) = 3{,}045.2$$

STEP 6:
Subtract $1/kn \cdot T^2$ from the step 5 result:

$$\text{SSA} = 3{,}045.2 - \left(\frac{1}{20} \cdot 246^2\right) = 3{,}045.2 - 3{,}025.8 = 19.4$$

STEP 7:
Subtract SSA from SST to form SSW:

$$\text{SSW} = \text{SST} - \text{SSA}$$
$$= 52.2 - 19.4 = 32.8$$

In the present case there are two sets of degrees of freedom to compute instead of one as we had in χ^2 analysis. The following general table shows how we complete the analysis-of-variance hypothesis test:

SOURCE TABLE

Source of Variation	Degrees of Freedom	Sum of Squares	Mean Square	F
Among	$k-1$	SSA	$\text{MSA} = \dfrac{\text{SSA}}{k-1}$	$\dfrac{\text{MSA}}{\text{MSW}}$
Within	$k(n-1)$	SSW	$\text{MSW} = \dfrac{\text{SSW}}{k(n-1)}$	
Total	$nk-1$	SST		

Inserting the data from our problem:

Source of Variation	Degrees of Freedom	Sum of Squares	Mean Square	F
Among trucks	3	19.4	6.47	3.16
Within trucks	16	32.8	2.05	
Total	19	52.2		

By using Appendix H we can determine the appropriate df. For the numerator of our F ratio we have $k - 1 = 3$ df, and for the denominator we have $k(n - 1) = 4(4) = 16$ df. In Appendix H the critical F value is found by moving across the top to the appropriate df for the numerator and down to the appropriate df for the denominator. We find $F 3, 16$ at the .05 level to be 3.24 and at the .01 level to be 5.29. Because our computed F is less than the critical F value, we accept the null hypothesis that $\mu_1 = \mu_2 = \mu_3 = \mu_4$. In our specific problem, we conclude that the differences among sample means are due to sampling error. As a result, we conclude that there is not a significant difference among the 4 brands of trucks. We still face the α and β error possibilities discussed before. Even though there were differences among the sample means, these differences were not sufficient as compared to the sampling error.

Two-Way Analysis of Variance

We have found that there is no difference among trucks; however, it is possible that influences other than brand of truck had an impact on the data. Because of the way in which analysis of variance works, the only place where the impact of this outside influence can be felt is in SSW, which is the error component of the analysis. Therefore, it is possible that an outside influence could have "inflated" SSW.

Upon further examination of the data, our firm found that there were differences among drivers of the trucks. The data matrix actually looked like the following:

Drivers	Truck 1 (mpg)	Truck 2 (mpg)	Truck 3 (mpg)	Truck 4 (mpg)
Sheet-metal men	14	15	10	13
Salesmen	11	13	12	11
Plumbers	12	12	11	12
Installers	11	16	10	15
Repairmen	11	12	12	13

Because these different types of drivers used the trucks for different purposes and under different conditions, it was felt the type of driver could have an impact on the mpg. We now have two factors, trucks in the columns and drivers in the rows, which makes two-way analysis of variance the appropriate technique.

The procedures for two-way ANOVA are very similar to one-way methods, except that we must deduct the impact of the second factor from SSW and the number of degrees of freedom change. The following table describes the situation:

Source of Variation	Degrees of Freedom	Sum of Squares	Mean Square	F
Sum of squares columns	$k-1$	SSC	$\dfrac{SSC}{k-1}$	$\dfrac{MSC}{MSW}$
Sum of squares rows	$n-1$	SSR	$\dfrac{SSR}{n-1}$	$\dfrac{MSR}{MSW}$
Sum of squares within	$(n-1)(k-1)$	SSW	$\dfrac{SSW}{(n-1)(k-1)}$	
Total	$nk-1$	SST		

We are testing two hypotheses now:

$$\mu_1 = \mu_2 = \mu_3 = \mu_4 \quad \text{trucks}$$
$$\mu_1 = \mu_2 = \mu_3 = \mu_4 \quad \text{drivers}$$

We calculate SSR for drivers in a manner similar to the previous truck calculations, except that we sum the data from the rows, square each sum, add the squared sums, and divide the result by the number of observations in each row:

	Truck				
Drivers	1	2	3	4	Σ
Sheet metal	14 +	15 +	10 +	13 =	52
Salesmen	11	13	12	11 =	47
Plumbers	12	12	11	12 =	47
Installers	11	16	10	15 =	52
Repairmen	11	12	12	13 =	48

$$\frac{1}{n}\sum_{i=1}^{k} T^2 = \frac{1}{4}(52^2 + 47^2 + 47^2 + 52^2 + 48^2)$$

$$= \frac{1}{4}(2{,}704 + 2{,}209 + 2{,}209 + 2{,}704 + 2{,}304) = 3{,}032.5$$

Then,

$$\text{SSR} = 3{,}032.5 - \frac{1}{kn} \cdot T^2 = 3{,}032.5 - 3{,}025.8 = 6.7$$
$$\text{SSW} = \text{SST} - \text{SSC} - \text{SSR}$$
$$= 52.2 - 19.4 - 6.7 = 26.1$$

and the resulting source table using our data is:

Source of Variation	Degrees of Freedom	Sum of Squares	Mean Square	F
Columns: trucks	3	19.4	6.47	2.97
Rows: drivers	4	6.7	1.68	.77
Within	12	26.1	2.18	
Total	19	52.2		

Therefore, as we test the hypotheses at the .05 level and .01 level for 3, 12 and 4, 12 df, we find the following critical values in Appendix H:

$$F.05(3, 12\ df) = 3.49 \qquad F.05(4, 12\ df) = 3.26$$
$$F.01(3, 12\ df) = 5.95 \qquad F.01(4, 12\ df) = 5.41$$

Because both of our computed F values are less than the tabled values, we accept the null hypothesis in both cases. The differences between means of trucks and means of drivers, respectively, are assumed to be due to sampling error at the .05 and .01 levels.

Summarizing the procedures followed in analysis of variance, these steps apply:

1. Establish the null hypothesis.
2. Establish the level of significance.
3. Compute the elements of the source table (SST, SSA, SSW, or SST, SSC, SSR, SSW) as appropriate.
4. Determine degrees of freedom for F ratio(s) and find the correct critical values.
5. Compare the computed and tabled F values.
6. If computed F is greater than the critical F, reject the null hypothesis. If the value is equal to or less than the critical value, accept the null hypothesis.

There is a considerable body of analysis of variance procedures that are beyond the scope of this text. For example, we have ignored *interaction*, which examines the relation between trucks and drivers simultaneously. We also have not considered third and fourth sources of variation. Our presentation assumed that there were an equal number of observations per treatment. Many situations fit our examples, which serve as a basis for introduction to analysis of variance.

Assumptions

The analysis of variance procedures are much like that of χ^2 analysis and previous hypothesis-testing methods. Because ANOVA is a unique method, there are a few conditions that must be met so that the application is statistically correct.

1. The individual observations must be independent of one another. For example, the mpg of one truck cannot be influenced by data for another truck.
2. The variation within each cell must be approximately equal. This means that the variance for truck 1 through truck 4 should be about equal.
3. Samples should be random and assignments of people or objects to cells should be random. Random sampling is important in most statistical evaluations.

Exercises

8.8 A manager of a retail store has been uncertain where to put his bread display. To check certain theories he has moved the bread to various store locations and recorded weekly sales. The data are as follows (in hundreds of dollars):

Aisle 1	Aisle 4	Back	Front
89	90	74	96
94	94	79	88
90	96	103	97
79	91	91	81
91	86	90	86

(a) Perform a one-way analysis of variance on these data.
(b) What cautions regarding the data and results would you make to management?

8.9 A truck picking up shelled corn for a grain elevator has to make a daily run from the north of town to the south of town. There are essentially 3 routes to be taken. Recorded times along the routes are as follows:

Route 1	Route 2	Route 3
27	24	28
29	23	29
26	21	26
28	26	27
30	29	30

(a) Perform one-way analysis of variance on the data.
(b) What cautions are important?

8.10 Assume that in problem 8.9 the data were collected for 5 days of the week. For example, row 1 is Monday, row 2 Tuesday, and so forth. Perform two-way analysis of variance and draw the appropriate conclusions.

8.11 Labor and management at a local job shop are having a disagreement over production with regard to labor time and machines. They have collected the following data regarding output time in minutes:

| Machine | Operator | | | |
	1	2	3	4
A	15	13	15	12
B	14	12	13	13
C	12	13	14	11
D	16	15	13	12

(a) Perform two-way analysis of variance.
(b) Make recommendations to management.

8.12 A university athletic department has been collecting data regarding a new synthetic surface. As their football strategy has emphasized speed, they want a surface that will allow the fastest acceleration for running. Running backs in full equipment ran 100 yards on each of 4 surfaces in the following times (in seconds):

| | Brand Name | | | |
Running Backs	Titan	Beechmont	Scott	Nastro-Turf
1	11.3	10.9	11.2	10.9
2	11.0	10.7	11.0	10.8
3	10.8	11.2	11.6	10.9
4	10.6	10.6	11.0	10.5
5	10.9	10.8	11.4	10.9
6	11.1	11.0	10.9	11.0

(a) Perform one-way analysis of variance and test the appropriate hypothesis.
(b) Run two-way analysis of variance and test the appropriate hypotheses.
(c) As the data were collected at other universities around the nation prior to scheduled football games, what other comments do you have for the decision makers?

Appendix

Often in practice when we are hand-calculating a chi-square value from a 2 × 2 table where there is 1 df, or when we are dealing with machine output where a chi-square value has been calculated from a 2 × 2 table, there is a correction factor used. This correction, which is called a *correction for continuity*, is

$$\chi^2 \text{ corrected} = \frac{(|f_{o_1} - f_{e_1}| - .5)^2}{f_{e_1}} + \frac{(|f_{o_2} - f_{e_2}| - .5)^2}{f_{e_2}} + \cdots + \frac{(|f_{o_k} - f_{e_k}| - .5)^2}{f_{e_k}}$$

where k = the total number of cells. Typically, this correction is used only for df = 1 situations; however, if the sample is small and many expected frequencies are 5 or close to 5, it should be utilized even when df > 1. Generally, if the corrected and uncorrected χ^2 values lead to different conclusions, a larger sample should be used or methods treated elsewhere should be employed.

Learning Objectives

Introduction
   Variables
   Other Terms

Data Collection and Use

Bivariate Regression

Interpretation of Regression Estimates
   Constant Term a
   Regression Coefficient or Slope b
   Other Interpretations
   Standard Errors of a and b

A Numerical Example
   Formation of Specific Y_c and e Values
   Standard Error of the Estimate
   Sampling Error of b
   Confidence Interval for b
   Confidence Interval Around Y_c

Correlation Analysis

Assumptions of Regression Analysis
   Assumption 1: Linearity
   Assumption 2: Random Error Terms
   Assumption 3: Uniform Variance
   Assumption 4: Normality

Cautions

Appendix — Derivation of a and b

Learning Objectives

The primary learning objectives for this chapter are:

1. To be able to hand-calculate an elementary simple regression analysis problem.
2. To understand how regression analysis is similar to, yet different from, material covered in Chapters 6 to 8.
3. To understand how to interpret regression-analysis results once the computations are completed.
4. To be able to perform a hypothesis test on results of regression analysis.
5. To demonstrate an understanding of the assumptions underlying regression analysis.

Regression and Correlation Analysis

Introduction

Regression analysis and *correlation analysis* relate closely to many of the techniques we used in hypothesis testing. Measures of differences between means, differences between proportions, chi-square analysis, and analysis of variance were discussed in prior chapters. Through use of various methods we made decisions regarding whether a hypothesis was accepted or rejected. In these tests we confined ourselves to stating whether or not a difference existed. Examples of conclusions were (1) $\bar{X}_1$ is significantly different from $\bar{X}_2$, (2) two things are or are not independent of one another, or (3) three or more means are or are not different from one another.

In regression analysis, our approach is somewhat different in that we are seeking the *magnitude* of the *relation* between two variables. As an example, our firm is considering an increase in its advertising budget by $1,000, and logically asks, "How much are sales expected to go up?" (Don't forget that when the test is made we may find that sales go up $0, although we hope not.) A general discussion follows of regression (or regression analysis; we shall drop the word analysis from now on). This will be followed by a specific business example.

Variables

Regression involves two types of variables. The first is a *dependent variable,* which is the variable about which we are trying to explain change. Examples are sales volume, miles per gallon, or yield per acre. *Independent variable(s)* are the second type and are used to explain the change in the dependent variable. Examples are:

Independent Variable	Dependent Variable
Advertising expenditure	→ Sales volume
Weight of car	→ Miles per gallon
Amount of fertilizer	→ Yield per acre

As a specific example, assume that a landscaping firm had taken a random sample of 12 new-home buyers and collected the data shown in Table 9-1. The data relate the buyers' annual income before taxes to the amount of money they spent on landscaping work. It is important to recognize at this point that the independent variable *may or may not cause* the change. The buyer's income may not directly cause landscape expenditures, as these expenditures may be related to a buyer's attitudes, esthetic needs, and other factors. However, we may find that the level of expenditure is *associated* with the level of income. We would *a priori* expect landscape expenditures to rise with income levels.

In some cases we are merely attempting to *establish a relationship,* such as between past shopping behavior and future shopping behavior, where we could never prove that past behavior *caused* future behavior. In other cases causality is more nearly evident. For example, tempera-

TABLE 9-1. Regression Landscape Example

Home Buyer	Landscaping Expenditures, Y	Annual Income, X
1	$2,700	$18,600
2	3,600	20,400
3	1,800	19,400
4	5,400	24,200
5	5,200	24,000
6	6,300	28,400
7	1,300	37,200
8	8,400	56,800
9	4,600	26,400
10	5,900	23,600
11	7,100	45,400
12	4,100	24,600

ture and a town's use of natural gas are causally related. However, do not forget that other factors, such as severity of winds, switches to other methods of heating, and changes in price structure, can also have an impact on the use of natural gas. An assumption implicitly built into regression, though, is that of *one-way causation*. By this we mean that the independent variable explains the dependent, but *not* vice versa. When we run into a situation where the independent variable explains the dependent variable, but the dependent variable also explains the independent variable, it is called *bilateral causation*. This problem causes complications in regression.

Other Terms

Before showing how to calculate the exact relationship between two variables, it is necessary to discuss a few very important points.

In our example, landscaping expenditure is the dependent variable, while annual income is the independent variable. There are 12 observations for both variables. As we view these data, there are a number of things we could do that we have already learned. For example, we could calculate the mean and standard deviation, the median, the mode, and other descriptive measures for each variable. Yet, will such measures provide what the landscaping company wants? Partially the answer is yes, but the company also wants to know *how* annual income and landscaping expenditures are related. As we closely examine these data, we get the feeling that as income increases, landscaping expenditures also increase, but we are not sure how much the increase is. As a matter of fact, we are not completely sure that there is an actual increasing relation, although we strongly suspect the increase.

In Table 9-1 the landscaping column has been headed by Y, the dependent variable, and the income column by X, the independent variable. Many independent variables could have been used to explain changes in Y, or, as in our example, only one. When only one independent variable is used, the process of analysis is called *bivariate regression analysis* (or in some cases, *simple regression analysis*). If two or more independent variables are used, such as when we use income and amount of home mortgage, it is termed *multiple regression analysis*. Multiple regression will be discussed in the next chapter.

In bivariate regression the function being tested is $Y = f(X)$ where, for example, Y is landscape expenditures and X is income. One way of expressing the relation between X and Y is a linear function. This is commonly expressed as $Y = a + bX$ where a is the intercept of the line with the Y axis and b is the slope of the line. Once again, although one variable (Y) is said to be "dependent" and the other variable (X) "independent," this relationship need not imply a cause–effect relationship. In some cases, Y can be predicted from X; however, neither variable truly causes a change in the other.

Data Collection and Use

Two additional points need to be made. First, in many cases the decision maker and/or person doing the data collection has two options regarding collection of data for X and Y. In the first option, a sample might be collected by recording each variable at set intervals over a period of time. For example, if we wanted to test the relation between the Gross National Product (GNP) and automobile sales, our data might be collected quarterly for 7 years. Data collected in such a fashion are called *time-series data,* or *longitudinal data.* The landscaping example used a different option, as data were collected for the 12 households at one point in time and the observations for each household were recorded. Data collected in this way are called *cross-section* data, or *latitudinal data.* Keep in mind, though, that in both examples one sample was taken. The GNP example had 2 variables with 28 observations, while the landscaping example had 2 variables and 12 observations.

Both methods of data collection (time series and cross section) are commonly used. Market surveys are good examples of cross-section data, while much economic forecasting is done using time-series data. Either procedure yields a similar set of data. The computational procedures do not "know how the data were collected" and, therefore, all statistics are calculated in the same fashion. The distinction between time-series and cross-section data becomes important at the interpretation stage of regression analysis when we are trying to reach conclusions.

Bivariate Regression

In the landscape example we have data for only 12 households. These households represent a subset, or sample, of the households the firm has as customers. Examine Figure 9-1, which is called a *scatter diagram.* As we can see, there is a scatter of points rather than a simple series of points all following a line. The important question becomes: How do we best estimate the relationship between X and Y shown in this scatter of

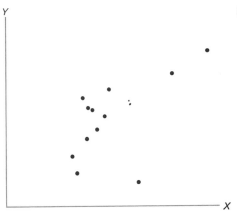

FIGURE 9-1. Bivariate Regression Example

points? We could run many different straight lines through these points, assuming that the relationship is linear. Think for a minute about how you would run a line that represented the best fit between X and Y. Using our standard deviation and analysis-of-variance background, we want to run a line that has the minimum sum of squared deviations away from it as compared to all other possible lines. In more general terms, we want to run a line that has the minimum error as compared to all other possible lines.

From Figure 9-2 the equation we can use is

$$Y_c = a + bX$$

where

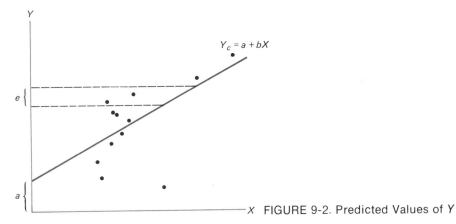

FIGURE 9-2. Predicted Values of Y

Y_c = an estimate of Y (called "Y sub-c" or "Y calculated")
a = the Y intercept of the line
b = the slope of the line

In fact, we are really dealing with three types of situations, which can be explained through reference to Figure 9-2 and our previous hypothesis-testing work.

Situation 1. The underlying parameters we want to estimate can be found from the following model for the population

$$Y = \alpha + \beta X + \epsilon$$

where

Y = landscape expenditures
α = the true but unknown *population* intercept
β = the true but unknown *population* slope
ϵ = the deviation of the actual value of Y from the true regression line

The underlying relationship that we are attempting to estimate is represented in situation 1. The population values α and β are those which would be derived if we had data for all the firm's landscape customers, not just the 12.

Situation 2. Rarely do we have the luxury of having population data; rather, we have sample data. In this case we have data for 12 households. We therefore must *estimate* the α and β values through the previously explained equation

$$Y_c = a + bX$$

Situation 3. If once again we have only sample data but are making reference to the values of Y, we say that

$$Y = a + bX + e$$

where

Y = actual value of Y (one of the 12)
a = estimated intercept
b = estimated slope
$e = Y - Y_c$ (actual Y value $-$ calculated Y value)

In summary, we have

$$Y = \alpha + \beta X + \epsilon \quad \text{population relationship}$$
$$Y = a + bX + e \quad \text{sample relationship}$$
$$Y_c = a + bX \quad \text{sample regression equation used for estimation}$$

We can now make a few additional observations from an examination of Figure 9-2. First, we notice that the line does not hit all points. As a result, there are "misses" in the attempt of this line to express the relationship between X and Y. These misses are called "e's," as they are calculated from sample data; the e is measured along the Y axis. The Y_c represents the predicted value of Y. One use of the line will be to estimate values of Y from values of X, and Y_c would be the expected value the line would estimate Y to be for a given value of X.

Also from these statements we can see that for any value of Y for which you have data, there will also be the corresponding error term e:

$$Y = a + bX + e$$

and for Y_c is

$$Y_c = a + bX$$

Therefore, we can see that

$$e = Y - Y_c$$

We need to deal with e's because the fit between X and Y is not perfectly represented by the line. For any given value of X, we find an estimate of Y from the line which is called Y_c. These relationships may seem fairly "visible" at this point as we only have 12 observations, but imagine how Figure 9-2 would appear if there were 10,000 observations for X and Y, respectively.

Interpretation of Regression Estimates

At this point we shall discuss a number of the terms that have been introduced. Special attention will be placed on interpretation.

FIGURE 9-3. Intercept Example

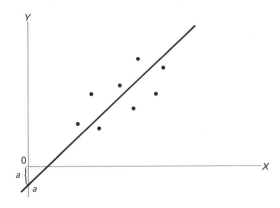

Constant Term a

The estimated value of a in bivariate regression tells us the point where the regression line intersects the Y axis. In other words, a has a major influence in positioning the regression line. This term is not necessarily positive, as we can have a situation such as shown in Figure 9-3, where the regression line actually intersects the Y axis at a point below zero. In some cases the a may appear to make no sense as it is well outside the range of the measured observations. The estimate a, a mathematical constant, generally serves the purpose of placing or positioning the regression line.

Regression Coefficient or Slope b

The *regression coefficient* b where $Y_c = a + bX$ is interpreted as the expected change in the dependent variable Y with respect to a one-unit change in the independent variable X. More simply, it is called the *slope* of the regression line. The regression coefficient b is one of the most useful concepts of regression in terms of interpretation and decision making. For example, assume that we run a linear regression on the relation between advertising and sales and find that every dollar spent on advertising yields an $8.00 return in sales, or a $1.20 return in profits. Remembering that there are other marketing and organizational components contributing to the ratio, with hope we get back more than we expend. Decisions as to how to allocate money can then be made based on how much return we estimate we can make. This parameter estimate b, which is used to estimate β, constitutes the crucial difference between regression and other statistical techniques we have discussed before in that we can find out not only *if* two variables are related but more importantly *how* they are related.

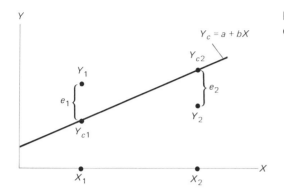

FIGURE 9-4. Standard Error of the Estimate

Other Interpretations

Error Term. The examination of Figure 9-4 indicates that error exists in our model as represented by $e = Y - Y_c$. We also know from prior experience with standard deviations that we need a measure of the dispersion of these errors so as to gain some insight into the accuracy of the regression line. From our example, in Figure 9-4 each observed value of X has a corresponding observed value of Y, as well as a predicted Y_c, and an individual e value.

For purposes of illustration only, two values of Y and X each have been selected from the 12 we had previously. The other values will be distributed around the line in a positive upsloping relationship. The true Y associated with a particular X is not necessarily or usually the value Y_c falling on the regression line, although a true value of Y *could* fall on this line. In that case, the e would be zero. Each Y_c value represents the expected value of Y for a given value of X. When we examine Figure 9-4, we should recall similar dispersion as when we were calculating standard deviations in Chapter 3.

Standard Error of Estimate. Similarly, another type of standard deviation called the *standard error of the estimate* can be calculated such that it represents the standard deviation of points around the regression line. This statistic will be labeled S_{YX} and is represented by the formula

$$S_{YX} = \sqrt{\frac{\Sigma (Y - Y_c)^2}{n - 2}} \quad \text{or} \quad S_{YX} = \sqrt{\frac{\Sigma e^2}{n - 2}}$$

As we saw in Chapter 7, we lose a degree of freedom for each parameter estimate. As a result, estimating α and β with a and b causes us to lose 2 df. This calculation of S_{YX} will have as many e terms as there are observations of Y.

Standard Errors of a and b

Recalling that b is just an estimate of the true but unknown population parameter β, there also exists a probable sampling error for that estimate as well as for a. The calculation procedures for the standard error of a (S_a) are covered elsewhere, and for the standard error of b, (S_b), exact procedures are shown in the following illustration.

A Numerical Example

Table 9-1 presented data compiled for a landscaping firm interested in finding the relation between annual income (X) and landscaping expenditures (Y). Prior to conducting the analysis, the managers of the firm were not sure what form the outcome might take. Any one of the situations shown in Figure 9-5 could have resulted from plotting the data in a scatter diagram.

If the data resulted in scatter diagram A, there appears to be some increasing landscape expenditures as income increases. Situation B is the exact opposite, or inverse, relationship. In situation C, expenditures ap-

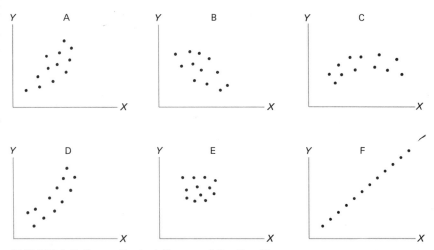

FIGURE 9-5. Some Typical Forms of Scatter Diagrams

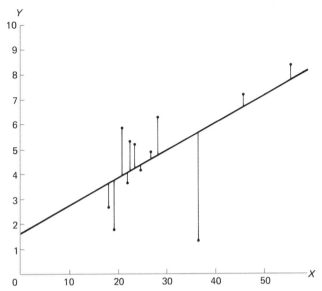
FIGURE 9-6. Scatter Diagram and Regression Line Income and Landscape Expenditure Example

pear to increase with income for a while and then beyond a certain point the expenditures level off and finally decrease. When data form a diagram such as situation D, expenditures increase at an increasing rate with income. In E there appears to be no relation between X and Y, as a line that minimized error could be drawn anywhere through the cluster of points. In F the fit between X and Y is perfect. As proprietors of the landscaping firm, we would assume that B, E, and F are not realistic; however, A and D would be very likely situations. As we examine linear methods of regression analysis, our attention will be concentrated on situation A, with the curvilinear situation D being left to more advanced treatments of regression. Figure 9-6 presents the actual scatter diagram of points for our example as exactly plotted from the data shown in Table 9-1. The plot resembles diagram A in Figure 9-5, although the fit does not appear to be as close.

The next step in the regression analysis becomes one of calculating the relevant a and b so that the regression equation is known. The method used for calculation of these statistics is called the *method of least squares,* or *ordinary least squares* (OLS).* Following are the principles upon which this method is based:

---

* A proof shown in the appendix to this chapter shows how the method of OLS works from a calculus standpoint. It can be skipped without loss of continuity.

1. The method minimizes the sum of the squared deviations $\Sigma (Y - Y_c)^2$ away from the line as compared to any other possible line. This is the reason it is called the least-squares line.

2. If we added all the unsquared deviations (e's) from the regression line, the sum would be zero. This is true because any summation of deviations from a mean will equal zero. Recall that, as shown in Figure 9-6, the deviations from the line are measured vertically in units of Y.

3. The regression line represents the best-fitting straight line for the sample data, because it minimizes error. This may or may not be the best fit if all the population data were available; however, we do not have data from the whole population.

The computational procedures for regression analysis can be accomplished by utilizing the following formulas:

$$b = \frac{\Sigma XY - n\bar{X}\bar{Y}}{\Sigma X^2 - n\bar{X}^2}$$
$$a = \bar{Y} - b\bar{X}$$

For later calculations of the standard error of the estimate, we shall also need to calculate ΣY^2 and $\bar{Y} \Sigma Y$. Therefore, these figures are also shown in Table 9-2, where the first two columns represent the data taken directly from Table 9-1 for Y and X, respectively. The other computations are those necessary to calculate a and b for the relationship $Y_c = a + bX$. To

TABLE 9-2. Regression Analysis Calculations, Landscape Example

Home Purchaser	Y (000)	X (000)	XY	Y^2	X^2
1	2.7	18.6	50.22	7.29	345.96
2	3.6	20.4	73.44	12.96	416.16
3	1.8	19.4	34.92	3.24	376.36
4	5.4	24.2	130.68	29.16	585.64
5	5.2	24.0	124.80	27.04	576.00
6	6.3	28.4	178.92	39.69	806.56
7	1.3	37.2	48.36	1.69	1,383.84
8	8.4	56.8	477.12	70.56	3,226.24
9	4.6	26.4	121.44	21.16	696.96
10	5.9	23.6	139.24	34.81	556.96
11	7.1	45.4	322.34	50.41	2,061.16
12	4.1	24.6	100.86	16.81	605.16
	56.4	349.0	1,802.34	314.82	11,637.00

$\bar{Y} = 4.7 \quad \bar{X} = 29.08 \quad \Sigma XY = 1802.34 \quad \Sigma Y^2 = 314.82 \quad \Sigma X^2 = 11,637.00$

facilitate computations, the numbers are expressed as thousands; decimals have been inserted and the (000) omitted. The equation values are calculated as follows:

$$b = \frac{\Sigma XY - n\bar{X}\bar{Y}}{\Sigma X^2 - n\bar{X}^2} = \frac{1{,}802.34 - [12(29.08)(4.7)]}{11{,}637 - 12(29.08)^2} = .1089$$

$$a = \bar{Y} - b\bar{X} = 4.7 - .1089(29.08) = 1.53$$

Substituting, we find the regression line to be

$$Y_c = a + bX = 1.53 + .1089X$$

An immediate conclusion is that for each $1,000 increase in income, landscaping expenditures increase $108.90.

Formation of Specific Y_c and e Values

We can see that the previous equation represents the relationship of X and Y across the 12 observations. As we will recall from the general discussion of regression, there are as many Y_c's or e's as there are Y's. How then do we form these values? They are found by substituting the original values of X for each of the 12 households into the regression equation. For example, the X values for the first two households were 18.6 and 20.4 so that

$$Y_{c_1} = 1.53 + .1089(18.6) = 3.56$$
$$Y_{c_2} = 1.53 + .1089(20.4) = 3.75$$

Using this procedure, we can construct a table such as Table 9-3, showing a Y_c and e for each household.

Standard Error of the Estimate

We can see from Table 9-3 that there are differences between the calculated values (Y_c) as compared to the original actual values of Y. The standard error of the estimate is a measure of deviation of the original points from the regression line. This is a measure of how well the line fits the data. We know that in unsquared form the average deviation from the line is zero. We can utilize the numbers from Table 9-2 to calculate the standard error of the estimate S_{YX}:

$$S_{YX} = \sqrt{\frac{\Sigma Y^2 - \bar{Y}\Sigma Y - b(\Sigma XY - \bar{X}\Sigma Y)}{n-2}} = \sqrt{\frac{49.74 - .1089(162.228)}{10}}$$

$$= \sqrt{3.21} = 1.79$$

TABLE 9-3. Estimated Values of Y and Error Terms, Landscape Example

Household	Y (000)	Y_c (000)	e $(Y - Y_c)$ (000)
1	2.7	3.56	−.86
2	3.6	3.75	−.15
3	1.8	3.64	−1.84
4	5.4	4.17	1.23
5	5.2	4.14	1.06
6	6.3	4.62	1.67
7	1.3	5.58	−4.28
8	8.4	7.72	.68
9	4.6	4.40	.20
10	5.9	4.10	1.80
11	7.1	6.47	.63
12	4.1	4.21	−.11

You will recall that previously we examined two other formulas for S_{YX} which can be used. The formula presented here is designed to allow calculation from a worksheet such as that shown in Table 9-2. In most cases calculations using this method can be made more rapidly, although the results will be the same as derived through the previous equations.

The next question is: What does this 1.79 represent? As we discussed before, the standard error of the estimate is a standard deviation. Therefore, the 1.79 represents the standard deviation of points around the regression line. Given that we have a large sample of 30 or more observations, we can say that we would expect approximately 68.26 per cent of the points to fall within ±1 standard deviation of the regression line, or ±1.79. The same could be said for ±2S_{YX}, including approximately 95 per cent of the points. In the case of small samples, the appropriate values from the t distribution must be utilized.

Sampling Errror of b

Many times when we deal with regression analysis we may wonder if the true population regression coefficient β actually is 0, and whether our regression analysis showed a relationship only due to chance or sampling error. The process of testing whether the true population regression coefficient might be 0 is very similar to our hypothesis-testing procedure discussed in Chapter 7. Just as we did before when we calculated $S_{\bar{x}}$, we have to calculate the standard error for b. This standard error, called S_b (S sub-b), can be calculated from the data found in Table 9-2 and from our calculated standard error of the estimate:

$$S_b = \frac{S_{YX}}{\sqrt{\Sigma X^2 - \bar{X} \Sigma X}} = \frac{1.79}{\sqrt{1{,}488.08}} = .046$$

The hypothesis-testing steps are the same as before when we tested a null hypothesis. In this case we are testing the null hypothesis whether the population β is actually zero. Remember that b is only an estimate of β and could be off owing to sampling error. The steps are:

1. H_0: $\beta = 0$
   H_a: $\beta \ne 0$
2. Establish a level of significance. For the bivariate case the t distribution is used at $n - 2$ degrees of freedom. For large samples, normal distribution values are sometimes used. In this case, $n - 2 = 10$.
3. Calculate S_b:

$$S_b = \frac{S_{YX}}{\sqrt{\Sigma X^2 - \bar{X} \Sigma X}} = .046 \text{ here}$$

4. Compute t:

$$t = \frac{b - \beta}{S_b} = \frac{b - 0}{S_b} = \frac{b}{S_b} \quad \text{or} \quad \frac{.1089}{.046} = 2.37 \text{ here}$$

5. Compare the absolute value of the computed t to the tabled t.
6. If it is greater, reject H_0. (Recall that the reason a t value may come out negative is because $t = b/S_b$ and b may be negative.)

Because we are dealing with a small sample and an unknown population standard deviation, you will recall that it is necessary to use the t distribution. In bivariate regression t tests, we are dealing with $n - 2$ degrees of freedom and we shall therefore go to the t table at $12 - 2 = 10$ df. At 10 df, $t_{.025} = 2.23$. Recall that in running a two-tailed test at the .05 level, we will have .05 of the total area in both tails combined, and .025 under a single tail. Because our value of $2.37 > 2.23$, we reject the null hypothesis and accept the alternative that there is a significant relationship between income and landscape expenditures at the .05 level. Note that in this case we are dealing with a positive regression coefficient—that is, the expenditure increases with the increase in income. The coefficients in some regressions will be negative so the comparison made in step 5 is made on the basis of absolute value of the computed t only.

Confidence Interval for b

For a large sample we could set up a 95 per cent confidence interval for β as $b \pm 1.96\,(S_b)$ just like the confidence intervals around a mean $\bar{X}$.

In the current case of a small sample where $n = 12$ and $df = 10$, the interval becomes

$$b \pm 2.23 \, (S_b) = .1089 \pm 2.23 \, (.046) \text{ or}$$

$$.00632 \text{ to } .21148$$

In other words, 95 per cent of the time this procedure yields an interval within which β lies. Translating back to dollars, the relation between landscape expenditures and income lies between \$6.32 spent per \$1,000 income and \$211.48 spent per \$1,000 income. Following this procedure, we shall be correct 95 per cent of the time. As we noted in previous work, this interval can be narrowed by increasing the sample size or decreasing the confidence level. Also, note that if we would have accepted H_0 the confidence interval constructed would include 0. Substantiating our original rejection of the null hypothesis, the interval does not include 0.

Confidence Interval Around Y_c

In some cases we wish to use the value of Y_c to predict the expected population value of Y, given a particular X. In other terms, we are interested in expanding our sample predictions of Y to what we would estimate would be true for the population. Y_c represents an estimated value of Y for a given X as calculated from sample data. Therefore, let μ_{YX} equal the corresponding expected value of Y for a given X for the population. Because Y_c has associated sampling error, we generally form an interval around Y_c much as we did for b. For a specific value of X, we have to calculate an estimate of the variance of Y_c. The variance of Y_c will be called $\sigma^2(Y_c)$ and the estimate of this variance is $S^2(Y_c)$. The estimated variance of Y_c can be computed from the following formula:

$$S^2(Y_c) = \frac{S^2_{YX}}{n} + (X - \bar{X})^2 \frac{S^2_{YX}}{\Sigma X^2 - \bar{X} \Sigma X}$$

The S^2_{YX} figure can be found by squaring the S_{YX} figure computed earlier; the other necessary computations have also been done previously.

$$S_{YX} = 1.79$$
$$S^2_{YX} = 3.20$$
$$\Sigma X^2 - \bar{X} \Sigma X = 1{,}488.08$$

Therefore, if X were 25, for example,

$$S^2(Y_c) = \frac{3.20}{12} + (25 - 29.08)^2 \frac{3.20}{1{,}488.08} = .267 + 16.65(.002) = .30$$

Using the estimated variance figure, we can set up the 95 per cent confidence interval around Y_c:

$$Y_c - t_{.025}\, S(Y_c) < \mu_{YX} < Y_c + t_{.025}\, S(Y_c)$$

So if $X = 25$,

$$Y_c = a + bX = 1.53 + (.1089)(25)$$
$$= 4.25$$

and the confidence interval at $n - 2 = 12 - 2 = 10$ degrees of freedom is

$$4.25 - 2.23(\sqrt{.30}) < \mu_{YX} < 4.25 + 2.23(\sqrt{.30})$$
$$4.25 - 1.22 < \mu_{YX} < 4.25 + 1.22$$
$$3.03 < \mu_{YX} < 5.47$$

We can then make statements about μ_{YX}. For example, if we took 100 samples of size 12 and constructed confidence intervals, such as above when $X = 25$, we would expect approximately 95 of them to contain the true $E(Y|X)$, which is μ_{YX}. Correspondingly, we could make up confidence intervals for any other given value of X, which fell into our range of X values observed in the regression problem.

Correlation Analysis

The correlation coefficient r is a measure of the *degree* of relationship between two variables. It can range from -1 to $+1$ where both extremes represent a perfect relationship or correlation. The plus one indicates perfect direct, or positive, linear correlation, and the minus one indicates a perfect inverse, or negative, linear relationship. The total variation in Y about its mean is

$$\Sigma\, (Y - \bar{Y})^2$$

Note that this is simply the numerator in the formula for the variance of Y. This total variation can be broken down, as will be explained after examination of Figures 9-7, 9-8, and 9-9. In Figure 9-7 we see a relationship where there is a perfect positive correlation. In other words, the

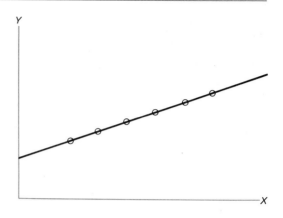

FIGURE 9-7. Example of Perfect Positive Correlation

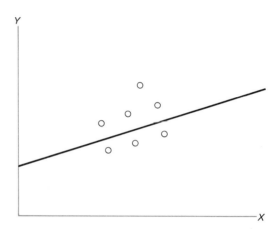

FIGURE 9-8. Example of Imperfect Positive Correlation

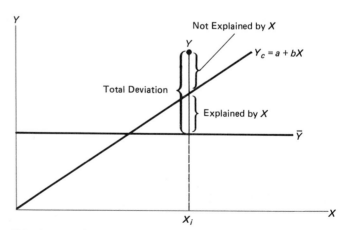

FIGURE 9-9. Explained and Unexplained Variation

correlation coefficient equals 1.0. In Figure 9-8 we can see that even though the relation between X and Y is positive, it is not a perfect 1.0.

From Figure 9-9, we can see that the total variation is made of two components: explained variation and unexplained variation. Expressing this by equation:

$$\Sigma (Y - \bar{Y})^2 = \Sigma (Y_c - \bar{Y})^2 + \Sigma (Y - Y_c)^2$$
$$\text{total} \quad\quad = \text{explained} \quad + \text{unexplained}$$

This leads us to a valuable statistic called r^2, which is a measure of the relative explanatory power of the regression equation or model. The r^2 value is called the *coefficient of determination*.

$$r^2 = \frac{\text{explained variation of } Y}{\text{total variation of } Y} = \frac{\Sigma (Y_c - \bar{Y})^2}{\Sigma (Y - \bar{Y})^2}$$

This statistic can vary from 0 to 1 inclusive. The zero represents the case where a model is associated with none of the variation in the dependent variable. An r^2 of 1 is where the model is associated with all variation in the dependent variable and there is no unexplained variation. For example, $r^2 = .92$ means 92 per cent of the variation in the dependent variable is related to variation in the independent variable. There are a variety of ways to calculate r^2. As it is the square of the correlation r between X and Y, let us calculate r and square it.

$$r = \frac{\Sigma XY - \bar{X} \Sigma Y}{\sqrt{(\Sigma X^2 - \bar{X} \Sigma X)(\Sigma Y^2 - \bar{Y} \Sigma Y)}} = \frac{162.23}{\sqrt{1,488.08 \times 49.74}} = \frac{162.23}{272.06}$$
$$= .5963 = .60$$

where the numerical values are taken from the landscape problem. The figure .60 represents the degree of relation between X and Y. Thus we have a relatively large positive relation between income and landscape expenditures. Perhaps more revealing from a decision-making standpoint, $r^2 = (.6)^2 = .36$. This means that 36 per cent of the variation in Y is accounted for by variation in X. The r statistic and r^2 value serve as very useful guides to management. They are indices to use in telling us how well our model and equation are doing. In the present case, 64 per cent of the variance is still left unexplained. This result is useful to management since it may want to consider other variables to help predict landscape expenditures. For example, in the case where a $r^2 = .05$, they may wish to discard the model entirely and seek other variables and/or estimation methods.

Assumptions of Regression Analysis

For any statistical method, there are certain assumptions made while making calculations and interpreting results. This is also true in the use of bivariate regression. These assumptions generally must be met in order to properly perform regression and use the findings. Violating these assumptions brings in certain ramifications which are well treated in advanced courses. These ramifications will be discussed briefly.

Assumption 1: Linearity

When we established the equation $Y_c = a + bX$, we were making the basic assumption that the relation between X and Y was linear rather than taking some curvilinear form. If the relationship were actually curvilinear, an assumption of linearity will result in an increase in the standard error of the estimate. This will produce a resulting increase in S_b, which then may lead to inability to reject the null hypothesis. In other words, the hypothesis test may indicate a poor fit. Plotting a scatter diagram will help if the data are likely to be representable by something other than a linear equation.

Assumption 2: Random-Error Terms

When the population relationship $Y = \alpha + \beta X + \epsilon$ was established and we used $Y_c = a + bX$ to estimate the parameters, we were saying in essence that X is the factor which we feel most important in predicting Y. If there are other outside variables important in predicting Y, the only place their presence can be noticed is in the e's. In other words, there may be an important variable omitted that then has its impact on the error terms. The consequences of this problem (called *autocorrelation of residuals*) are that the t tests and calculations of S_b are seriously biased and, as a result, the planned hypothesis test may well be jeopardized.

There are two rough rules of thumb that can be used to judge if autocorrelation of residuals exists. First, as e is expected to be a random variable, we would expect about as many positive e terms as there are negative e terms. (Remember, though, that as a result of sampling error this balance may not be exactly equal due to chance alone.) For example, in our landscape problem we had 5 negative e's and 7 positive e's. This does not appear to be seriously out of line with what might be expected.

The second rule for data collected on a time-series basis (only) is to

FIGURE 9-10. Clustering of Negative and Positive Error Terms

examine for "clustering" of negative and positive error terms. For example, if there were negative error terms for 3 or 4 years and then positive error terms for 3 or 4 years, the results might look like Figure 9-10, which indicates that there appear to be outside influences causing these clusters.

In cross-section data a simple rearrangement of the order of the data, because they do not extend over time, will generally alleviate this problem. There are other tests, such as the Durbin–Watson test treated elsewhere, which can aid in testing for autocorrelation.

Assumption 3: Uniform Variance

The assumption of *uniform variance*, sometimes also called *homoscedasticity*, means that the points are fairly uniformly spread around the regression line, as they are in Figure 9-11. Examples that illustrate violations of this assumption are shown in A, B, C, and D in Figure 9-12. Violation of this assumption makes it very difficult to predict Y from X because the accuracy of the prediction rests upon where we are in the scatter diagram. For example, in scatter diagram A, we have small error for small values of X, no experience for intermediate values of X, and

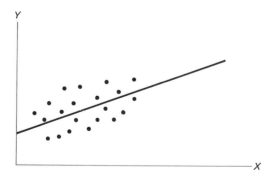

FIGURE 9-11. Uniform Variance Along the Regression Line

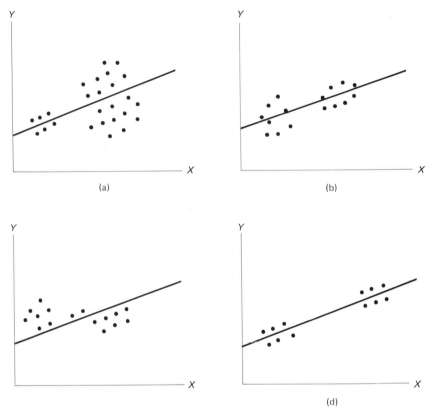

FIGURE 9-12. Violations of Uniform Variance

large error for larger values of X. Other problems can be seen in diagrams B, C, and D. The illustrations contain ranges of X where we are not sure what the relationship between X and Y is, and in addition it makes a difference at what point we are on the X axis in terms of our prediction of Y. This is very hazardous because in many business situations, we may not have scatter diagrams of all the observations. We can take a sampling of points of X and Y, though, and plot these points to observe their distribution around the regression line. The plotting of scatter diagrams can reveal this problem and possibly help in identifying alternative estimation methods. Because the data often appear in clusters that are separate from one another when nonuniform dispersion exists, it may be better to fit two or more separate regression lines rather than use a single "misfit" line. In other words, you should separate out those observations which are clustered together and run regression analyses on them individually.

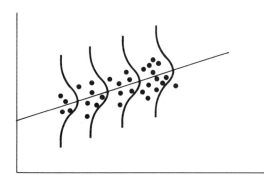

FIGURE 9-13. Normal Distribution Along the Regression Line

Assumption 4: Normality

It is generally assumed that if S_{YX} and S_b are to be used for hypothesis testing, the points are normally distributed along the regression line as in Figure 9-13. As we learned before, this essentially means for large samples that about 95 per cent of the points are distributed in the region bounded by two lines at $\pm 1.96 S_{YX}$ from the regression line. A close-to-normal distribution is especially important for small samples if inferences are going to be drawn regarding α and β. When very small samples are used, results are generally unstable and subject to a good deal of sampling error. For large samples, the central limit theorem allows us to make the inferences without the necessity of the normality assumption. Scatter diagrams can be studied and, if necessary and/or possible, additional data can be collected to help detect if this problem may exist.

Cautions

Regression analysis can be a very valuable tool; however, like any tool it can also be misused. Therefore, the following precautions should be heeded:

1. *Extrapolation error.* When predictions of Y are made using regression analysis which depended on historical data, an error is often made through predicting a Y value by use of an X value that is out of the range of the original data that were collected. In other words, the regression line has been extrapolated as shown in Figure 9-14. The prediction of Y_c is not based along a range of X values where we have had prior experience. This should be avoided by identifying the range of X values over which predictions are valid.

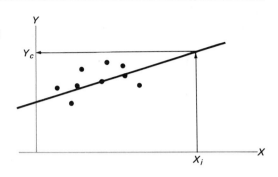

FIGURE 9-14. Extrapolation Error

2. *Spurious results.* Regression results can often be unstable in nature. As a result, what you feel may be true may not actually be the case. It is often a good idea to do a follow-up regression and compare the results to the original results or perform other validation procedures, some of which will be discussed in the Chapter 10.

3. *Implied cause.* Often when results turn out to be significant, the decision maker makes the assumption that X caused Y. Regression within itself does not imply this; regression only shows an association between X and Y. The only way that cause and effect can be inferred is through knowledge and business judgment, which served as the underpinnings for construction of the model to begin with. The computer can process almost any numbers, and the printouts look impressive. However, the printouts and results are only as good as the data utilized and the judgment used.

4. *Brute force of the computer.* Electronic data-processing equipment has been a marvel with regard to processing regression analysis programs. However, with the computer has come the tendency for some users to just "throw in a batch" of data and see what comes out. The computer is no substitute for sound business judgment. As a result, the user should be careful to have well-thought-out ideas prior to submission of data for analysis. Good judgment will help to avoid violations of the assumptions discussed earlier as well as avoidance of other pitfalls related to use of regression.

Exercises

9.1 (a) What is the difference between a regression analysis model and a regression equation?

(b) What is the difference between ϵ and e?
(c) What is a regression coefficient?
(d) What differences are there between regression analysis and other methods of hypothesis testing?

9.2 Explain:
(a) the meaning of "least squares" as applied to regression analysis
(b) how Y_c and Y differ
(c) how a confidence interval is formed around Y_c
(d) how a confidence interval is formed around b
(e) how the coefficient of determination helps in examining the outcome of regression analysis

9.3 The landscape company used as an example in this chapter has collected additional information for the same 12 households regarding the amount of a mortgage taken on the new house. The data for landscape expenditures and mortgage value are as follows:

Household	Landscape Expenditures	Mortgage Amount
1	$2,700	$24,200
2	3,600	32,100
3	1,800	28,200
4	5,400	36,100
5	5,200	30,400
6	6,300	40,100
7	1,300	48,200
8	8,400	80,300
9	4,600	20,100
10	5,900	32,600
11	7,100	54,700
12	4,100	48,200

(a) Compute the parameter estimates for the equation $Y_c = a + bX$ where Y (landscape expenditures) is the dependent variable.
(b) Test the hypothesis H_0: $\beta = 0$ at the .05 level.
(c) What conclusions can you draw?

9.4 Corn Research Incorporated has been attempting to find the relationship between acres of corn planted and average price of corn per bushel paid the farmer in the prior year. They have the following data:

Year	Acres Planted	Average Price per Bushel
1	71,156,000	$1.24
2	65,126,000	1.03
3	64,264,000	1.08
4	66,849,000	1.16
5	74,055,000	1.33
6	66,972,000	1.08
7	71,912,000	1.57
8	77,746,000	2.55

(a) Calculate the linear regression equation relating acres planted (Y) to average price per bushel (X).
(b) Test the hypothesis that $\beta = 0$, at the .05 level.
(c) Establish a confidence interval around b. Interpret the results.
(d) Establish a confidence interval around Y_c if the price is $1.30 per bushel. Interpret the results.

9.5 In a search for other variables of importance in predicting acres planted, Corn Research Incorporated has found the following data for corn exports in billions of bushels:

Year	Corn Exports (billions of bushels)
1	.471
2	.620
3	.525
4	.601
5	.506
6	.790
7	1.258
8	1.243

(a) Calculate the linear regression equation relating acres planted (Y) to corn exports (X).
(b) Test the hypothesis that $\beta = 0$.
(c) Establish a confidence interval around b. Interpret the result.
(d) How do the results in problems 9.4 and 9.5 differ?

9.6 For year 9, U.S. Department of Agriculture estimates for average price and exports are $3.51 per bushel and 1.243 billion bushels, respectively.
(a) Develop estimates of Y_c using the two equations found in problems 9.4 and 9.5.

(b) Form confidence intervals around Y_c at the 95 per cent level. Interpret these estimates.
(c) Why the difference between the estimates formed in (b)?

9.7 The personnel department at Richmond Incorporated is very interested in predicting manufacturing employee absenteeism. When an employee is hired, a personality and physical background test is given which will hopefully predict such factors as absenteeism. Data have been collected from 15 employees and their absentee rates have been recorded for a one-year period. The data are:

Employee	Personality and Physical Test Score	Days of Absence
1	90	1
2	68	4
3	70	3
4	59	6
5	74	4
6	88	2
7	83	3
8	80	0
9	90	1
10	95	1
11	78	5
12	81	3
13	86	2
14	75	8
15	77	4

(a) Calculate the linear regression equation relating test scores (X) and days of absence (Y).
(b) What is the sampling error associated with b?
(c) Is b significantly different from 0 at the 95 per cent level?
(d) Calculate and interpret the correlation coefficient.
(e) Assume that an incoming employee has a test score of 80, what would you predict his absenteeism to be?

9.8 The following data were obtained from a regression analysis computer run:

$$Y_c = 42{,}805.87 + 6.50X$$
$$S_{YX} = 4{,}356.04$$
$$S_b = 3.96$$
$$r = .527$$

where

Y = demand for a model of a typewriter in units
X = direct mail advertising expenditures measured in thousands of dollars

(a) What is the meaning of the b of 6.50?
(b) What statements do you have regarding the usefulness of the equation?
(c) Assuming that you have no further information, what questions should be raised regarding the outcome?

9.9 The following set of 10 numbers was drawn from a random-number table:

Y	X
02	94
69	65
20	81
88	15
62	47
17	62
34	74
41	27
82	19
18	45

(a) Calculate the parameter estimates for $Y_c = a + bX$.
(b) What is the meaning of b in this case?
(c) Calculate and interpret the correlation coefficient.
(d) Test $H_0: \beta = 0$, at .05 level.
(e) How can this type of effect have an impact on other regression analyses?

Appendix — Derivation of a and b

Let $Y = \alpha + \beta X + \epsilon$ be our stochastic model as described in this chapter.

$$Y = a + bX + e \tag{1}$$

where $e = Y - Y_c$ and $Y_c = a + bX$, which in this case is the regression line. Therefore,

$$e = Y - (a + bX) \tag{2}$$

Following then,

$$\Sigma e^2 = \Sigma [Y - (a + bX)]^2 \tag{3}$$

which represents the sum of all the squared errors around the regression line. The magnitude of the summed error term (Σe^2) is dependent upon the choice of estimate values a and b. It is the objective of OLS to minimize the sum of these squared errors. If we minimize (calculus) Eq. 3 with respect to a and b, the solutions are as follows:

$$\Sigma e^2 = \Sigma Y^2 - 2 \Sigma Y(a + bX) + \Sigma (a + bX)^2 \tag{4}$$

The necessary condition for minimization is

$$\frac{\partial \Sigma e^2}{\partial a} = 0 \quad \text{and} \quad \frac{\partial \Sigma e^2}{\partial b} = 0 \tag{5}$$

Differentiating (4) with respect to a, we obtain

$$\frac{\partial \Sigma e^2}{\partial a} = -2 \Sigma Y + 2 \Sigma (a + bX) = 0$$
$$= -2[\Sigma (Y - a - bX)] = \Sigma e = 0 \tag{6}$$

or

$$na + b \Sigma X = \Sigma Y$$

Differentiating (4) with respect to b yields

$$\frac{\partial \Sigma e^2}{\partial b} = -2 \Sigma YX + 2 \Sigma (a + bX)X = 0 \tag{7}$$

or

$$a \Sigma X + b \Sigma X^2 = \Sigma YX$$

Then solve the two normal equations (6 and 7) simultaneously for values of b and a.

The values obtained in the simultaneous solution are the OLS estimates of the true (but unknown) population parameters α and β.

10

Learning Objectives

Introduction

Interpretation

Assumptions

Evaluation
   Net Regression Coefficients
   Significance of the Independent Variables
   Other Interpretations
   Collinearity
   Further Analysis of the Results
   Interpretation of Beta Weights and Y_c Values

General Cautions in Multiple Regression

A Case Example
   Problem
   Output
   Steps in Interpretation
   Remedy

Categorical Data

Validation
   Forecasting With Regression
   Summary

Exercises

Appendix

Learning Objectives

Because strong emphasis is put on interpretation in this chapter, the primary learning objectives also stress this approach:

1. To gain an understanding of the similarities and differences between simple and multiple regression.
2. To gain a general understanding of how to interpret regression analysis output from a computer.
3. To become aware of some of the problem areas associated with doing multiple regression.
4. To recognize some ways to avoid or eliminate the problems referred to in objective 3.

Multiple Regression Analysis

Introduction

Multiple linear regression analysis is a logical extension of simple (bivariate) regression analysis. Multiple regression is a commonly used technique for analysis of data and business forecasting. In bivariate regression analysis there was one independent variable and one dependent variable. The term *multiple regression* means that there are two or more independent variables used. For example, we might desire to understand the relationship between expenditures on landscaping (the dependent variable Y) and two independent variables, say family income before taxes (X_1) and amount of new home mortgage (X_2). This relationship is estimated through use of the following equation:

$$Y = a + b_1 X_1 + b_2 X_2 + e$$

where

Y = expenditures on landscaping
X_1 = family income before taxes
X_2 = amount of new home mortgage

From Chapter 9 we know that Y_c is the value of Y estimated from the sample data. The underlying model in this case that is under examination is

$$Y = \alpha + \beta_1 X_1 + \beta_2 X_2 + \epsilon$$

where the values α, β_1, and β_2 are the population intercept and population regression coefficients, respectively. Therefore, a, b_1, and b_2 are sample estimates of these parameters. Once again, we are in the position of testing the significance of the specific parameter estimates. That is, do they differ more from zero than we would expect from chance alone?

Compared to the example used in the last chapter, a new variable, amount of new home mortgage (X_2), has been added to the analysis. This equation gives a relationship showing the impact of both X_1 and X_2 on the dependent variable Y. We should note that there is still only one dependent variable, and the form of the equation specified is still linear (there are no squared terms). More advanced treatments of multiple regression deal with curvilinear regression and logarithmic transformations.

Multiple linear regression analysis generally involves rather tedious mathematical computations; however, these can be accomplished readily by a computer. In keeping with the practical orientation of this text, the stress in this chapter will be on interpretation of the meaning of the results rather than on the mathematical calculations. The computational procedures for a multiple regression involving two independent variables can be found in the appendix to this chapter. In addition to placing stress on interpretation of results, special attention will be given to some of the possible pitfalls that relate to use of multiple regression and how to avoid them.

Interpretation

Table 10-1 presents the data from our landscaping example with observations included for the mortgage amount. In this example we can see that we have supplemented the data from Chapter 9 with one additional variable (mortgage amount). The data as given can be processed through the use of standard programs available with many computer systems.

TABLE 10-1. Linear Multiple Regression, Landscape Example

Household	Landscape Expenditures, Y (000)	Income, X_1 (000)	Mortgage Value, X_2 (000)
1	2.7	18.6	24.2
2	3.6	20.4	32.1
3	1.8	19.4	28.2
4	5.4	24.2	36.1
5	5.2	24.0	30.4
6	6.3	28.4	40.1
7	1.3	37.2	48.2
8	8.4	56.8	80.3
9	4.6	26.4	20.1
10	5.9	23.6	32.6
11	7.1	45.4	54.7
12	4.1	24.6	48.2

Figure 10-1 represents a computer printout of the pertinent multiple regression analysis results. We will first go through a rather detailed interpretation and explanation of these results and then proceed to a discussion of how to eliminate some of the problems inherent in the analysis.

Assumptions

The major difference between multiple and bivariate regression is that, in the former, the results are somewhat harder to evaluate since we are dealing with a "multivariate" or three-dimensional situation (or even more than three dimensions) rather than one which can be portrayed on a two-dimensional graph. There are still a few things that can be done from a practical business standpoint to examine for evidence of potential violation of assumptions.

First, we can examine the "table of residuals" in Figure 10-1 to determine if the autocorrelation problem (nonrandom error terms) exists. In the examination of this table, we first need to become acquainted with the column headings. There are still 12 households, and the Y values, landscape expenditures, are exactly duplicated. The column labeled "Y estimate" represents our Y_c values, which the computer has calculated. The residual column contains the e (define $Y - Y_c$) values for each

LANDSCAPE EXPENDITURES (Y) VERSUS INCOME (X)

VARIABLE NO.	MEAN	STANDARD DEVIATION	CORRELATION X VS Y	REGRESSION COEFFICIENT	STD. ERROR OF REG. COEF.	COMPUTED T VALUE
2	29.08328	11.62644	0.59583	0.10898	0.04645	2.34612
DEPENDENT						
1	4.69999	2.12646				

INTERCEPT	1.53059
MULTIPLE CORRELATION	0.59583
COEFFICIENT OF DETERMINATION	0.35502
STD. ERROR OF ESTIMATE	1.79113

ANALYSIS OF VARIANCE FOR THE REGRESSION

SOURCE OF VARIATION	DEGREES OF FREEDOM	SUM OF SQUARES	MEAN SQUARES	F VALUE
ATTRIBUTABLE TO REGRESSION	1	17.65858	17.65858	5.50430
DEVIATION FROM REGRESSION	10	32.08147	3.20815	
TOTAL	11	49.74005		

TABLE OF RESIDUALS

CASE NO.	Y VALUE	Y ESTIMATE	RESIDUAL
1	2.70000	3.55756	−0.85756
2	3.60000	3.75372	−0.15372
3	1.80000	3.64474	−1.84474
4	5.40000	4.16783	1.23217
5	5.20000	4.14603	1.05397
6	6.30000	4.62553	1.67447
7	1.30000	5.58453	−4.28453
8	8.40000	7.72047	0.67953
9	4.60000	4.40758	0.19242
10	5.90000	4.10244	1.79756
11	7.10000	6.47814	0.62186
12	4.10000	4.21142	−0.11142

FIGURE 10-1. Computer Analysis Landscape Example

household. Essentially, each e term represents how much the regression equation missed in predicting each of the 12 values of Y. The calculation methods for the Y_c and e values will be shown later in the chapter.

An examination of the column labeled "residuals" shows that we once again have a balance of 5 negative and 7 positive error terms. Further examination shows that there is one very large error term for household 7 as compared to the rest of the households. These results are somewhat supportive of not having an autocorrelation problem. As discussed elsewhere, a Durbin–Watson test should also be made, or an ordinary length-of-runs test as shown in Chapter 11.

LANDSCAPE EXPENDITURES (Y) VERSUS
INCOME (X_1) AND MORTGAGE VALUE (X_2)

BETA WT. (2) = 0.53227806E 00

BETA WT. (3) = 0.70552826E-01

VARIABLE NO.	MEAN	STANDARD DEVIATION	CORRELATION X VS Y	REGRESSION COEFFICIENT	STD. ERROR OF REG. COEF.	COMPUTED T VALUE
2	29.08328	11.62644	0.59583	0.09735	0.11268	0.86401
3	39.59995	16.46532	0.55003	0.00911	0.07956	0.11452
DEPENDENT						
1	4.69999	2.12646				

INTERCEPT	1.50783
MULTIPLE CORRELATION	0.59662
COEFFICIENT OF DETERMINATION	0.35596
STD. ERROR OF ESTIMATE	1.88664

ANALYSIS OF VARIANCE FOR THE REGRESSION

SOURCE OF VARIATION	DEGREES OF FREEDOM	SUM OF SQUARES	MEAN SQUARES	F VALUE
ATTRIBUTABLE TO REGRESSION	2	17.70523	8.85262	2.48709
DEVIATION FROM REGRESSION	9	32.03482	3.55942	
TOTAL	11	49.74005		

TABLE OF RESIDUALS

CASE NO.	Y VALUE	Y ESTIMATE	RESIDUAL
1	2.70000	3.53910	−0.83910
2	3.60000	3.78631	−0.18632
3	1.80000	3.65343	−1.85343
4	5.40000	4.19270	1.20730
5	5.20000	4.12129	1.07870
6	6.30000	4.63803	1.66197
7	1.30000	5.56854	−4.26854
8	8.40000	7.76914	0.63086
9	4.60000	4.26109	0.33891
10	5.90000	4.10240	1.79760
11	7.10000	6.42606	0.67394
12	4.10000	4.34190	−0.24190

FIGURE 10-1.

The assumptions of linearity, normality, and uniform variance are somewhat harder to test since we are dealing with a multivariate space (more than one independent variable). We can, however, plot individual scatter diagrams with X_1 versus Y and X_2 versus Y, respectively. The scatter diagram of X_1 versus Y was presented in Chapter 9. The scatter diagram of X_2 versus Y is shown in Figure 10-2. Because we do not have much else to go on at this point, we should make a visual inspection of the relation between X_2 and Y. The points appear to be somewhat closely

LANDSCAPE EXPENDITURES (Y) VERSUS
MORTGAGE VALUE (X)

VARIABLE NO.	MEAN	STANDARD DEVIATION	CORRELATION X VS Y	REGRESSION COEFFICIENT	STD. ERROR OF REG. COEF.	COMPUTED T VALUE
3 DEPENDENT	39.59995	16.46532	0.55003	0.07104	0.03411	2.08270
1	4.69999	2.12646				

INTERCEPT	1.88700
MULTIPLE CORRELATION	0.55003
COEFFICIENT OF DETERMINATION	0.30253
STD. ERROR OF ESTIMATE	1.86258

ANALYSIS OF VARIANCE FOR THE REGRESSION

SOURCE OF VARIATION	DEGREES OF FREEDOM	SUM OF SQUARES	MEAN SQUARES	F VALUE
ATTRIBUTABLE TO REGRESSION	1	15.04810	15.04810	4.33764
DEVIATION FROM REGRESSION	10	34.69194	3.46919	
TOTAL	11	49.74004		

TABLE OF RESIDUALS

CASE NO.	Y VALUE	Y ESTIMATE@	RESIDUAL
1	2.70000	3.60605	−0.90605
2	3.60000	4.16723	−0.56723
3	1.80000	3.89020	−2.09020
4	5.40000	4.45137	0.94863
5	4.20000	4.04647	1.15353
6	6.30000	4.73551	1.56448
7	1.30000	5.31090	−4.01090
8	8.40000	7.59113	0.80887
9	4.60000	3.32481	1.28519
10	5.90000	4.20275	1.69725
11	7.10000	5.77263	1.32737
12	4.10000	5.31090	−1.21090

FIGURE 10-1.

grouped as was the case for X_1 and Y, and the relation appears to be linear. There is also generally uniform dispersion around the line as was the case for X_1 and Y.

Evaluation

From a somewhat visual and intuitive standpoint, there are no serious violations of assumptions. Advanced works present far more complex

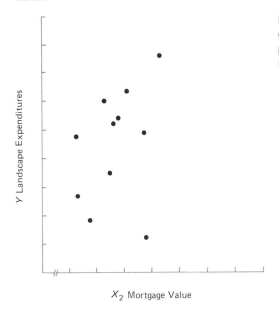

FIGURE 10-2. Scatter Diagram of Mortgage Expenditures and Landscape Expenditures

and accurate methods of examination. For our purposes and many practical business purposes, we base our conclusions on the visual examination. As we scan the column headings in Figure 10-1, we can see some terms that should be familiar to us now. For variables 2 (X_1) and 3 (X_2), the means and standard deviations are presented. These were calculated by the computer exactly as we did our hand computations in Chapter 3.

In the column headed "correlation X versus Y," we observe a measure of the closeness of the relationship between the dependent variable and each of the independent variables. Where there is no relationship between the dependent variable and each of the independent variables, each correlation coefficient will be zero. The higher the absolute value of the correlation coefficient, the closer the relationship between variables. Although these correlation coefficients are commonly used, a more readily understandable and useful measure of the value of the regression equation is found in the coefficient of determination, which will be discussed later in the chapter.

Net Regression Coefficients

In multiple regression the regression coefficients are called *net regression coefficients*. They are interpreted in a fashion somewhat similar to bivariate regression coefficients in Chapter 9. For example, $b_1 = .097$ and indicates that for a $1,000 increase in income, landscape expenditures increase $97. This assumes that we are *holding variable X_2 constant*. For X_2, where $b_2 = .009$, a $1,000 increase in mortgage expenditures is associated with a $9 increase in landscape costs (assuming that variable X_1 is held constant). In this case we are dealing with positive net regression coefficients, which means that as one variable increases the other also increases. We often obtain negative net regression coefficients, which indicate that for a one unit increase in X, Y decreases b units. Again, this assumes holding the other independent variables constant. The standard errors are similar to those found in bivariate regression. In this case they are .112 and .079, respectively.

Significance of the Independent Variables

Our next step is to ascertain the significance of our results. In other words, what is the likelihood that our results occurred by chance alone? Upon examination of the t values (which are found by taking b/S_b), we note that they are .864 and .114, respectively. To interpret significance of a t value, we again go to the t table shown in Appendix B. To determine the appropriate degrees of freedom (df), we calculate $n - k - 1$, where

n = total number of observations per variable
k = total number of independent variables

For our example, $n - k - 1 = 12 - 2 - 1 = 9$.

In moving to the t table at the .05 level (remember that a test at the .05 level means that .025 of the area is under each tail for a total of .05; therefore, when we use Appendix B we shall go to the column headed by $t_{.025}$), for example, and 9 df we find the appropriate t value to be 2.26. We follow past procedures in hypothesis testing. Since the calculated t is less than the tabled t, we accept the null hypothesis that $\beta = 0$ (no relationship). Given our results in Chapter 9, this is somewhat puzzling as in that example income *was* related to landscape expenditures. The reason for this curious outcome is due to a special problem in multiple regression called *collinearity*, to which we will return later.

Other Interpretations

Continuing with our analysis of the printout, we observe that the estimated intercept a is 1.507. This can no longer be interpreted as the Y intercept, as we are dealing in a three-dimensional space and it actually represents the intersection of planes. For our purposes, let us simply consider this to be a mathematical constant.

The "multiple correlation," R (capital R in multiple regression), represents the correlation between Y and X_1 and X_2, simultaneously. Its square, R^2, (once again called the coefficient of determination) represents the proportion of variation in Y "explained" by X_1 and X_2 together. In the multiple regression we have still only accounted for about .36 of the variation in Y, which was about the same explained variation we had for the bivariate example discussed in Chapter 9. The standard error of the estimate is interpreted much as it was in Chapter 9.

Collinearity

Collinearity is a special problem in multiple regression and refers to the degree of association between the *independent* variables. Table 10-2 represents the correlation matrix of all variables.

Table 10-2 shows a series of 1.0's down the "main diagonal." This stands to reason since a variable is completely correlated (1.0) with itself. The "off-diagonal" positions contain correlations of variables taken one pair at a time. They show the now familiar correlations of Y and X_1 as well as Y and X_2, and a new correlation, .901, representing the correlation between X_1 and X_2. This means that 90.1 per cent of the variation in X_1 is associated with variation in X_2. These two variables are nearly perfectly correlated.

Collinearity is a high degree of association between *independent* variables of the model. It means that the independent variables are so closely interrelated that their separate effects on the dependent variable Y are not easily detectable. From an intuitive standpoint, we can examine what has actually happened. We have told the computer that there are two independent variables; however, because of these variables' very close relationship, there is really just one separate independent effect. Some very excellent studies have shown the impact of collinearity and have revealed two basic results. When collinearity exists, (1) the standard errors of the regression coefficients tend to rise and (2) the resulting

TABLE 10-2. Correlation of Variables, Landscape Example

	Variable		
Variable	1	2	3
1. Landscape expenditure	1.0	.596	.550
2. Income	.596	1.0	.901
3. Mortgage amount	.550	.901	1.0

t ratios decline; this is exactly what happened in our case. Two additional impacts are (3) signs can change from positive, which they should be, to negative, which they should not be, and vice versa, and (4) the regression results tend to be very unstable, meaning that a rerun using the same equation on a different set of data might yield vastly different results.

Two general questions come up in practical business multiple regression problems with regard to collinearity. First, how do we know if we have collinearity, and second, if we do have it, how do we get rid of it or compensate for it?

To detect collinearity, there are few published guidelines. One rough rule is that if the correlation coefficient between independent variables is higher than the coefficient of determination R^2, collinearity exists. To rid the model of collinearity there are two general procedures. First, one or more of the offending variables can be dropped from the analysis. For example, mortgage amount might be dropped. The decision as to which variables to drop generally rests with the judgment of the person conducting the analysis since he may be able to judge which variable provides the least useful value. For example, if mortgage amounts are found to be insignificant in a bivariate regression run against landscape expenditures, they should be dropped. It may also be wise to drop a variable if the data are difficult to obtain. If one variable were dropped in our example, we are back to the simple (bivariate) regression form of one variable.

A second potential solution to collinearity problems is to add more observations. In this case we might decide to collect data from 100 households rather than 12. Often this will eliminate the problem since both X_1 and X_2 (as well as Y) tend to vary more.

Further Analysis of the Results

In many problems there will be no evidence of collinearity. Disregarding the problem of collinearity for now (which partially negates this particular analysis), we can proceed with analysis of the computer printout. The next portion contains the "analysis of variance for the regression." Essentially, this section deals with how well both X_1 and X_2 (taken in conjunction with one another) explained the variation in Y.

The total variance in Y is equal to $\Sigma (Y - \bar{Y})^2/n$, while the unexplained variation is equal to $\Sigma (Y - Y_c)^2/n$. The deviation from the regression that is explained equals

$$(Y - \bar{Y}) - (Y - Y_c) = Y_c - \bar{Y}$$

It follows then that the explained variance is $\Sigma (Y_c - \bar{Y})^2/n$. Putting these terms in equation form, we have

$$\frac{\text{total}}{\text{variance}} = \frac{\text{explained}}{\text{variance}} + \frac{\text{unexplained}}{\text{variance}}$$

or

$$\frac{\Sigma (Y - \bar{Y})^2}{n} = \frac{\Sigma (Y_c - \bar{Y})^2}{n} + \frac{\Sigma (Y - Y_c)^2}{n}$$

Because the denominator in each of the terms is the same, the numerators are often used alone. When used alone the numerators are called *sum of squares*.

The analysis of variance section of the printout can be interpreted in much the same way we treated F tests in Chapter 8. The *attributable to regression section* refers to

$\Sigma (Y_c - \bar{Y})^2 =$ sum of squares attributable to (or explained by) regression

while *deviation from regression* is

$$\Sigma (Y - Y_c)^2 = \text{sum of squares error}$$

and *total* is our now familiar

$$\Sigma (Y - \bar{Y})^2 = \text{total sum of squares}$$

To compute the appropriate degrees of freedom there will be k degrees of freedom for the numerator, $n - k - 1$ degrees of freedom for the denominator, and $n - 1$ total degrees of freedom. The mean squares and F values are found exactly as in Chapter 8. In interpreting the F value of 2.487, we are testing the hypothesis:

H_0: The independent variables when taken as a whole did not explain a significant proportion of the dependent variable.

As we move to Appendix H, we find that the appropriate F value at 2,9 df is 4.26 at the .05 level. As has been the case previously, we accept H_0 and conclude that the two independent variables taken simultaneously did not explain a significant portion of the landscape expenditures.

In passing, we should note that F and R^2 are very closely tied together. For our example,

$$R^2 = \frac{\text{explained variation}}{\text{total variation}} = \frac{17.705}{49.740} = .3559$$

and

$$F = \frac{R^2/k}{(1-R^2)/n-k-1} = \frac{.3559/2}{.6441/9} = \frac{.1780}{.0716} = 2.486$$

Tying this material into what we previously examined, we can say that treatment sum of squares is essentially the same as sum of squares attributable to regression. In total, the components are:

$$\Sigma(Y-\bar{Y})^2 = \Sigma(Y_c - \bar{Y})^2 + \Sigma(Y - Y_c)^2$$
$$\text{total} = \text{attributable} + \text{deviation}$$
$$\text{to regression} \quad \text{from regression}$$

which would be synonomous with

$$\text{SST} = \text{SST}_r + \text{SSE}$$
$$\text{total} = \text{treatment} + \text{error}$$

which we examined before. The method for determining degrees of freedom is also very similar. In summary, it is important to remember that the intent is to determine if the regression independent variables "explain" a significant portion of the dependent variable's variation, or whether the result could simply be due to sampling error.

Interpretation of Beta Weights and Y_c Values

We can recall from examination of the table of residuals that there were as many e terms and Y_c's as there were Y's. A Y_c is calculated by

$$Y_c = a + b_1 X_1 + b_2 X_2$$

For the first household, using the data from Table 10-1,

$$Y_c = 1.50783 + .09735(18.6) + .00911(24.2) = 3.539$$

which is the figure given as the Y estimate on the printout. We should recall in this example that our independent variables were both expressed in units of thousands of dollars. *Assuming that both variables came out significant,* which they did not, and that they were in differing units, the question might arise as to which is more important than the other. If they are not in like units, for example if they were in units of dollars and thousands of pounds, we cannot make a direct comparison of the regression coefficients, because a coefficient is interpreted as the change in Y for a *one unit* change in X (holding all other X's constant). To make the comparison, we can make use of the "beta weights," which are the regression coefficients expressed in standard deviation units (SDU). From these a direct comparison can be made as to relative importance by simply ranking them in terms of absolute value. Recall that when you see the notation E−01 on the computer printout, this means to move the decimal over one place to the left making the beta weight for $X_2 = .070552826$. If a beta weight is followed by a E+01, for example, this would mean to move the decimal one place to the right.

General Cautions in Multiple Regression

Problems discussed in this and in Chapter 9 often get decision makers in trouble when using regression analysis. Table 10-3 lays out the impacts and solutions regarding collinearity and autocorrelation in summary form. We can observe from the solution components of the table that the potential solutions for autocorrelation and collinearity could be the exact opposite. In the case where both problems exist, obtaining more observations appears to be the most productive approach to obtaining a useful solution.

TABLE 10-3. Impacts and Solutions for Problems in Regression Analysis

	Impact	Solution
Autocorrelation	S_b decreases t ratio increases artificially	(a) Include another variable(s) (b) More observations
Collinearity	Causes S_b to increase t ratio declines	(a) Drop offending variables (b) More observations

A Case Example

To this point we have been examining multiple regression analysis with two independent variables. In many instances, it is necessary or desirable to have more than two independent variables. The following case situation examines some of the methods and problems associated with this extended form of regression analysis.

Problem

Standard Farm Seeds, Inc., is trying to examine variables that are related to how many acres of corn are being planted nationally. It is important to them to be able to predict corn acreage in order to more effectively plan corn seed production. They have collected data for a 9-year period on corn acreage and have established the following model:

$$Y = \alpha + \beta_1 X_1 + \beta_2 X_2 + \beta_3 X_3 + \beta_4 X_4 + \beta_5 X_5 + \epsilon$$

where

Y = acres of corn planted, millions of acres
X_1 = prior-year corn production, millions of bushels
X_2 = prior-year average price of corn per bushel, dollars
X_3 = prior-year carryover of corn stored in bins, silos, and elevators, millions of bushels
X_4 = prior-year corn exports, millions of bushels
X_5 = prior-year domestic consumption of corn, millions of bushels

The first step in problem analysis was to run a linear multiple regression analysis with the 5 independent variables, the dependent variable, and 9 observations per variable.

Output

The data were processed on the computer and the results are shown in Figure 10-3. As was the case for our previous work, the correlation matrix is followed by the calculation of coefficients, t values, analysis of variance, and the table of residuals.

Steps in Interpretation

Once again we shall proceed through the interpretation of the regression analysis output and in the next section deal with any remedies to problems discussed.

CORRELATIONS OF VARIABLES

VARIABLE	1	2	3	4	5	6
1	1.000	0.136	0.751	-0.439	0.620	0.527
2	0.136	1.000	0.157	-0.228	0.692	0.709
3	0.751	0.157	1.000	-0.656	0.763	0.700
4	-0.439	-0.228	-0.656	1.000	-0.426	-0.354
5	0.620	0.692	0.763	-0.426	1.000	0.947
6	0.527	0.709	0.700	-0.354	0.947	1.000

MULTIPLE REGRESSION

SELECTION 1

BETA WT. (2) = -0.67687988E-01

BETA WT. (3) = 0.64420414E 00

BETA WT. (4) = 0.79542160E-01

BETA WT. (5) = 0.62921810E 00

BETA WT. (6) = -0.44377422E 00

VARIABLE NO.	MEAN	STANDARD DEVIATION	CORRELATION X VS Y	REGRESSION COEFFICIENT	STD. ERROR OF REG. COEF.	COMPUTED T VALUE
2	4.86955	0.60852	0.13638	-0.53334	10.49592	-0.05081
3	1.61667	0.85226	0.75128	3.62420	9.86579	0.36735
4	0.88256	0.23833	-0.43865	1.60021	15.32766	0.10440
5	0.80633	0.34357	0.61987	8.78116	25.88251	0.33927
6	4.24233	0.38887	0.52705	-5.47168	16.11729	-0.33949
DEPENDENT						
1	70.37434	4.79470				

Figure 10-3. Regression Output: Standard Farm Problem

1. As we examine the correlation matrix, we note that variable number 1 is the dependent variable, corn production. For diagnosis of collinearity we want to examine the correlations of variables 2 through 6. Once again we shall use the rough rule of thumb that a correlation coefficient r that is greater than the coefficient of determination R^2 is cause for concern. The following list presents the outcomes of those r's greater than R^2.

Variables	r		R^2
2 and 5	.692	>	.596
2 and 6	.709	>	.596
3 and 5	.763	>	.596
3 and 6	.700	>	.596
5 and 6	.947	>	.596

```
INTERCEPT                         81.83217
MULTIPLE CORRELATION              0.77201
COEFFICIENT OF DETERMINATION      0.59600
STD. ERROR OF ESTIMATE            4.97665
```

ANALYSIS OF VARIANCE FOR THE REGRESSION

SOURCE OF VARIATION	DEGREES OF FREEDOM	SUM OF SQUARES	MEAN SQUARES	F VALUE
ATTRIBUTABLE TO REGRESSION	5	109.61172	21.92233	0.88514
DEVIATION FROM REGRESSION	3	74.30118	24.76706	
TOTAL	8	183.91290		

MULTIPLE REGRESSION.

SELECTION 1

TABLE OF RESIDUALS

CASE NO.	Y VALUE	Y ESTIMATE	RESIDUAL
1	71.15599	69.35278	1.80321
2	65.12599	68.48167	-3.35568
3	64.26399	68.15355	-3.88956
4	66.84900	67.68210	-0.83310
5	74.05499	68.72906	5.32593
6	66.97198	66.73795	0.23404
7	71.91199	71.59941	0.31258
8	77.74599	74.77228	2.97371
9	75.28999	77.85976	-2.56976

FIGURE 10-3.

From the very beginning of our interpretation we are alerted that we may have problems with collinearity.

2. Next we can examine the "table of residuals," where we note that there are 4 negative and 5 positive values, which appears to be OK. Because these are time-series data, there does appear to be a problem in that 3 of the negative residuals are "bunched" together. This means that in years 2, 3, and 4 we overestimated Y. This overestimation could be caused by some force not contained in our model that is restricting the size of corn acreage, such as weather and/or prices for other crops, such as wheat or soybeans.

3. As we examine the regression coefficients and computed t values, we note that some of the problems shown in Table 10-3 appear to have surfaced. Looking at the t values at 3 degrees of freedom ($n - k - 1$), none is greater than the tabled $t_{.025}$ of 3.18. This is strange as some of the variables logically appear to be significant. The cause for the nonsignifi-

cant variables could be due to the previously discussed problem of collinearity.

4. If serious collinearity exists, the interpretation of other results becomes less important at this point; however, we can see that $R^2 = .596$, we had an F value that was insignificant, meaning that the model taken as a whole did not explain a significant portion of the variation in the dependent variable.

Remedy

Once the results were obtained, it was noted that none of the variables was significant and that collinearity could be a likely cause. The decision at this point was to drop some of the independent variables that were highly correlated with one another. As it turned out, because price was highly correlated with all other independent variables as well as with the dependent variable, it was selected as one of the independent variables to remain, as was prior-year carryover. A multiple regression run was made with the equation used for estimation being $Y_c = a + b_1 X_1 + b_2 X_2$ where

Y = acres planted
X_1 = prior-year average price of corn per bushel, dollars
X_2 = prior-year carryover of corn stored in bins, silos, and elevators, millions of bushels

In this analysis only price was significant. Finally, a regression analysis was run using price as the sole independent variable.

In Figure 10-4 the computer results from this simple regression analysis run are shown. Briefly:

1. There is still a relative balance of 5 negative versus 4 positive residuals, although the bunching still exists.
2. The regression coefficient is significant at the .05 level, $t_{.025}$, because $4.227 > 2.36$. As a result, we can conclude that price and corn acreage are related.
3. The sign of the coefficient is positive, meaning that as price goes up 1 unit, corn acreage goes up approximately 4.227 units.
4. The coefficient of determination R^2 is very nearly as large as when all 5 independent variables were used. However, there still is about .44 of the variation in corn acreage *not* explained by price.
5. Finally, the decision was made to perhaps use more observations and to see if additional variables, not examined before, could add to the explanatory power of price.

STANDARD FARM SEEDS, INC.
CORN ACREAGE (Y) VERSUS PRICE (X)

VARIABLE NO.	MEAN	STANDARD DEVIATION	CORRELATION X VS Y	REGRESSION COEFFICIENT	STD. ERROR OF REG.COEF.	COMPUTED T VALUE
3	1.61667	0.85226	0.75128	4.22660	1.40336	3.01176
DEPENDENT 1	70.37434	4.79470				

INTERCEPT   63.54134

MULTIPLE CORRELATION   0.75128

COEFFICIENT OF DETERMINATION   0.56443

STD. ERROR OF ESTIMATE   3.38289

ANALYSIS OF VARIANCE FOR THE REGRESSION

SOURCE OF VARIATION	DEGREES OF FREEDOM	SUM OF SQUARES	MEAN SQUARES	F VALUE
ATTRIBUTABLE TO REGRESSION	1	103.80515	103.80515	9.07073
DEVIATION FROM REGRESSION	7	80.10776	11.44396	
TOTAL	8	183.91290		

TABLE OF RESIDUALS

CASE NO.	Y VALUE	Y ESTIMATE	RESIDUAL
1	71.15599	68.78232	2.37367
2	65.12599	67.89473	-2.76874
3	64.26399	68.10606	-3.84207
4	66.84900	68.44418	-1.59518
5	74.05499	69.16270	4.89229
6	66.97198	68.10606	-1.13408
7	71.91199	70.17709	1.73489
8	77.74599	74.31917	3.42682
9	75.28999	78.37671	-3.08672

Figure 10-4. Regression Output: Standard Farm—Corn Acreage Versus Price

Categorical Data

In our previous discussion of multiple regression analysis, we have made use of examples and cases in which all our variables were intervally scaled (see definition in Chapter 2). In many cases a firm collects data or uses data that are nominally scaled (see definition in Chapter 2), such as war or peace, strike or no strike, marital status, occupation, and religion. All these variables might be useful to use as independent variables in some situations. Multiple regression analysis can easily handle nominally scaled variables that are expressed as dichotomies or multichotomies.

To accomplish this application, use is made of "dummy" variables which are binary coded into zero or one. Assume a situation in which we are trying to predict the number of days of work lost per year as a result of sickness by employees in a factory. We have narrowed the variables of interest down to age and marital status of the employee. Age is coded in years, and marital status includes married, single, divorced or separated, and widowed.

A c-state categorical variable can be coded into $c-1$ dummy variables. For example, the 4-state marital status variable can be coded into 3 dummy variables (represented by 3 columns of digits) as follows:

```
100  married
010  single
001  divorced or separated
000  widowed
```

For each of the columns a single 1 appears. When it does appear, it indicates that that employee falls into that category. Each employee will have only one 1 as he or she can fall into only one category. The absence of a 1 in the 3 columns indicates that the employee is in the fourth class, widowed. Carefully note that the outcomes are not coded 0, 1, 2, 3, because this implies an order in the data that is not present.

We could then run regression analysis on our data set with the data looking as follows:

Employee	Data				
	Days Lost	Age	Marital Status Dummy Codes		
1	1.5	39	1	0	0
2	2.6	47	0	0	1
3	3.0	40	0	0	0
.					
.					
.					
200	10.0	57	1	0	0

Assume that all variables were significant for the equation

$$Y_c = a + b_1 X_1 + b_2 X_2 + b_3 X_3 + b_4 X_4$$

where

Y_c = estimated number of days missed
X_1 = age

$$X_2 = \text{married}$$
$$X_3 = \text{single}$$
$$X_4 = \text{divorced or separated}$$

and respective parameter estimates were

$$a = 1.2$$
$$b_1 = 2.3$$
$$b_2 = 2.2$$
$$b_3 = 3.8$$
$$b_4 = 4.1$$

Note that these parameter estimates are simply presented as examples here. Actual figures would be derived from hand calculations or computer output. Therefore,

$$Y_c = 1.2 + 2.3(\text{age}) + 2.2(\text{married}) + 3.8(\text{single}) + 4.1(\text{divorced or separated})$$

The estimated lost time due to sickness for widowed employees is found from

$$Y_{cW} = 1.2 + 2.3(\text{age})$$

where the impact of being widowed is in the a term. More generally, the a term of 1.2 becomes the "base category" for the marital-status dummy variables. Widowed employees have the lowest sickness rate with married employees next, followed by single and then divorced or separated employees. In terms of interpreting the specific coefficients, married employees are sick 2.2 days more per year than widowed employees; single employees 3.8 more days per year than widowed employees; and divorced or separated 4.1 days more than widowed employees.

Validation

Because we are constantly faced with the problem of sampling error, there is always the possibility of obtaining results that appear to be significant but are actually random (or spurious). There are basically two ways to guard against this possible problem. The first is to take a set of data and run a regression; then, on a separate set of matched data, run another analysis and compare results to determine if they are similar.

A second, "more statistical," method of guarding against these spurious or random results is called *split-half reliability analysis*. This method is gaining increased favor in business usage and involves the following steps:

1. Take a total sample of, say, 500 observations (very reasonable for a cross-section data collection method).
2. Split the sample randomly in half. The split does not necessarily have to be into two equal halves, but let us assume for now that we have two groups, $n = 250$ each. For example, from the list of 500 we could start at a random point and place every other observation into a second group and retain the remainder for the first group.
3. Run the regression analysis on one group of 250 and obtain the a and b values. These values represent the estimates of how each independent variable is related to Y and include the mathematical constant a.
4. Take the data from the second group of 250 and substitute in the a and b estimates for each unit, for example a household, of the 250. From this we can form Y_{c1} (the estimate from the second sample). For example, if from the *first* analysis $a = 1.5$, $b_1 = 2.0$, and $b_2 = 3.0$, and the observations for the first unit in the *second* sample are

$$X_1 = 10$$
$$X_2 = 20$$

then

$$Y_{c1} = 1.5 + 2.0(10) + 3.0(20) = 81.5$$

For each household in this second sample of 250 form a Y_{c1}.

5. When we have completed this we shall have 250 Y_{c1} and 250 values of what Y actually was in the second sample of 250, as assumed in the following:

	Y_{c1}		Y
1	81.5	1	84
2	86.1	2	89
.	.	.	.
.	.	.	.
250	94.3	250	90.6

6. We then take the two variables Y_{c1} and Y and run a correlation as shown in Chapter 9. If the correlation coefficient is high, as for ex-

ample .80 would be, we can say that the results are stable. If r is small, we should investigate the use of additional variables and/or households.

7. Given stable results in the split-half test, we may wish to combine the data for all 500 households and perform a regression analysis and interpret the results as shown previously in the chapter. When we obtain these results, we will be more sure of the underlying stability of the analysis.

We can see that the above procedure will work well on a cross-section type of sample but is more difficult and often impossible for a time-series analysis, where data are collected for consecutive periods, such as years, quarters, or weeks.

Forecasting with Regression

Once the parameter estimates a and b have been obtained, they can be used for business forecasting and other purposes. The procedures for accomplishing this are very similar to calculation of a Y_c. For example, assume that we have the equation

$$Y_c = a + b_1 X_1 + b_2 X_2$$

where

Y = factory sales of automobiles
X_1 = Gross National Product (GNP)
X_2 = Consumer Price Index (CPI)

The multiple regression analysis gives us the estimates of $a = 116.8$, $b_1 = 4.2$, and $b_2 = -.80$ based upon an assumed analysis of quarterly data collected over the last 16 quarters. Our next task would be to obtain estimates of GNP and CPI for the coming quarter. These estimates could then be substituted into the equation to obtain a sales estimate as follows:

$$\text{predicted Sales } (Y_c) = a + b_1 X_1 + b_2 X_2$$

where

X_1 = estimated GNP
X_2 = estimated CPI

In making this prediction of sales, we are making one very important implicit assumption. This assumption is that history (or the factors governing sales over the last 16 quarters) is going to hold true in the future, at least through the next quarter. We are making this assumption because we have based our predicted value upon analysis of past data. As long as business conditions and the market of interest are fairly stable, this assumption is not excessively hazardous. When conditions are very unstable, however, estimates obtained in this manner may be unreliable. Even given these risks, the regression estimates are often better than simple guesswork.

Summary

To review once again the procedures for examination of multiple regression analysis results, the following steps should be helpful:

1. Examine the table of residuals for autocorrelation. This is found from checking the balance of positive and negative values, or making an ordinary length-of-runs test, as shown in Chapter 11.
2. Examine the pairwise correlations of variables for traces of collinearity. A rough rule of thumb was given that the correlation between independent variables should not exceed the R^2 of the model.
3. Take necessary steps to remedy problems detected in steps 1 and 2.
4. Examine the t values for significance of the independent variables individually. If the calculated t is equal to or less than the tabled t, accept $H_0: \beta_i = 0$. If not, reject.
5. Examine the analysis of variance results to detect the impact of the combined effects of the independent variables on the dependent variable. If the computed F value is less than or equal to the tabled F, accept the null hypothesis that the independent variables when taken in conjunction with one another do not explain a significant proportion of the variation in Y.
6. Examine R^2, beta weights, and other estimates as interested.
7. Make necessary stability tests.
8. Beyond all else, examine the results to make sure that they make sense. Regression results that defy sound business judgment may be subject to the pitfalls mentioned above.

Exercises

10.1 (a) How does linear multiple regression analysis differ from simple regression analysis?
(b) How does the interpretation of the b's differ between simple and multiple regression?
(c) Why is the analysis of variance section important in multiple regression when it is not in simple regression?
(d) What are examples of multiple regression analysis models?
10.2 Interpret the following computer results. (Follow steps shown in chapter.)

```
CORRELATIONS OF VARIABLES

    VARIABLE    1         2         3

        1     1.000     0.817     0.731

        2     0.817     1.000     0.898

        3     0.731     0.898     1.000


MULTIPLE REGRESSION

    SELECTION..... 1

        BETA WT. ( 2) =  0.83024406E 00

        BETA WT. ( 3) = -0.14762878E-01


VARIABLE   MEAN      STANDARD    CORRELATION   REGRESSION    STD. ERROR     COMPUTED
  NO.                DEVIATION   X VS Y        COEFFICIENT   OF REG.COEF.   T VALUE
   2      14.40833   5.88750     0.81699        0.38344      0.20158        1.90222
   3      49.49159  20.59654     0.73064       -0.00195      0.05762       -0.03382
DEPENDENT
   1       7.00833   2.71912


INTERCEPT                         1.57999

MULTIPLE CORRELATION              0.81702

COEFFICIENT OF DETERMINATION      0.66751

STD. ERROR OF ESTIMATE            1.73336


              ANALYSIS OF VARIANCE FOR THE REGRESSION

   SOURCE OF VARIATION       DEGREES     SUM OF        MEAN        F VALUE
                             OF FREEDOM  SQUARES       SQUARES
   ATTRIBUTABLE TO REGRESSION    2       54.28862      27.14430    9.03442
   DEVIATION FROM REGRESSION     9       27.04088       3.00454
        TOTAL                   11       81.32950
```

Exercise 10-2.

```
MULTIPLE REGRESSION

    SELECTION..... 1

            TABLE OF RESIDUALS

CASE NO.    Y VALUE     Y ESTIMATE      RESIDUAL
   1        3.60000      5.08717        -1.48717
   2        4.80000      5.41297        -0.61297
   3        2.40000      5.23060        -2.83060
   4        7.20000      5.90170         1.29830
   5        6.90000      6.10726         0.79274
   6        8.40000      6.92726         1.47274
   7       10.70000      8.59473         2.10527
   8       11.20000     12.27414        -1.07414
   9        6.10000      6.59254        -0.49254
  10        7.90000      5.64187         2.25813
  11        9.50000     10.15087        -0.65087
  12        5.40000      6.17884        -0.77884


    SELECTION..... 2

        BETA WT. ( 2) =  0.81698996E 00

VARIABLE    MEAN     STANDARD     CORRELATION   REGRESSION    STD. ERROR    COMPUTED
  NO.                DEVIATION      X VS Y      COEFFICIENT   OF REG.COEF.  T VALUE
   2      14.40833    5.88750       0.81699       0.37732       0.08422     4.48026
DEPENDENT
   1       7.00833    2.71912

INTERCEPT                            1.57173

MULTIPLE CORRELATION                 0.81699
```

Exercise 10-2.

```
COEFFICIENT OF DETERMINATION    0.66747

STD. ERROR OF ESTIMATE          1.64452

              ANALYSIS OF VARIANCE FOR THE REGRESSION
     SOURCE OF VARIATION       DEGREES      SUM OF        MEAN        F VALUE
                              OF FREEDOM    SQUARES       SQUARES
     ATTRIBUTABLE TO REGRESSION    1        54.28520      54.28520    20.07269
     DEVIATION FROM REGRESSION    10        27.04430       2.70443
     TOTAL                        11        81.32950


     SELECTION..... 2

              TABLE OF RESIDUALS
     CASE NO.    Y VALUE      Y ESTIMATE    RESIDUAL
        1        3.60000       5.08084      -1.48084
        2        4.80000       5.42043      -0.62043
        3        2.40000       5.23177      -2.83177
        4        7.20000       5.91095       1.28905
        5        6.90000       6.09961       0.80039
        6        8.40000       6.92972       1.47028
        7       10.70000       8.58994       2.11006
        8       11.20000      12.28771      -1.08771
        9        6.10000       6.55240      -0.45240
       10        7.90000       5.64682       2.25318
       11        9.50000      10.13697      -0.63697
       12        5.40000       6.21281      -0.81281


     SELECTION..... 3

          BETA WT. ( 3) =  0.73064286E 00
```

Exercise 10-2.

VARIABLE NO.	MEAN	STANDARD DEVIATION	CORRELATION X VS Y	REGRESSION COEFFICIENT	STD. ERROR OF REG.COEF.	COMPUTED T VALUE
3	49.49159	20.59654	0.73064	0.09646	0.02850	3.38405
DEPENDENT						
1	7.00833	2.71912				

INTERCEPT   2.23447

MULTIPLE CORRELATION   0.73064

COEFFICIENT OF DETERMINATION   0.53384

STD. ERROR OF ESTIMATE   1.94712

ANALYSIS OF VARIANCE FOR THE REGRESSION

SOURCE OF VARIATION	DEGREES OF FREEDOM	SUM OF SQUARES	MEAN SQUARES	F VALUE
ATTRIBUTABLE TO REGRESSION	1	43.41684	43.41684	11.45181
DEVIATION FROM REGRESSION	10	37.91266	3.79127	
TOTAL	11	81.32950		

SELECTION..... 3

TABLE OF RESIDUALS

CASE NO.	Y VALUE	Y ESTIMATE	RESIDUAL
1	3.60000	5.14750	-1.54750
2	4.80000	6.10243	-1.30243
3	2.40000	5.63943	-3.23944
4	7.20000	6.58472	0.61528
5	6.90000	5.89987	1.00013
6	8.40000	7.06701	1.33299
7	10.70000	8.04124	2.65876
8	11.20000	11.91885	-0.71885
9	6.10000	4.65556	1.44444
10	7.90000	6.16031	1.73969
11	9.50000	8.83220	0.66780
12	5.40000	8.05089	-2.65089

Exercise 10-2.

10.3 Using data for factory sales of automobiles and various potential independent variables a correlation matrix was formed from the following data, and a subsequent computer run was made.
  (a) Interpret the correlation matrix.
  (b) What problem is suggested if all independent variables are used simultaneously?
  (c) What impact can this problem have on regression analysis results?
  (d) After analysis, population and unemployment rate were selected as independent variables. Why was this decision made?
  (e) Interpret the regression analysis output.

Year	Y Dependent Factory Sales	X Independent Variables						
		Pop	Un Rate	Scrap Cars	$\bar{X}$ Family Income	Auto Credit Instal	GNP	CPI For Trans
1970	6546817	201,722	4.9	3,773	10,289	35,490	4,769	112.7
1969	8223715	199,145	3.5	4,327	10,423	36,602	4,590	107.2
1968	8822158	195,264	3.6	3,703	10,049	34,130	4,305	103.2
1967	7436764	193,420	3.8	3,639	9,683	30,724	3,995	100
1966	8598326	191,605	3.8	4,413	9,360	30,556	3,815	97.2
1965	9305561	187,141	4.5	4,200	8,932	28,619	3,525	95.9
1964	7751822	186,493	5.2	4,401	8,579	24,934	3,296	94.3
1963	7637728	183,677	5.7	4,773	8,267	22,254	3,120	93
1962	6933240	181,143	5.5	5,174	7,975	19,381	3,004	92.5
1961	5542707	178,140	6.7	6,034	7,765	17,135	2,831	90.6
1960	6674796	175,277	5.5	6,163	7,668	17,658	2,788	89.6
1959	5591243	174,521	5.5	6,056	7,524	16,420	2,731	89.6
1958	4257812	171,485	6.8	6,433	7,126	14,152	2,569	86
1957	6113344	168,400	4.3	6,266	7,138	15,370	2,576	83.3
1956	5816109	165,311	4.1	5,972	7,122	14,459	2,492	78.8

Y = Factory Sales of Automobiles
X_1 = Population
X_2 = Unemployment Rate
X_3 = Scrap Cars
X_4 = Mean Family Income
X_5 = Auto Credit Installments
X_6 = Gross National Product
X_7 = Consumer Price Index for Transportation

Correlations of Variables

Variable		Sales 1	Population 2	Unemployment Rate 3	Scrap Cars 4	$\bar{X}$ Family Income 5	Auto Installment Credit 6	GNP 7	CPI For Transportation 8
Sales	1	1.000	0.694	−0.661	−0.789	0.702	0.750	0.627	0.554
Population	2	0.694	1.000	−0.467	−0.921	0.980	0.973	0.960	0.973
Unemployment	3	−0.661	−0.467	1.000	0.596	−0.604	−0.638	−0.603	−0.382
Scrap Cars	4	−0.789	−0.921	0.596	1.000	−0.918	−0.928	−0.880	−0.837
$\bar{X}$ Family Income	5	0.702	0.980	−0.604	−0.918	1.000	0.994	0.986	0.952
Installment Credit	6	0.750	0.973	−0.638	−0.928	0.994	1.000	0.980	0.933
GNP	7	0.627	0.960	−0.603	−0.880	0.986	0.980	1.000	0.959
CPI for Transportation	8	0.554	0.973	−0.382	−0.837	0.952	0.933	0.959	1.000

Output

Variable	Computed t	Beta Weight	Correlation	Standard Error
Population	2.476	.49332	.69446	24.79
Unemployment	−2.163	.43099	.66122	264.32

$b_1 = 61.41$
$b_2 = 571.91$
$F = 10.11$
$R^2 = .627$
Residuals: 8(+), 7(−)

10.4 Interpret the following computer results. (Follow steps shown in chapter.)

EXERCISE 10-4. Correlation of Variables

	Variable					
Variable	1	2	3	4	5	6
1 (Dependent variable)	1.00000					
2	.78581					
3	.79036	.95593				
4	−.12344	−.13100	−.10916			
5	.77071	.98895	.97147	−.14415		
6	.25300−01	.26296	.31302	.15205	.31544	
7	.77981	.99183	.96384	−.13770	.99723	.31475

Regression Analysis

Variable	Mean	Standard Deviation	Correlation X vs. Y	Regression Coefficient	Standard Error of Regression Coefficient	Computed t Value	Beta Weight
1 (Dependent variable)	6,065.78906	1,802.31201					
2	295.48364	96.49286	.78581	−.11486	24.54091	−.00468	−.61492920E−02
3	95.96307	12.03638	.79037	155.47504	100.97235	1.53978	.10383091E 01
4	4.52632	2.79620	−.12344	−8.90046	103.65297	−.08587	−.13808608E−01
5	170.26831	18.01198	.77071	−313.13428	241.78409	−1.29510	−.31294098E 01
6	4.92105	1.05703	.02530	−421.82617	307.72827	−1.37077	−.24739456E 00
7	22.00000	5.62732	.77981	954.98730	802.87500	1.18946	.29817352E 01

Intercept: 25,603.10938
Multiple correlation: .85121
Coefficient of determination: .72455
Error of estimate: 1,158.49048

Analysis of Variance for the Regression

Source of Variation	Degrees of Freedom	Sum of Squares	Mean Squares	F Value
Attributable to regression	6	42,364,736.00000	7,060,789.00000	5.26100
Deviation from Regression	12	16,105,200.00000	1,342,100.00000	
Total	18	58,469,936.00000		

Residuals

Case	Y Value	Y Estimate	Residual
1	2,149.00000	2,597.25391	−448.25391
2	3,558.00000	4,273.76563	−715.76563
3	3,909.00000	5,402.06250	−1,493.06250
4	5,119.00000	4,494.87500	624.12500
5	6,666.00000	5,042.38672	1,623.61328
6	6,117.00000	6,011.70703	105.29297
7	5,559.00000	4,975.99219	583.00781
8	7,920.00000	5,753.00781	2,166.99219
9	5,816.00000	5,891.05078	−75.05078
10	6,113.00000	6,322.05469	−209.05469
11	4,258.00000	5,732.38672	−1,474.38672
12	5,591.00000	6,464.08203	−873.08203
13	6,675.00000	6,560.52344	114.47656
14	5,543.00000	6,177.15625	−634.15625
15	6,963.00000	6,881.88281	81.11719
16	7,638.00000	7,122.23828	515.76172
17	7,752.00000	7,702.62891	49.37109
18	9,306.00000	8,437.16797	868.83203
19	8,598.00000	9,407.41797	−809.41797

10.5 Interpret the following computer results. (Follow steps shown in chapter.)

CORRELATIONS OF VARIABLES

VARIABLE	1	2	3	4	5
1	1.0000	0.423	0.849	0.238	0.416
2	0.423	1.000	0.225	0.576	0.121
3	0.849	0.225	1.000	0.082	0.089
4	0.238	0.576	0.082	1.000	0.257
5	0.416	0.121	0.089	0.257	1.000

MULTIPLE REGRESSION

SELECTION 1

BETA WT. (2) =   0.23594934E 00

BETA WT. (3) =   0.77039087E 00

BETA WT. (4) = −0.46739757E−01

BETA WT. (5) =   0.33083701E 00

VARIABLE NO.	MEAN	STANDARD DEVIATION	CORRELATION X VS Y	REGRESSION COEFFICIENT	STD. ERROR OF REG. COEF.	COMPUTED T VALUE
2	14.27739	3.50940	0.42252	0.45096	0.26270	1.71667
3	102.14658	16.14339	0.84897	0.32009	0.04691	6.82354
4	20.68665	1.09926	0.23756	−0.28520	0.84256	−0.33849
5	8.00000	4.47214	0.41576	0.49620	0.17053	2.90980
DEPENDENT						
1	37.05992	6.70743				

INTERCEPT	−0.14465
MULTIPLE CORRELATION	0.93818

EXERCISE 10-5.

COEFFICIENT OF DETERMINATION   0.88018

STD. ERROR OF ESTIMATE   2.74716

ANALYSIS OF VARIANCE FOR THE REGRESSION

SOURCE OF VARIATION	DEGREES OF FREEDOM	SUM OF SQUARES	MEAN SQUARES	F VALUE
ATTRIBUTABLE TO REGRESSION	4	554.38672	138.59668	18.36472
DEVIATION FROM REGRESSION	10	75.46899	7.54690	
TOTAL	14	629.85571		

MULTIPLE REGRESSION

   SELECTION 1

TABLE OF RESIDUALS

CASE NO.	Y VALUE	Y ESTIMATE	RESIDUAL
1	34.20000	32.20367	1.99632
2	38.00000	37.18149	0.81851
3	38.39999	38.71030	−0.31030
4	36.70000	37.87263	−1.17264
5	37.20000	39.12448	−1.92448
6	35.29999	37.27962	−1.97963
7	33.29999	32.98203	0.31796
8	23.89999	23.99776	−0.09776
9	25.59999	28.02509	−2.42509
10	33.29999	33.62601	−0.32602
11	39.39999	36.74951	2.65048
12	46.39999	41.52034	4.87965
13	47.70000	45.61409	2.08591
14	43.70000	43.69176	0.00824
15	42.79999	47.32042	−4.52043

DURBIN-WATSON STATISTIC = 0.87065E 00

EXERCISE 10-5.

Appendix

Computations of Estimates for Linear Multiple Regression Landscape Example

Family	Landscape Expenditures, Y	Income, X_1	Mortgage, X_2	X_1Y	X_2Y	X_1X_2	Y^2	X_1^2	X_2^2
1	2.7	18.6	24.2	50.22	65.34	450.12	7.29	345.96	585.64
2	3.6	20.4	32.1	73.44	115.56	654.84	12.96	416.16	1,030.41
3	1.8	19.4	28.2	34.92	50.76	547.08	3.24	376.36	795.24
4	5.4	24.2	36.1	130.68	194.94	873.62	29.16	585.64	1,303.21
5	5.2	24.0	30.4	124.80	158.08	729.60	27.04	576.00	924.16
6	6.3	28.4	40.1	178.92	252.63	1,138.84	39.69	806.56	1,608.01
7	1.3	37.2	48.2	48.36	62.66	1,793.04	1.69	1,383.84	2,323.24
8	8.4	56.8	80.3	477.12	674.52	4,561.04	70.56	3,226.24	6,448.09
9	4.6	26.4	20.1	121.44	92.46	530.64	21.16	696.96	404.01
10	5.9	23.6	32.6	139.24	192.34	769.36	34.81	556.96	1,062.76
11	7.1	45.4	54.7	322.34	388.37	2,483.38	50.41	2,061.16	2,992.09
12	4.1	24.6	48.2	100.86	197.62	1,185.72	16.81	605.16	2,323.24
	56.4	349.0	475.2	1,802.34	2,445.28	15,717.28	314.82	11,637.00	21,800.10

To solve for values of a, b_1, and b_2 we have to solve the following set of equations using the values calculated above:

$$\Sigma Y = na + b_1 \Sigma X_1 + b_2 \Sigma X_2$$
$$\Sigma X_1Y = a(n)(\Sigma X_1) + b_1 \Sigma X_1^2 + b_2 \Sigma X_1X_2$$
$$\Sigma X_2Y = a(n)(\Sigma X_2) + b_1 \Sigma X_1X_2 + b_2 \Sigma X_2^2$$

Substituting into the equations, we have

$$56.4 = 12a + 349b_1 + 475.2b_2$$
$$1{,}802.34 = 4{,}188a + 11{,}637b_1 + 15{,}717.28b_2$$
$$2{,}445.28 = 5{,}702.4a + 15{,}717.28b_1 + 21{,}800.10b_2$$

Solving these equations simultaneously, we get the values

$$a = 1.51$$
$$b_1 = .097$$
$$b_2 = .009$$

11

Learning Objectives

Introduction

Meaning and Advantages of Nonparametric Statistics and When To Use Them

Test for Significance of Difference Between Two Proportions

Sign Test

Mann–Whitney *U* Test

Ordinary Length-of-Runs Test
  Cautions and Comments

Tests for Goodness of Fit

Rank-Order Correlation
  Spearman's Rho Coefficient
  Tests of Significance of Spearman's Rho

Learning Objectives

The primary learning objectives for this chapter are:

1. To be able to detect situations where nonparametric methods are appropriate, and to select the appropriate method for a particular situation.
2. To become aware of the advantages and disadvantages of nonparametric methods.
3. To recognize the ease with which most nonparametric methods can be applied.
4. To understand the types and importance of nonparametric methods in their applications to business problems.
5. To understand and be able to use the test procedures.

Introduction

Nonparametric Statistics

As we have progressed through the text, we have been introduced to many terms that were called parameters. Examples are μ, σ, and β. In most cases assumptions regarding these parameters were made. For instance, the central limit theorem was a key factor in our hypothesis-testing experiences, assumptions were made about the equality of variances in analysis of variance, and other sometimes stringent assumptions were made about the form of the parent population from which a sample was drawn. We have consistently noted that, when some of the important assumptions are violated, this violation may impair the validity of a statistical test. In some instances, we are dealing with very small samples, such as cases where destructive tests are run. In other cases, we have very limited experience, such as in a new-product introduction; or, finally, we do not have sufficient information to allow us to speculate on what the nature of the population is. This chapter presents some methods that can be used when the necessary assumptions required for other techniques do not appear to be appropriate. In general, the term *nonparametric statistics* relates to use of methods not concerned with population parameters or which do not depend on rigid assumptions concerning the population distribution. The user's knowledge about the data or the nature of the data themselves will govern our choice in selecting the appropriate procedure.

Meaning and Advantages of Nonparametric Statistics and When To Use Them

When we are not dealing with speculation regarding a population parameter, and make no assumptions about the shape of the underlying population, we can often use nonparametric methods. We can still make descriptions, but we do not make statements pertaining to the shape of population distribution. Also, the way the data are scaled can have an impact on the type of methods we use. For example, ordinal or nominally scaled data sometimes require use of nonparametric procedures.

Generally when we still want to do statistical analysis, but our data do not meet the necessary parametric assumptions, we can apply nonparametrics. For example, if the sample size is small, the data consist merely of ranks, or assumptions required by parametric tests are not otherwise met, nonparametrics are often appropriate. We should be careful to note that because we are not making (meeting) the set of assumptions, nonparametric methods are not as "powerful" as parametric methods. In the terms of statistics, "power" refers to the probability of making an error such as rejecting the null hypothesis when it is false. The power of a nonparametric test is hard to discuss because no "true" correct situation is known.

Nonparametric tests have several factors in their favor. They can be referred to as *robust* because they are generally best able to withstand violation of assumptions necessary for parametric tests. In general, nonparametric tests are relatively efficient if the sample size is large, and when it is not large they are better than doing no statistical analysis at all. What follows is a discussion of some of the more widely used nonparametric tests.

Test for Significance of Difference Between Two Proportions

Assume that a firm desires to evaluate a television advertisement. They are going to test the ability of groups of consumers who have seen a television show to recall a television advertisement for Duraply Radial Tires. The advertising agency has stated that a single message of their new advertisement will be remembered as well 1 week after its airing as it was 2 days after airing. To test this hunch, or hypothesis, an experiment was run where two separate groups were tested. For each group member, the 2-day and the 1-week group, a "+" was recorded if he or she recalled the ad in a telephone interview and a "−" if the message was not recalled.

This is a clear case of data that are dichotomous (i.e., there are only two possible responses), and we wish to test the significance of the difference between the proportion from group 1 who recall the ad to the proportion of group 2 who recall. The procedure that is generally used in this situation is the same as the one we examined in Chapter 7.

$$z = \frac{p_1 - p_2}{s_d}$$

where

$$\hat{\pi} = \frac{n_1 p_1 + n_2 p_2}{n_1 + n_2} \quad \text{and} \quad s_d = \sqrt{\frac{\hat{\pi}(1 - \hat{\pi})(n_1 + n_2)}{n_1 + n_2}}$$

where

p_1 = proportion of group 1 correctly recalling
p_2 = proportion of group 2 correctly recalling
n_1 = number of people in group 1
n_2 = number of people in group 2

Data from a sample are shown in Table 11-1. Respondents 3, 5, 6, 11, 16, 17, 18, and 20 of group 1 did not recall the advertising message, while respondents 21, 22, 27, 30–34, 36, 38, 39, and 40 of group 2 did not recall the ad. Therefore,

$$p_1 = \frac{12}{20} = .60$$
$$p_2 = \frac{8}{20} = .40$$
$$\hat{\pi} = \frac{20(.60) + 20(.40)}{20 + 20} = \frac{12 + 8}{40} = .50$$

We should note that .50 is the same figure we would get if we took an arithmetic mean of .60 and .40. This will only be the case when the two sample sizes are identical.

$$s_d = \sqrt{\frac{\hat{\pi}(1-\hat{\pi})(n_1+n_2)}{n_1+n_2}} = \sqrt{\frac{.50 \times .50\ (40)}{40}} = \sqrt{\frac{10}{.40}} = \sqrt{.25} = .5$$

$$z = \frac{.60 - .40}{.50} = .40$$

If we are testing the hypothesis at the .05 level that

$$H_0: \pi_1 - \pi_2 = 0$$
$$H_a: \pi_1 - \pi_2 \neq 0$$

when H_0 indicates the two samples (2 days, and 1 week) come from the same population, the critical z value would be 1.96. Because $.40 < 1.96$, we would accept H_0. This procedure, although not limited to nonparametric situations, serves as an excellent lead into the next section.

TABLE 11-1. Recall Data for Duraply Advertisement

Group 1: 2 Days	Group 2: 1 Week
R_1 +	R_{21} −
R_2 +	R_{22} −
R_3 −	R_{23} +
R_4 +	R_{24} +
R_5 −	R_{25} +
R_6 −	R_{26} +
R_7 +	R_{27} −
R_8 +	R_{28} +
R_9 +	R_{29} +
R_{10} +	R_{30} −
R_{11} −	R_{31} −
R_{12} +	R_{32} −
R_{13} +	R_{33} −
R_{14} +	R_{34} −
R_{15} +	R_{35} +
R_{16} −	R_{36} −
R_{17} −	R_{37} +
R_{18} −	R_{38} −
R_{19} +	R_{39} −
R_{20} −	R_{40} −

Sign Test

A procedure similar to previous material involves examination of rank orders and is commonly called the "sign test" of two groups. Assume that a group of 35 drivers have been asked to ride blindfolded in two large-sized cars. Upon completion of the ride, they are to assign a rank of 1 to 5 for each car, based upon a number of criteria. For example, one criterion is smoothness of ride where 1 represents a very smooth ride and 5 represents a very rough ride. The results for the two cars, Marx 18 and Coup De Neville, are as shown in Table 11-2. A $-$ sign of difference indicates that Marx 18 is preferred to Coup De Neville, while a $+$ score indicates that Coup De Neville is preferred, and a zero indicates a tie.

We can view this problem as a test of a null hypothesis that there is no difference in the ranks of the smoothness of ride for the two cars. Under the null hypothesis, we expect the proportion to prefer Marx 18 to equal the proportion preferring Coup De Neville.

$$H_0: \pi_1 = \pi_2 = .50$$
$$H_a: \pi_1 \neq \pi_2 \neq .50$$

Given that the sample is relatively large, we can use the normal approximation to the binomial where

$$s_p = \sqrt{\frac{\pi(1-\pi)}{n}}$$

In the case of the sign test, ties are excluded, so we had

```
12   + scores
18   − scores
 5   ties
35   observations = n
```

Therefore, using the 30 untied scores

$$s_p = \sqrt{\frac{.50 \times .50}{30}} = .09$$

TABLE 11-2. Rank Order Scores for Ride of Two Cars

Rider	Score for Marx 18	Score for Coup De Neville	Sign of Difference
1	2	3	−
2	1	3	−
3	1	2	−
4	3	4	−
5	1	1	0
6	2	2	0
7	3	1	+
8	2	1	+
9	4	2	+
10	3	4	−
11	2	1	+
12	1	3	−
13	2	3	−
14	2	3	−
15	3	4	−
16	2	3	−
17	2	2	0
18	1	1	0
19	5	1	+
20	3	3	0
21	2	1	+
22	1	2	−
23	3	2	+
24	2	4	−
25	1	4	−
26	3	2	+
27	1	2	−
28	2	1	+
29	1	3	−
30	2	1	+
31	2	1	+
32	3	4	−
33	2	4	−
34	4	3	+
35	4	5	−

Because we are examining pluses, although minus values could alternatively be used,

$$p = \frac{12}{30} = .40$$

$$z = \frac{p - \pi}{s_p} = \frac{.40 - .50}{.09} = \frac{.10}{.09} = 1.11$$

At the .05 level, $z = 1.96$ and thus we accept the null hypothesis that $\pi_1 = \pi_2 = .50$.

Mann–Whitney U Test

The Mann–Whitney U test for differences between independent samples goes one step further than the sign test just presented. In the Mann–Whitney U test, the specific ranks are taken into account and are summed. This differs from the sign test in that the exact ranks are taken into account rather than solely the sign of the difference in scores. As a result, this test is often called a "rank sum test." Take the following instance where data have been collected for random samples of right-handed golfers and left-handed golfers regarding how far they can hit a drive off the tee. The data shown in Table 11-3 were found.

If the ranks of drives were not substantially different for the left-handed or right-handed golfers, we would expect the totals of the ranks

TABLE 11-3. Length of Tee Shots for 15 Left- and 15 Right-Handed Golfers

Left-Handed	Right-Handed
175	181
177	182
180	186
183	189
185	194
187	195
188	198
190	199
191	200
192	201
193	226
196	237
197	241
225	246
228	257

to be about the same with differences due only to sampling error. To test this, we utilize the following statistic:

$$U = n_1 n_2 + \frac{n_1(n_1 + 1)}{2} - R_1$$

We can also make use of a standard deviation formula:

$$\sigma_U = \sqrt{\frac{n_1 n_2 (n_1 + n_2 + 1)}{12}}$$

where

$n_1 =$ the number of observations in sample 1
$n_2 =$ the number of observations in sample 2
$R_1 =$ the sum of the ranks in sample 1
$R_2 =$ the sum of the ranks in sample 2

From Table 11-4 we can obtain the figures necessary to test the hypothesis that there is no difference between ranks of left- and right-handed drives, under the assumption that samples were random and golfers were matched by golf handicap. From this table we can find:

$$R_1 = 4 + 5 + 8 + \cdots + 30 = 286$$

$$U = (15)(15) + \frac{(15)(15 + 1)}{2} - 286$$

$$= 225 + \frac{240}{2} - 286 = 59$$

$$\sigma_U = \sqrt{\frac{(15)(15)(15 + 15 + 1)}{12}} = 24.1$$

It also can be shown that $\mu_U = (n_1 n_2)/2 = 112.5$ where μ_U equals the expected mean of U. Therefore, to test the hypothesis we can use methods which are very similar to our previous work.

$$z = \frac{U - \mu_U}{\sigma_U} = \frac{59 - 112.5}{24.1} = -2.22$$

If we are conducting the test at the .05 level (critical value = 1.96), we would conclude that the two ranks are significantly different from one

TABLE 11-4. Rank Orders for Golf Example

Rank	Drive Length (yards)	Left- or Right-Handed	Right-Handed Ranks	Left-Handed Ranks
1	175	L	4	1
2	177	L	5	2
3	180	L	8	3
4	181	R	11	6
5	182	R	16	7
6	183	L	17	9
7	185	L	20	10
8	186	R	21	12
9	187	L	22	13
10	188	L	23	14
11	189	R	25	15
12	190	L	27	18
13	191	L	28	19
14	192	L	29	24
15	193	L	30	26
16	194	R	286	179
17	195	R		
18	196	L		
19	197	L		
20	198	R		
21	199	R		
22	200	R		
23	201	R		
24	225	L		
25	226	R		
26	228	L		
27	237	R		
28	241	R		
29	246	R		
30	257	R		

another because 2.22 (absolute value) is greater than 1.96. In other words, we have sufficient evidence to say with 95 per cent confidence that left- and right-handed golfers have differing lengths of tee shots.

Before proceeding with additional nonparametric tests, a few points need to be noted.

1. It would appear that a simple t test could be run on the last set of data. On occasion the data collected in an experiment are badly skewed and there is a small sample such as in the golf example. The Mann–Whitney U test serves as a substitute for the t test for the evaluation of two independent groups of data.

2. It is especially useful to use the Mann–Whitney U test when there are small samples and little is known about the population because the appropriate sampling distribution can be shown to approach normality when n_1 and n_2 are each 10 or greater.

3. We have been testing hypotheses that the proportions, signs, or sums of the ranks are expected to be about equal. Other one-tailed and/or more specific hypotheses can also be tested.

4. In the previous example of the Mann–Whitney U test, there were no ties. In the case of a tie, say for number 6 and 7, each would be assigned $(6 + 7)/2 = 6.5$ as its rank. If there were a tie for 6, 7, and 8 where they all had the same value, the appropriate value for each would be $(6+7+8)/3 = 7$. When the sample is relatively large and there are not a great number of ties, they will have little impact on the outcome of the test.

Exercises

11.1 Suppose that we examined two groups for their ability to identify if a wine were domestic. Group 1 represented those people who drink nondomestic wines exclusively, with the other group drinking domestic wine exclusively. A + indicates a correct identification; a − indicates an incorrect identification.

Group 1 Nondomestic Wine Drinkers	Group 2 Domestic Wine Drinkers
−	+
−	+
+	+
+	−
+	−
−	+
−	+
+	+
+	−
−	+
−	+
+	−
+	+
−	+

Test the significance of difference in the proportions.

11.2 At a meeting of purchasing agents, a test has been conducted concerning what their evaluations are of two sales approaches.

On a 1 to 10 scale with 1 very good and 10 very bad, they have evaluated each of the presentations. The scores are as follows:

Purchasing Agent	Presentation 1	Presentation 2
1	2	5
2	3	4
3	8	6
4	5	5
5	3	2
6	4	3
7	3	1
8	8	7
9	10	8
10	3	4
11	4	3
12	5	4
13	6	5
14	3	6
15	4	4
16	2	3
17	1	2
18	4	2
19	2	3
20	4	3

Assuming the use of z values from the normal distribution, perform a sign test on these data and test the hypothesis that the two methods are equally as effective.

11.3 The following data have been collected regarding scores received for a pilot television show which is under consideration by a major network. Data were collected from cities of 500,000 people and over and cities under 500,000. Each score is on a 100-point basis with 0 being bad and 100 being good.

Less Than 500,000	500,000 or More
45	38
55	40
58	46
60	73
47	59
53	61
68	63
70	66
65	49
62	74

Establish the correct hypothesis and perform a Mann–Whitney U test at the .05 level.

11.4 An experiment was conducted to see if consumers could identify Big Blue gum when it was compared in a taste test to its competitor. The results showed:

$$p_1 = .60 \quad p_2 = .43$$
$$n_1 = 50 \quad n_2 = 75$$

where

$p_1 =$ proportion of regular Big Blue users able to identify Big Blue

$p_2 =$ proportion of occasional Big Blue users able to identify Big Blue

Is there a difference in ability to identify between regular and occasional Big Blue users?

11.5 A group of 26 students was ranked on a test for manual dexterity. The students were classified on the basis of whether they had participated in a voluntary dexterity achievement program. The scores for the 26 students are as follows on the basis of 100 points maximum:

Participated	Did Not Participate
79	81
84	89
89	78
76	53
94	59
63	92
58	86
76	83
75	78
83	77
84	76
88	64
90	70

On the basis of a Mann–Whitney U test at the .01 level, are there differences between the group who participated and those who did not?

Ordinary Length-of-Runs Test

In the areas of marketing, production, and accounting, it is often important to examine the sequence in which events happen. For example, a quality-control engineer may be very interested in analyzing the sequence of defective parts coming off a production line. It is often an expensive and/or time-consuming job to test the quality of items in production, but it may be much less costly to run a test than to take a chance that a large part of a production run may be defective. If defective parts are coming off the line in somewhat random fashion, this will have an impact on the engineer's sampling procedure since he will have to take fairly large samples throughout the day. If defective parts tend to be clustered, there can be relatively small samples taken frequently throughout the day.

The same type of analogy can be drawn for an auditor who is examining accounts receivable for their accuracy or a market researcher who is examining brand purchase patterns. Assume that the general pattern for the characteristic examined turns out to be as follows:

$$111 \quad 0 \quad 11 \quad 000 \quad 1 \quad 0 \quad 1 \quad 0 \quad 11 \quad 0000 \quad 1$$

where we might assign 1 to a "good" result and 0 to a "bad" result. As we examine the sequence of events, we can see that there are some clusters of bad as well as good results.

The type of hypothesis tested is very similar to those we have discussed before:

H_0: The sequence is random.
H_a: The sequence is not random.

The null hypothesis means that the ordered sequence of defective and correct parts can be considered a random process, or that the arrangements of outcomes are all equally likely. The alternative hypothesis implies that there is some outside nonrandom influence causing the outcomes to be clustered in some "biased" fashion.

As another example, consider a product manager for ECCO gasoline. Almost all product managers are interested in "brand loyalty" toward their particular product. They design advertising messages to persuade people to be loyal as well as to entice them to shift away from other brands. As we examine a car owner's purchases of gasoline for a one-year period, we find the following pattern where a 1 denotes purchase of ECCO when gasoline is purchased and 0 denotes purchase of an alternative brand:

1111 00 11111 00 111111 000 111111 000 11111111 00 1111 0 1 0 11

If we examine the clusters of purchases, we can let c denote the total number of clusters or runs of any type. This is indicated by the total number of clusters separated by a space. In this case we have 15 such clusters.

The distribution for c has a mean

$$\mu_c = \frac{2n_1 n_2}{n_1 + n_2} + 1$$

and a standard deviation

$$\sigma_c = \sqrt{\frac{2n_1 n_2(2n_1 n_2 - n_1 - n_2)}{(n_1 + n_2)^2(n_1 + n_2 - 1)}}$$

where n_1 is the number of occurrences of one outcome and n_2 is the number of occurrences of the other, and c refers to the total number of runs (clusters). As was the case for some of our earlier nonparametric tests, when the sample size is large, the sampling distribution is close to a normal distribution. The same applies to the sampling distribution of c. A general rule of thumb is that either n_1 or n_2 should be at least 20, with neither being below 5, for the normal approximation to hold. As was the case in many of our previous hypothesis tests, we can calculate a z or standard deviation unit (SDU) value:

$$z = \frac{c - \mu_c}{\sigma_c}$$

In analyzing our purchase concentrations for ECCO, $n_1 =$ purchase of ECCO $= 36$, and $n_2 = 14$. From this,

$$\mu_c = \frac{2(36)(14)}{36 + 14} + 1 = 21.16$$

$$\sigma_c = \sqrt{\frac{(2)(36)(14)[(2)(36)(14) - 36 - 14]}{(36 + 14)^2(36 + 14 - 1)}} = 2.81$$

$$z = \frac{15 - 21.16}{2.81} = -2.19$$

If we are testing for significance at the .05 level, our calculated absolute value (2.19) is greater than the critical value at the .05 level (1.96) and we reject the null hypothesis. In this case the number of runs c is different enough from the expected number of runs μ_c that we can state that purchase patterns of gasoline for this household are nonrandom. We state this at the .05 level. From the product manager's standpoint, he could conclude that other factors such as loyalty are operative in the purchase sequence. Carefully note that the test does not show what other factors exist, but just whether or not the sequence is random.

Cautions and Comments

The length-of-runs test is not considered to be one of the more powerful tests because no strong underlying assumptions are required and many things can have an impact on the sampling of a sequence of events. However, this type of test is virtually the only one that examines the sequence of events. Because of this uniqueness, these length-of-runs tests are used fairly widely in a variety of practical applications.

Tests for Goodness of Fit

In previous chapters we have examined various distributions such as the binomial, normal, Poisson, F, χ^2, and t. Generally, there are assumptions made pertaining to the use of these distributions in practical situations, but often these applications do not exactly comply with the stated assumptions. Chi-square analysis is often used to test how well a sample frequency distribution corresponds to a hypothetical population distribution. The procedures used for this *goodness of fit* test are very similar to the chi-square (χ^2) procedures discussed in Chapter 8.

In the use of the χ^2 goodness-of-fit test, we generally make an assumption about the population distribution, or have some knowledge about it from past studies. The null hypothesis, then, is that there is no significant difference between a frequency distribution from a sample and the assumed population distribution. In other words, we are testing to determine if there is a difference.

EXAMPLE

As an example of an application of this test, consider a sports magazine editor who wishes to test some format changes to determine how readers will react. Considerable expense is involved in test marketing,

TABLE 11-5. National Income Breakdowns of Subscribers to a Sports Magazine

Income Category National Subscribers	Percentage of Subscribers
$ 0– 4,999	4.5
5,000– 9,999	7.0
10,000–14,999	28.6
15,000–19,999	39.4
20,000–24,999	15.5
25,000 and over	5.0
	100.0

so the editor wants to select a test city that is a good representation of the national market. One of the factors important to the company is family income. The editor subscribes to a syndicated national readership service that provides accurate income data for the national subscribers to the sports magazine. These breakdowns are shown in Table 11-5.

One of their important considerations is to have a good match between the test market respondents and the national market in terms of income. For the test market, 2,200 magazine readers are selected through random sampling procedures and a questionnaire is sent out prior to the format change; another questionnaire is to be sent out after the format change, which is to be done only in the test market area. On the prechange questionnaire, respondents are asked their yearly family income before taxes. After appropriate followup and verification procedures, 2,000 usable responses are obtained. The data pertaining to income for the test sample are shown in Table 11-6.

TABLE 11-6. Sample Income Breakdowns of Magazine Readership

Income Category Sample Subscribers	Percentage
$ 0– 4,999	5.3
5,000– 9,999	8.7
10,000–14,999	31.2
15,000–19,999	37.2
20,000–24,999	14.0
25,000 and over	3.6
	100.0

TABLE 11-7. Calculations of χ^2 for National and Sample Comparison

Income	Sample Frequency, f_o	National Frequency, f_e	$f_o - f_e$	$(f_o - f_e)^2$	$\dfrac{(f_o - f_e)^2}{f_e}$
$ 0– 4,999	106	90	16	256	2.84
5,000– 9,999	174	140	34	1156	8.26
10,000–14,999	624	572	52	2704	4.73
15,000–19,999	744	788	−44	1936	2.47
20,000–24,999	280	310	−30	900	2.90
25,000 and over	72	100	−28	784	7.84
	2,000	2,000			$\chi^2 = 29.04$

We assume that the national figures are accurate and can be used in calculating our theoretical frequencies. Therefore, the national and sample percentages can be used in conjunction with the 2,000 responses to calculate the expected and actual frequencies, respectively, as shown in Table 11-7.

Prior to accepting or rejecting our null hypothesis, we have to determine the degrees of freedom. In this situation we cannot use the $(r - 1)(c - 1)$ formulation that we used for contingency table analysis. Generally, the number of degrees of freedom when the population data are known, as shown above, can be found by taking $r - 1$ where r is equal to the number of class intervals (also the number of rows in this case). The reason why we use $r - 1$ is because we know the total expected frequencies must equal 2,000 and after we calculate the first 5 expected frequencies the last (100) is not free to vary. In other words, it must equal 100 to make the total equal 2,000. In this case, since there are 6 intervals, the degrees of freedom are $(6 - 1) = 5$. We calculate χ^2 as

$$\chi^2 = \sum \frac{(f_o - f_e)^2}{f_e}$$

where

f_o = observed frequency
f_e = expected frequency

From Appendix G, using 5 degrees of freedom, we find the appropriate χ^2 value to be 11.1 at the .05 level. The computed χ^2 is 29.04. As a result,

TABLE 11-8. Golf Ball Preferences

Ball Preferred	Number of Professionals Selecting
Topmark	68
Special X	42
Title Flight	51
Distance Ultra	79
Flight Power	60
	300

we conclude that the sample frequency distribution and the national frequency distribution are significantly different.

As another example, consider a preference test where professional golfers were asked to rate 5 alternative golf balls for how they "felt" when hit with a 3 iron. Each golfer was instructed to select the ball that felt best when hit. Names were left off the balls and the rotation in which golfers hit them was randomly varied. After the experiment the preferences were noted, as shown in Table 11-8.

From examination of Table 11-8, we can see that there was substantial variation in preferred golf balls. The key question is whether this variation is due to sampling error or to actual differences in preferences. In short, the null hypothesis is:

H_0: Preferences for all golf balls are the same.
H_a: Preferences for all golf balls are not the same.

Once again we can make use of the chi-square test, where

$$\chi^2 = \sum \frac{(f_o - f_e)^2}{f_e}$$

where

f_o = frequency observed
f_e = frequency expected

Under the null hypothesis we would expect all golf balls to have an equal number of selections of first choice. Therefore, f_e for each ball would be $.20 \times 300 = 60$. The key question is whether the actual frequencies differ

TABLE 11-9. Calculations of χ^2 Values for Golf Ball Problem

Ball Preferred	Observed Frequency, f_o	Expected Frequency, f_e	$f_o - f_e$	$(f_o - f_e)^2$	$\dfrac{(f_o - f_e)^2}{f_e}$
Topmark	68	60	8	64	1.07
Special X	42	60	−18	324	5.4
Title Flight	51	60	−9	81	1.35
Distance Ultra	79	60	19	361	6.02
Flight Power	60	60	0	0	0
	300	300			$\chi^2 = 13.84$

from the expected frequencies more than would be due to chance. The appropriate calculations are shown in Table 11-9.

In this case the calculated $\chi^2 = 13.84$ and since we have 5 brands of balls, we have $r - 1$, or 4 degrees of freedom. If we were conducting this test at the .05 level, the critical χ^2 would be 9.5. Therefore, since 13.84 is greater than 9.5, we would reject the null hypothesis and conclude that there were differences in golf ball preferences across the 5 balls.

We should once again note certain cautions in using χ^2 tests. First, the data should be in frequencies; second, the sample should be random; finally, the sample size should be adequate with at least 50 total observations and no less than 5 expected observations in any cell.

Rank-Order Correlation

Frequently, we collect data about two or more characteristics in the same sample. If these variables can be considered as independent of each other, then there is no relationship or association and one variable's value has no effect on the value of the other variable. We are often interested in the degree of association or strength of association between two rank orders.

Spearman's Rho Coefficient
Consider the case where a firm is attempting to analyze the degree of accuracy that a personnel test has in predicting the success of machine operators. The part being produced is very expensive and the major

criterion of success used to evaluate an operator is the total number of good parts turned out on a weekly basis. A test is used by the personnel department to attempt to predict how well a person will perform on the job. The test score consists of data from each person for manual dexterity, past experience, and intelligence. The objective is to be able to rank the persons taking the test in order of expected job performance. The rank correlation check will attempt to determine if there is correlation between the test performance and the job performance. To test the difference in the personnel rank and the performance rank, we can utilize the Spearman *rho* rank-order correlation coefficient:

$$\text{rho} = 1 - \frac{6 \Sigma D^2}{n(n^2 - 1)}$$

where

rho = the rank-order correlation coefficient
D = the difference score between X (personnel) and Y (performance)
n = the number of pairs of scores

Data for 12 new employees are presented in Table 11-10. Calculations of the rho coefficient are presented in Table 11-11.

TABLE 11-10. Rank Order of Employees in Machine Operator Problem

Employee	Personnel Test Rank X	Production Performance Rank Y
A	1	4
B	2	3
C	3	6
D	4	5
E	5	1
F	6	2
G	7	7
H	8	9
I	9	10
J	10	8
K	11	12
L	12	11

TABLE 11-11. Calculation of Rho Coefficient for Machine Operator Problem

Employee	Personnel Test Rank, X	Production Performance Rank, Y	Difference in Ranks, $D = X - Y$	$D^2 = (X - Y)^2$
A	1	4	−3	9
B	2	3	−1	1
C	3	6	−3	9
D	4	5	−1	1
E	5	1	4	16
F	6	2	4	16
G	7	7	0	0
H	8	9	−1	1
I	9	10	−1	1
J	10	8	2	4
K	11	12	−1	1
L	12	11	1	1
				60

$$\text{rho} = 1 - \frac{6 \sum D^2}{n(n^2 - 1)} = 1 - \frac{6(60)}{12[(12)^2 - 1]} = 1 - (.21) = .79$$

We can interpret this rho coefficient in much the same way we interpreted a correlation coefficient in Chapters 9 and 10, except in this case it represents the degree of association between the two rank orders. In the case where rho $= +1$, there is a perfect positive correlation between the two ranks. This means that the test ranked machine operators exactly the same as their rank in performance. When rho $= -1$, the methods ranked operators *exactly* the opposite. A correlation coefficient (rho) of .79 means that there is a fairly high positive correlation between the two ranks, whereas a zero would indicate no association between the two ranks.

Tests of Significance of Spearman's Rho

There are two procedures for testing the significance of rho. For both procedures we are testing the null hypothesis:

H_0: rho $= 0$ (i.e., no correlation)
H_a: rho $\neq 0$

If n is 30 or larger, a z test can be used and when n is between 10 and 30, a t test should be used. When n is less than 10, the results can be very unstable, and it recommended that Spearman's rho not be calculated. When n is between 10 and 30, we compute a t value where

$$t = \text{rho} \sqrt{\frac{n-2}{1-\text{rho}^2}}$$

In our machine operator example, where $n = 12$ and rho $= .79$,

$$t = .79 \sqrt{\frac{12-2}{1-(.79)^2}} = 4.07$$

The degrees of freedom are $n - 2 = 10$. At the .05 level, $t_{.025}$ with 10 df $= 2.23$. Therefore, because $4.07 > 2.23$, our correlation coefficient of .79 is significant at the .05 level.

When n is 30 or greater,

$$z = \text{rho} \sqrt{n-1}$$

For example, if rho $= .58$ and $n = 50$,

$$z = .58 \sqrt{49} = 4.06$$

If z is greater than 1.96, then rho is significant at the .05 level for a two-tailed test. For the current example, $4.06 > 1.96$ and, therefore, rho is significant.

We should very carefully note in our personnel example that just because our rho coefficient is significant, this does not *mean* that our personnel test is a good one. It simply means that the correlation value of .79 is significant. We might, for example, find a test that yielded an even higher correlation with production performance of operators. The results do indicate, however, that our test provides a fairly accurate prediction of the ensuing production-performance rank order.

Exercises

11.6 U.S. Census data indicate that the profile of age groups for Scottsville are as follows:

Age Groups	Percentage
18–24	21
25–34	33
35–49	24
50–64	10
65 and over	12

In a recent sample that you took in Scottsville, you found the following age breakdowns for a sample of 500:

Age Groups	Percentage
18–24	18
25–34	35
35–49	27
50–64	10
65 and over	10

If the Census data are an accurate representation of the age breakdown in Scottsville, is there reason to suspect that our sample differs significantly from the Census breakdown?

11.7 Recently 6 brands of beer were tested in a blind taste test for preference. Sound procedures were followed with random assignment of tasters, and rotation. The results for a sample of 240 beer tasters were as follows:

Brand	Number Preferring
A	46
B	38
C	41
D	47
E	39
F	29

(a) What is the correct hypothesis to be tested?
(b) What are your conclusions regarding test of this hypothesis?

11.8 Recently a reading test was used to assign grade school students to classes. At the end of the school year, all students were again tested for reading ability. The question of interest is whether there were any significant shifts in the rank of students. There were 35 students tested with pre- and post-school-year ranks as follows:

Student	Pre-School-Year Rank	Post-School-Year Rank
1	1	02
2	2	17
3	3	24
4	4	14
5	5	32
6	6	27
7	7	19
8	8	29
9	9	26
10	10	30
11	11	31
12	12	34
13	13	08
14	14	15
15	15	35
16	16	11
17	17	10
18	18	04
19	19	05
20	20	20
21	21	23
22	22	25
23	23	09
24	24	18
25	25	07
26	26	06
27	27	03
28	28	13
29	29	01
30	30	12
31	31	16
32	32	21
33	33	22
34	34	28
35	35	33

(a) What is the appropriate hypothesis?
(b) At the .05 level, what are your conclusions regarding this hypothesis?

11.9 Some members of the swim team have been doing isometric exercises to improve coordination and strength. Prior to doing the exercises, they were ranked on how fast they swam the 200-meter freestyle and at the end of 6 months they were re-timed and ranked. The results were as follows:

Swimmer	Preexercise Rank	Postexercise Rank
1	1	3
2	2	4
3	3	1
4	4	2
5	5	6
6	6	7
7	7	5
8	8	9
9	9	8
10	10	10

(a) What is the appropriate hypothesis?
(b) At the .05 level, what do you conclude?
(c) Does your answer say anything conclusive about the value of the exercises? Why or Why not?

11.10 A group of preschool children were involved in an experiment on the effectiveness of the new "Silent Teacher" electronic calculator. This machine would ask a series of numerical questions and the student would answer. Following each response, the student was told by the calculator whether the answer was correct or not. A group of researchers were interested in testing a "frustration" hypothesis that a mistake or series of mistakes would cause a student to become frustrated and continue making mistakes. One student had the following pattern of responses for 30 responses where a 1 was a correct answer and a 0 was a mistake.

111001111000101011100001100110

(a) What hypothesis is to be tested?
(b) At the .05 level, is there reason to reject the hypothesis?
(c) How do you interpret your results?

11.11 A group of college students were tested by both the TAS and the TCA tests which were stated to test the same abilities. The results were:

Student	TAS	TCA
1	90	96
2	85	79
3	78	85
4	63	78
5	83	79
6	80	92
7	50	36
8	94	71
9	76	86
10	67	79
11	85	89
12	84	90
13	89	93
14	67	76
15	84	75

(a) Using the rank order correlation method, test the appropriate hypothesis at the .01 level.

(b) What can you say regarding the effectiveness of either of the tests?

11.12 A group of volunteers took an electronic driving test prior to consuming 4 ounces of whiskey. After the consumption of the alcohol and a 1-hour wait, the group was retested. The results of the two tests are as follows:

Volunteer	Test 1	Test 2
1	94	82
2	88	88
3	86	78
4	77	84
5	90	81
6	94	86
7	70	78
8	80	78
9	85	67
10	70	60
11	91	86
12	99	89
13	79	82
14	93	90
15	78	79

(a) Perform a sign test to test the hypothesis that alcohol has no impact on the driving test scores.

(b) Perform a rank order correlation to determine if alcohol had a differing impact on the volunteers' test scores. (Use .05 level.)

11.13 A group of 10 college professors have been ranked independently on teaching ability by the college dean and a panel of students. The ranks are as follows:

Professor	College Dean	Student Panel
1	6	5
2	3	3
3	4	2
4	1	1
5	9	8
6	10	7
7	5	9
8	7	10
9	2	4
10	8	6

(a) Calculate the coefficient of rank order correlation to measure the consistency of the two evaluations.
(b) What statements can you make about the consistency?

11.14 The Department of Health has been investigating the side effects of a recent inoculation program. One hypothesis is that the methods of administration, air gun or hypodermic needle, may have an effect on later reactions to the shots. Two samples were taken of 30 patients each with the following outcome, where 1 indicates that the patient had side effects and 0 indicates that he or she did not.

Group 1: Air Gun

100001100001101001000100000110

Group 2: Hypodermic Needle

001110010000110011000100111010

(a) Perform a length-of-runs test to see if patients tended to cluster in terms of having or not having side effects for both groups.
(b) Test to see if the proportion of patients having side effects in group 1 differs from the proportion of group 2 patients having side effects.
(c) What conclusions can you draw from the results in (a) and (b)?

12

Learning Objectives

Structure of Decision Making

Decisions and Probability
  Weighted Averages

Subjective Probability and Decisions under Uncertainty

Payoff Tables for Decision Making

Expected Opportunity Loss

Expected Value of Perfect Information

Bayesian Analysis
  Types of Bayesian Analysis

Example Using Bayesian Analyses
  Prior Analysis
  Posterior Analysis
  Bayes' Theorem: A Review
  Preposterior Analysis

Benefits of Bayesian Analysis

Limitations of Bayesian Analysis

Decision Making Under Conditions of Uncertainty

Learning Objectives

The primary learning objectives for this chapter are:

1. To review earlier topics in the text as they relate to decision making.
2. To understand payoff tables as they are used to help arrive at decisions.
3. To provide a more in-depth understanding of Bayesian analysis to supplement the introduction to the topic in Chapter 4.
4. To be able to use decision trees as logic diagrams to compile information and arrive at quantitative results.

Structure of Decision Making

As enterprises become more complex, the situations requiring decisions become more involved, and a greater number of variables and uncertainties bear upon each decision. The presence of more variables and a greater number of uncertainties brings in the need for more precise and systematic approaches to analysis of the data prior to making a decision. Some of these statistical approaches have become grouped into a category often called *decision theory*. Many of these techniques make use of *Bayesian statistics*. Although we introduced Bayesian statistics earlier in the text, in this chapter we extend the concepts and explore some further practical applications. For additional topics and approaches to decision making under uncertainty beyond the scope of this text, see the authors' *Quantitative Methods for Management*.*

If a decision-making situation were presented systematically, the framework of the decision would contain certain elements, which can be summarized as follows:

1. *Environmental Influences.* Environmental, or outside, influences affect the decision-making process, but they are not under the control of the decision maker. For example, in the marketing of a new product, the market demand would not be under the control of the decision maker. This is true, even though he may influence this

---

* Ross H. Johnson and Paul R. Winn, *Quantitative Methods for Management* (Boston: Houghton Mifflin Company, 1976).

demand partially through advertising and other promotional tools. A clear case where an influence is not under the control of the decision maker is when an outcome is dependent upon good weather. Sometimes such influences are referred to as *events,* or *states of nature,* and often they are defined in terms of probabilities.

2. *Alternative Courses of Action.* The decision maker usually has alternative strategies that he can employ in making a decision. Sometimes there may be only two choices of a strategy, such as whether to produce or not produce a new product. In another situation we may need to select which product or products to manufacture out of a number of available alternative products. The decision of how many of each item to produce would be a selection of a strategy from a number of alternative possible production quantities.

3. *Measure of Payoff and Decision Rule.* The use of statistical methods usually requires the quantification of values so that the relative worth of alternative strategies can be compared. One of the most common quantitative measures is dollars. In order to solve a problem, a rule must be formulated to express the decision selected in terms of the payoff values, such as "maximize dollar income" or "minimize dollar costs."

Some of the problems presented in this chapter utilize these three elements in the form of a table or series of tables. Tables often help us to arrive at a final decision in a systematic manner and reduce the likelihood of error. In some cases, the form in which the model is to be set up will allow for the use of a computer in the data analysis.

In setting up a decision model, we must recognize that there are many qualitative as well as quantitative factors that have an impact on the final decision. Under some conditions, the manager may have to estimate the probability of the occurrence of an influencing factor by assigning subjective probabilities. Care must be taken to differentiate between objective and subjective probabilities. *Objective probabilities* involve a known probability and distribution, such as the binomial, Poisson, or others discussed previously. *Subjective probabilities* are used when a probability or distribution is not known or not available. This does not imply that subjective probabilities are hit-or-miss estimates; they should be carefully thought out and be based upon the best information that can be found.

In decision-making situations, there is a varying degree to which the facts are known. When all facts regarding the influencing factors are known, the process is called *decision making under certainty.* Even though the system may be complex, the logic model can be completely defined and the best action can be selected. Traditionally risk and uncertainty have referred to different degrees of lack of knowledge. When the states are unknown but objective evidence exists which allows the decision maker to assign probabilities to the various states, the process has been called *decision making under risk.** When there is no objective information regarding the probabilities, the process has been referred to as *decision making under uncertainty.* We must recognize, however, that risk can also refer to the possible amount of loss if an outcome is unfavorable. Even when we are working under uncertainty, however, the decision maker may assign subjective probabilities to the state of nature. Once probabilities have been assigned to possible outcomes, regardless of how these probabilities were derived, the solution procedure is the same. Therefore, for all practical purposes, calculations involving risk and uncertainty are handled the same and will be referred to in this chapter as conditions of uncertainty. In this chapter, decisions made under varying conditions of uncertainty will be explored, and the use of various decision rules will be investigated.

Decisions and Probability

Typically, the decision maker is dealing with a decision framework that can be functionally stated as

$$R = f(A_j, S_i),$$

where

$R =$ the return to the decision maker, usually measured as a payoff such as dollars of sales or profits

$A_j =$ a set of alternative actions that are under the control of the decision maker

$S_i =$ a set of events or influences that can occur independently of the selection of a particular course of action; these events (or states of nature) are not under the control of the decision maker

The process for making a decision consists of specifying which alternative courses of action are feasible and what the reaction might be to each

---

* Jerome D. Braverman, *Probability, Logic, and Management Decisions* (New York: McGraw-Hill Book Company, 1972).

course of action. Along with this is some statement of potential return if combinations of A_j and S_i are selected or occur.

Weighted Averages

The decision maker frequently may know the true probability of occurrence of each state of nature. For example, in the situation illustrated in Table 12-1, assume that the probabilities of states of nature S_1 and S_2 are, respectively, .4 and .6. The payoff for each combination of alternatives is given in the body of the table. For example, if action A_1 were taken and event S_1 occurred, the payoff would be 12 (which may represent $12,000). We assume that money has a constant or unchanging value; in other words, the ten-thousandth dollar of a return is worth just as much as the first dollar. We are not dealing with the situation of diminishing returns, which is so common in economic theory. Our decision maker does realize that there is no guarantee of the occurrence of a state of nature. As a result, the selection of an appropriate strategy will be based upon the combination of the chances of each state of nature occurring and the payoffs (or losses) associated with each occurrences. Calculating the chance and payoff combinations, we find that if the decision maker selected A_1, the expected return R would be

$$R_1 = .4(12) + .6(-6) = 4.8 - 3.6 = 1.2$$

If alternative A_2 were selected, the expected return would be

$$R_2 = .4(-5) + .6(9) = -2.0 + 5.4 = 3.4$$

Since R_2 is larger, the decision maker would choose alternative course of action A_2. Under this decision rule the manager has used the chances for the different states of nature as *weighting* factors in the calculation. The resulting figures, 1.20 and 3.40, are *weighted averages* of the pertinent payoffs. The payoffs shown in Table 12-1 are called *conditional payoffs* because each is conditioned upon the occurrence of a particular state of nature.

TABLE 12-1. Payoff Matrix

Potential Outcomes	Alternative Strategies	
	A_1	A_2
S_1 (.4)	12	−5
S_2 (.6)	−6	9

Subjective Probabilities and Decisions Under Uncertainty

Executives who manage a business or other enterprise are becoming increasingly interested in decision making under conditions of uncertainty. They recognize the value of systematic procedures that tend to remove much of the guesswork from decisions. This field of technology, sometimes called *decision theory*, provides models to assist these managers in selecting the most appropriate course of action when the enterprise requires a decision and the information available is not complete.

Consider a situation in which an entrepreneur has developed a new type of machine to make soft ice cream, and he wants to make an evaluation of the market for the new product. The texture of the ice cream produced is different from that of other brands and the inventor does not have much to go on in terms of previous acceptance. A decision must now be made as to how many machines to manufacture and have ready for sale to potential customers.

The inventor assembles the machine himself and must therefore make the decision on how many to make prior to the start of the oncoming heavy sales period, since during this period his time will be largely taken up by delivery and installation of ordered items.

The market reaction to the new product is uncertain. It is projected through market analysis that there will either be a high demand (60 units) or low demand (20 units). Initial reactions from friends and business associates indicate that the probability of a high demand is .7 and the probability of a low demand is .3. The entrepreneur has made some rough estimates of the profits under each combination of decision and demand, which are shown in Table 12-2. The same information is portrayed as a decision tree in Figures 12-1 and 12-2. Each node on the

TABLE 12-2. Payoff Table for Ice Cream Machines

Market Demand	Possible Action (number to produce)	
	A_1: Produce 20	A_2: Produce 60
D_1 (High Demand: .7)	$11,000	$28,000
D_2 (Low Demand: .3)	$7,000	−$19,000

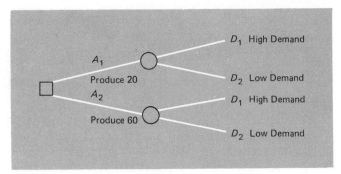

FIGURE 12-1. Decision Tree for Ice Cream Machine Problem

decision tree is represented by either a square (decision node) or a circle (probabilistic node). A square represents a point where the decision is under control of the decision maker. He has no control at the circle nodes and the branch taken is governed by chance according to the assigned probabilities.

We can now evaluate the weighted expected profits for each of the possible decisions A_1 and A_2. These are as follows:

$E(A_1) = .7(11,000) + .3(7,000) = 7,700 + 2,100 = \$9,800$
$E(A_2) = .7(28,000) + .3(-19,000) = 19,600 - 5,700 = \$13,900$

These calculations show that the greatest expected payoff is evident (\$13,900) if the decision is made to make 60 (A_2).

In this instance, we were required to use subjective probability estimates because no prior data existed. In another situation we could have

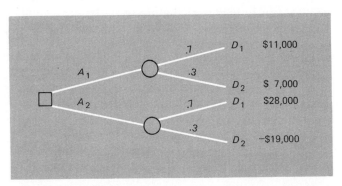

FIGURE 12-2. Ice Cream Machine Data Tabulated into Decision Tree

objective probabilities if the demand probabilities were based upon prior sales data.

In thinking about the ice cream problem, we may try to establish other alternatives. One might be to train an assistant to do some of the installations. This might enable the inventor to make a certain number of machines later (but probably at a higher cost). Alternatives such as this can also be evaluated by the approaches discussed here.

Payoff Tables for Decision Making

Payoff tables provide a systematic means of presenting data for evaluation. The results of the tabulations are in a form that enables a decision to be readily made. Suppose, for example, that the manager of a landscape firm wants to determine how many trees to have delivered each month to be planted in developments under his responsibility. The trees are fragile and the majority of those that are not planted within a few days after receipt do not survive. The developer, on the other hand, wants the trees installed when the grading of the lot is completed so that the homes can be shown and sold.

The objective is to maximize the profits of the landscape company on the trees. The manager has collected information on past sales for a 20-month period as shown in Table 12-3.

The cost per tree is $25.00 and the selling price is $48.00. Table 12-4 shows the profits for each combination of demand (sales) and amount ordered. The numbers ordered are given across the top of the table and are under the control of the decision maker. The demands listed down

TABLE 12-3. Tree Sales per Month for a 20-Month Period

Trees per Month	Frequency	Probability
43	2	.10
44	6	.30
45	10	.50
46	2	.10
	20	

TABLE 12.4. Payoffs (dollars) for Decisions on Number of Trees Ordered

Possible Demands	Alternative Actions (number ordered)			
	43	44	45	46
43	(989)	964	939	914
44	989	(1,012)	987	962
45	989	1,012	(1,035)	1,010
46	989	1,012	1,035	(1,058)

the left side of the table depend on chance or the number of sites ready, and this factor is not under the control of the landscape firm. The values in the body of the table represent the payoff (profit) for each decision-demand combination. The circled items across the diagonal represent optimal combinations, of course, since the best result is to sell the exact number of trees ordered.

How are these payoffs (or profits) determined? Take first the decision to order 44 trees. If all 44 are sold, the profit is $44 \times (\$48.00 - \$25.00)$, or $1,012.00. In this case the firm makes $23.00 on each tree. If 44 were ordered and only 43 used, it would lose the cost of one tree and its profit would be

$$43(23.00) - 25.00 = \$964.00$$

If the manager ordered 44 and the demand were 45 or 46, he would still make the $1,012.00 if no value is assigned to the loss of good will in the eyes of the developer by not having enough available. Table 12-4 is completed following this procedure.

The information presented so far still does not tell us how many trees to order each time. In order to arrive at this decision we shall prepare an expected profit table for each of the possible alternative actions (number ordered). The table for ordering 43 trees is given in Table 12-5. For each of the possible actual demands (from 43 to 46), the payoff is multiplied by its respective probability. These expected profits are then summed to obtain the total expected profit if that number of trees were sold. In the case of ordering 43 trees, the expected payoff is $989.00

TABLE 12-5. Expected Profits When 43 Trees Are Ordered

Actual Demand	Probability (from Table 12-3)	Payoff (from Table 12-4)	Expected Profit (probability × payoff)
43	.10	$989.00	$ 98.90
44	.30	989.00	296.70
45	.50	989.00	494.50
46	.10	989.00	98.90
	1.00		$989.00

Table 12-6 shows similar calculations if 44 trees are ordered. In this case, the total expected profit is $1,007.20, which is better than the alternative where 43 are ordered.

The alternative for ordering 45 trees is shown in Table 12-7. If 45 were ordered and the actual demand was 43, the payoff from Table 12-4 is $939.00. The probability of a demand of 43 is .10 as taken from Table 12-3. These two quantities are entered in Table 12-7 and multiplied to obtain the expected profit of $939.00. This procedure is continued to

TABLE 12-6. Expected Profits When 44 Trees Are Ordered

Actual Demand	Probability (from Table 12-3)	Payoff (from Table 12-4)	Expected Profit (probability × payoff)
43	.10	$ 964.00	$ 96.40
44	.30	1,012.00	303.60
45	.50	1,012.00	506.00
46	.10	1,012.00	101.20
	1.00		$1,007.20

TABLE 12-7. Expected Profits When 45 Trees Are Ordered

Actual Demand	Probability (from Table 12-3)	Payoff (from Table 12-4)	Expected Profit (probability × payoff)
43	.10	$ 939.00	$ 93.90
44	.30	987.00	296.10
45	.50	1,035.00	517.50
46	.10	1,035.00	103.50
	1.00		$1,011.00

TABLE 12-8. Expected Profits When 46 Trees Are Ordered

Actual Demand	Probability (from Table 12-3)	Payoff (from Table 12-4)	Expected Profit (probability × payoff)
43	.10	$ 914.00	$ 91.40
44	.30	962.00	288.60
45	.50	1,010.00	505.00
46	.10	1,058.00	105.80
	1.00		$990.80

complete Table 12-7. The total expected profit of $1,011.00 for ordering 45 trees is greater than either 43 or 44. The calculations for ordering 46 are given in Table 12-8. The expected profit of $990.80 is less than the two previous cases. Therefore, we can conclude that the optimum alternative is to order 45.

In our example, we assumed that leftover trees were a complete loss. What if we sold them for $20.00? This could be easily brought into our decision process by considering a leftover tree as a loss of $25.00 − $20.00, or $5.00. The same evaluation procedure could be followed using this figure.

Problems of this nature occur in a variety of business situations. An airline must order a certain number of meals to go on its planes. A hospital may purchase certain drugs that deteriorate if not used within a specified time. Baked goods only retain full value if sold within a limited time.

Expected Opportunity Loss

The example of the landscaper and the trees demonstrated how to select the best alternative where the objective was to maximize profit or payoff. Sometimes we may choose a different criterion for our decision. Rather than set our objective of maximizing expected payoff, sometimes referred to as largest expected monetary value (EMV), the decision maker could select a different objective of minimizing expected opportunity loss. The expected opportunity loss (EOL) is defined as the difference between the expected monetary value of a particular decision and the expected monetary value of the best decision that could have been made under the circumstances. In commonsense language, this means that a manager should try to make the best of a situation by keeping losses as small as possible. If a newsstand owner orders 100 Sunday papers and sells only 90 the opportunity loss is the difference between the profit if 90 were stocked and sold and the actual profit when 100 were stocked and 90 sold. The cost of the extra newspapers that had to be thrown away is equal to the opportunity loss. The owner could have avoided the loss by having made a different decision—that is, to stock only 90. As another example of opportunity loss, assume that the owner stocked 80 but could have sold 90. In this instance, the opportunity loss is the profit the owner did not make on the 10 extra papers he could have sold.

Take an example where a distributor of private-branded batteries in a large metropolitan area is attempting to estimate the probable lost profits for the coming month if a popular automobile battery style is out of stock. Owing to the large number of sizes and styles of batteries, the cost of carrying inventory has become sizable. Heavy demands have made it difficult for the factory to meet demand and special orders can no longer be filled expeditiously from factory inventory. An order is placed with the distributor by a service station, large retail chain store, or garage whenever a car owner has a battery failure. Under these circumstances a battery is needed immediately, and if the distributor is unable to fill an order for a particular battery, the customer will go elsewhere.

The distributor has kept records over a 100-day period of all cases where orders could not be filled and were therefore lost. Table 12-9 shows a tabulation of the orders lost on a day-by-day basis as a result of being out of stock. These values, together with the proportions of days (converted to probabilities), can be used to calculate the average number of lost orders per day (4.83) in Table 12-10. We are using these figures to predict the losses under similar conditions for the coming month.

TABLE 12-9. Orders Lost by Distributor over a 100-Day Period As a Result of Being Out-of-Stock

Number of Orders Lost	Frequency (number of days)	Proportion of Days, $P(S)$
0	2	.02
1	3	.03
2	5	.05
3	10	.10
4	16	.16
5	28	.28
6	20	.20
7	12	.12
8	4	.04

TABLE 12-10. Calculation of Average Number of Lost Orders per Day

Lost Orders, S	P(S)	S × P(S)
0	.02	0
1	.03	.03
2	.05	.10
3	.10	.30
4	.16	.64
5	.28	1.40
6	.20	1.20
7	.12	.84
8	.04	.32
		4.83

We then go to Table 12-11, which tabulates the expected opportunity loss for a day. The loss (L) is based on an average profit of $5.00 per battery.

As an example of the calculations consider an S (number of lost orders) of 4. From the table there is a probability of .16 that there will be 4 lost orders. If 4 orders are lost, the profit lost is $4 \times \$5.00$, or $20.00. The product of $P(S) \times L$ is then $.16 \times \$20.00$, or $3.20. These values are summed in the last column to obtain a total expected opportunity loss of $24.15 per day. Using a figure of 300 business days per year, this would represent a yearly value of $300 \times \$24.15$, or $7,245.00, for this one item alone. We can see that when all the hundreds of inventory items are considered, the values we are dealing with become substantial.

TABLE 12-11. Calculation of Expected Opportunity Loss

S	P(S)	Loss, L	Expected Opportunity Loss, P(S) × L
0	.02	$ 0	$0
1	.03	5.00	.15
2	.05	10.00	.50
3	.10	15.00	1.50
4	.16	20.00	3.20
5	.28	25.00	7.00
6	.20	30.00	6.00
7	.12	35.00	4.20
8	.04	40.00	1.60
			$24.15

Having illustrated the concept of opportunity loss, let us now return to the example of the landscaper and the trees. In that example we saw how to select the best alternative to obtain the maximum expected profit. Let us now consider this example using opportunity loss, to see if we arrive at a different decision with this changed criterion. In some situations we may desire to perform calculations by each method. If both indicate that the same alternative should be selected, we can be more sure of our decision. In other cases, we may select either one or the other of these two criteria.

Let us now see how to evaluate the opportunity loss table shown in Table 12-12. If we order 44 trees and then sell 44, we have done the best we can. An opportunity loss occurs if we select a certain amount to order, have a certain demand, but could have done better (given this demand) by ordering a different quantity. This quantity could have been either more or less than the quantity actually ordered. Suppose that we order 44 and the demand is actually 43. Our profit is $964.00 in Table 12-4. With an actual demand of 43 we could have done better if we had ordered only 43 instead of 44. The improvement would have been $989.00 − $964.00, or $25.00 more. Thus we can say that we lost the opportunity to make an additional $25.00 by ordering 44 instead of 43. Again, this covers only one type of tree for one period. If the landscape company were involved with several developments, the amount of money at stake would become sizable.

Now consider the case where 44 were ordered and the demand was 45. The actual profit from Table 12-4 was $1,012.00. Since the demand was 45, a better decision would have been to order 45. This would give a profit of $1,035.00, which is better by $23.00. This $23.00 is then the opportunity loss figure for the order-44–demand-45 position in Table 12-12. The rest of Table 12-12 is completed in this manner.

TABLE 12-12. Opportunity Loss Table for Tree-Ordering Decision

Possible Demands	Alternative Actions (number to order)			
	43	44	45	46
43	$ 0	$25.00	$50.00	$75.00
44	23.00	0	25.00	50.00
45	46.00	23.00	0	25.00
46	69.00	46.00	23.00	0

TABLE 12-13. Expected Opportunity Loss When 43 Trees Are Ordered Each Day

Actual Demand	Probability	Opportunity Loss (from Table 12-12)	Weighted Expected Opportunity Loss
43	.10	$ 0	$ 0
44	.30	23.00	6.90
45	.50	46.00	23.00
46	.10	69.00	6.90
			$36.80

We now must prepare opportunity loss tables for each order quantity (43, 44, 45, and 46) in order to calculate the expected opportunity loss for each potential decision. Table 12-13 shows the figures for the order quantity of 43. The calculations are similar to those for the payoff tables done previously.

With all four tables prepared (Tables 12-13 through 12-16) we are in a position to make a decision. The $14.80 for 45 trees (in Table 12-15) is the smallest weighted expected opportunity loss. Ordering 45 trees also gave us the maximum payoff in our previous calculations. Therefore, both these two methods have pointed to the same decision—order 45 trees. This is not a mere coincidence since the action with the highest expected monetary value also has the lowest expected opportunity loss.

TABLE 12-14. Expected Opportunity Loss When 44 Trees Are Ordered

Actual Demand	Probability	Opportunity Loss (from Table 12-12)	Weighted Expected Opportunity Loss
43	.10	$25.00	$ 2.50
44	.30	0	0
45	.50	23.00	11.50
46	.10	46.00	4.60
			$18.60

TABLE 12-15. Expected Opportunity Loss When 45 Trees Are Ordered

Actual Demand	Probability	Opportunity Loss (from Table 12-12)	Weighted Expected Opportunity Loss
43	.10	$50.00	$ 5.00
44	.30	25.00	7.50
45	.50	0	0
46	.10	23.00	2.30
			$14.80

TABLE 12-16. Expected Opportunity Loss When 46 Trees Are Ordered

Actual Demand	Probability	Opportunity Loss (from Table 12-12)	Weighted Expected Opportunity Loss
43	.10	$75.00	$ 7.50
44	.30	50.00	15.00
45	.50	25.00	12.50
46	.10	0	0
			$35.00

Expected Value of Perfect Information

Making decisions based on perfect information is the other extreme from being in a position to make decisions without any information. In the landscape example, perfect information would permit us to know ahead of time the exact number of trees to be used each month. If we had perfect information, we would then be able to order the exact quantity needed and there would never be any shortage or excess. We would always be on the diagonal of the payoff table (Table 12-4), where the optimum payoffs occur.

In a study of expected value of perfect information (EVPI), we recognize that in practical situations we will never achieve perfect information. Then why do we calculate EVPI? Assume that we have determined the best strategy in a decision-making situation based on available data. A natural question that might arise is: Will it pay us to collect more data? Better data would permit us to make closer predictions. In the case of the landscaping problem, a more sophisticated formula may yield different demands in different seasons of the year or as a function of other economic indexes. In this situation where we are considering paying for better data, the next question that naturally arises is: How much is this added data worth? We recognize that perfect information will not be achieved; however, the EVPI will give us an upper limit on what we should pay for information beyond the present information.

Let us proceed next to calculate the EVPI for the landscaping problem. If perfect information were available, the operator would always order a quantity of trees that would match the future demand, and the profits would be those on the diagonal of Table 12-4. In that case, the average expected profit per month over a period of months would be as shown in Table 12-17 as $1,025.80. As expected, this is higher than the expected profit when 45 trees were ordered; however, the difference is only $14.80 per month (or $177.60 for a 12-month year). Although we would not expect any type of estimating procedure to be perfect, this does give us an insight to how much can be spent logically to improve the data.

We can now apply this reasoning to the soft ice cream entrepreneur, where the data are shown in Table 12-18. Under conditions of perfect information we would always make the right decision. If the predicted and actual demand were high (60), the profit would be $28,000; whereas

TABLE 12-17. Expected Payoffs from Tree Sales When Perfect Information Is Available

Actual Demand	Probability	Payoff	Expected Profit
43	.10	$ 989.00	$ 98.90
44	.30	1,012.00	303.60
45	.50	1,035.00	517.50
46	.10	1,058.00	105.80
			$1,025.80

TABLE 12-18. Payoff Table for Ice Cream Machines

Market Demand	Possible Action (number to produce)	
	A₁: Produce 20	A₂: Produce 60
D_1 (High Demand: .7)	$11,000	$28,000
D_2 (Low Demand: .3)	$7,000	−$19,000

if predicted and actual demand were 20, the profit would be $7,000. Taking these probabilities into account, the expected profit is

$$.7(28,000) + .3(7,000) = 19,600 + 2,100 = \$21,700$$

This is the expected payoff if perfect information were available. If we did not have the perfect information available, our best strategy would be A_2 with an expected profit of $13,900. The difference here is

$$21,700 - 13,900 = \$7,800$$

This indicates that the entrepreneur could consider paying for better information, but certainly never more than the $7,800.

Exercises

12.1 A fruit stand operator receives deliveries of peaches twice a week. Peaches left at the end of 3 days must be discarded. The estimated sales for the 3-day period are as follows:

Estimated Sales (bushels)	Probability
4	.40
5	.50
6	.10

The cost per bushel is $4.00 and the selling price (sold in smaller quantities) is $9.00 per bushel. Construct a payoff table and determine the expected profits for each order quantity and select the best strategy.

12.2 In problem 12-1, what is the expected profit with perfect information? Determine the value of perfect information.

12.3 With reference to the example given in the text for the landscaper and the trees, assume that the selling price is $50.00 per tree and that leftover trees are sold for half price. Determine the optimum order quantity and the resulting expected profit.

12.4 Use an opportunity loss criterion and table to evaluate the best strategy in 12-3.

12.5 In problem 12-3, what is the value of perfect information?

12.6 A grain elevator operator receives grain from farmers, then either stores it or ships it in railroad carloads. His storage capacity is filled because of a bumper crop. Owing to a shortage of railroad cars, he must lease them by the week and one week in advance. It costs $200.00 per week to lease a car. If he can fill a carload and ship it, he makes $300.00 per carload beyond the lease fee. Current weekly demand for his services is running as follows:

Carloads	Probability
3	.2
4	.3
5	.4
6	.1

If a farmer brings in grain and the elevator operator cannot accept it, the farmer takes it elsewhere. What is the elevator operator's optimum strategy to maximize profits per week?

12.7 Evaluate problem 12-6 in terms of minimizing opportunity loss. What is the value of perfect information?

12.8 A magazine stand operator orders each magazine based on his history of sales. The following are his records for sale of *Newsweek*:

Demand (number of magazines)	Probability
40	.25
50	.40
60	.35

Construct a payoff table and determine how many magazines the owner should order. Unsold magazines at the end of the week have no value.

12.9 Evaluate problem 12-8 based on expected opportunity loss. What is the value of perfect information?

12.10 A department store has the chance to buy a lot of 5,000 toys for $15,000. The regular price is $4.00 each. The toys are sold before Christmas at $6.00 each. The following represents the expected demand before Christmas:

Demand	Probability
3,000	.10
4,000	.65
5,000	.25

Toys left over can be disposed of at $2.00 each. Construct a payoff table and determine the best action.

12.11 Evaluate problem 12-10 based on opportunity loss and determine the value of perfect information.

12.12 A magazine publisher is considering the publication of a special issue for the state fair. The fixed cost of preparing the special issue is $60,000. The cost of printing and handling is $.50 per copy. The issue is to be sold at $1.50 per copy. The attendance of the fair averages 400,000 and demand is estimated as follows:

Proportion Buying	Probability
.10	.30
.15	.45
.20	.25

Should the special issue be published?

12.13 Depending on conditions at the supplier, lots of material have been coming through at 2, 10, or 25 per cent defective. It costs $800.00 to inspect a lot to determine which of the 3 qualities are present, and to sort it out so that it averages the acceptable level of 2 per cent. If the lot is not checked, it later costs $400.00 for a 10 per cent defective lot and $2,800.00 for a 25 per cent defective lot. Prepare a payoff table and suggest a strategy.

12.14 In problem 12-13, the following probabilities are associated with the various per cent defective:

2%	.8
10%	.1
25%	.1

Determine the expected costs for various actions and decide on a strategy.

12.15 In problem 12-14, assume the following probabilities.

2%	.8
25%	.2

Solve for the best strategy. What would be the expected value of perfect information?

Bayesian Analysis

So far in this chapter we have considered decision-making situations in which the outcome was related to an influence or state of nature. We had information on the likelihood or probabilities that the influence would take different values. In these cases, we were placed in the position of trying to make a decision based on available information. We did not, however, consider the possibility that the person responsible for making the decision might seek additional information about the influencing factor. The value of perfect information was considered if it were possible to obtain it. We shall now examine more thoroughly the implications of Bayesian analysis, some aspects of which allow us to examine the merits of seeking further, or better, information. This information would be obtained by sampling to obtain additional information about the influencing factor, or state of nature. Once the decision maker has evaluated the presently available data, questions which should be raised are

1. Should we seek more or better information?
2. Is it possible to obtain better information than we currently have, considering cost of this information?

3. How much information should be obtained?
4. What decision rule should we employ and what are the associated risks?

Bayesian decision theory provides a scientific approach to selecting the best of these courses of action. Again, outside influences (or states of nature) are not under the control of the decision maker; however, it is possible to derive information about them from judgment, collection of more data, or experimentation.

If the chosen alternative is to defer the decision until additional information is obtained, it then becomes necessary to have a systematic method for utilization of the added information. We still have the original data which we shall call the *prior information*. The new information to be gained is called the *sampling information*. The combination of the prior and sampling information is called the *posterior information*. Before the sample is taken, the decision maker may wish to perform a *preposterior analysis* to specify a decision rule regarding whether the sample should be taken in the first place.

Types of Bayesian Analyses

Before starting to develop the various types of Bayesian Analyses, each will be defined further as it relates to the others.

Prior Analysis. In prior analysis, the decision maker specifies each state of nature and the associated probability and payoff for each alternative action–state of nature combination. The selection of the best alternative course of action is made on the basis of this information. Prior analysis can be considered as the "benchmark" from which we decide whether or not to collect further information. Often a decision will have to be made on this basis when time or resources do not permit the collection of additional information.

Preposterior Analysis. In some cases, the availability of time and/or the importance of the decision calls for the decision maker to delay his decision, pending the collection and interpretation of additional data. The firm must pay for the collection and analysis of this information, and it will seldom be entirely accurate. There will still be some risk in the results. The decision maker will have to trade off the cost of the information against its value to the firm, allowing for the risks or potential results of a decision based only on prior analysis. This type of analysis, known as preposterior analysis, involves determination of whether or not to collect the information needed for the next step, the posterior analysis.

Posterior Analysis. When the new information is available, the decision maker incorporates it into the analysis, calculates the expected payoffs of alternative actions, and selects that action leading to the highest expected payoff. This step, involving incorporation of new information and revision of prior probabilities, is called posterior analysis. The process of revision of prior conclusions based on inclusion of new information is the primary strength of Bayesian analysis.

Sequential Analysis. Sometimes the decision maker has the time available to weigh the value of additional information on a step-by-step basis. He may do this if it is expensive to collect more data and he wishes to expend funds only as necessary. He can collect some additional information, evaluate it, and then if the result is not significantly clear, it may be necessary to go on to the next step. He can stop at any step in the sequence where the evidence is sufficient to make one conclusion or the other.

Example Using Bayesian Analysis

Our firm is considering the production of a new type of pest control, Zapest, for insect control on farm crops. We have carried out a considerable amount of research and development on the product; however, there are still uncertainties as to Zapest's continued effectiveness as pests become acclimated to it, as well as possible carryover effects on crop growth. The firm is in the position of trying to decide whether to expend funds on this new product or to increase promotion and development of some of its older products. In this case, there are judged to be two states of nature

1. Product is effective.
2. Product is inadequate.

The company has determined that if Zapest is produced and marketed, and is effective, the profit will be $600,000 over a 3-year period. If it turns out to give inadequate performance, the net loss will be $900,000. On the other hand, if Zapest is not produced and competitors introduce something similar which is effective, the loss to us will be $200,000, stemming from loss of market share and cost already sunk into the development of Zapest. If the product is not effective, and we do not

TABLE 12-19. Conditional Payoff Matrix for Production of Zapest Pesticide

Product Effectiveness	Alternative Actions	
	A_1: Produce Zapest	A_2: Do Not Produce Zapest
Effective, S_1 (.7)	$600,000	−$200,000
Inadequate, S_2 (.3)	−$900,000	$800,000

produce it, our improved profits are estimated as $800,000, based on added promotion and improvement of present products. At this point, our evaluation shows a probability of S_1 (effective) to be .70, and the probability of S_2 (inadequate) to be .30. The conditional payoff matrix is shown in Table 12-19.

Prior Analysis

At this point, we can conduct a prior analysis to determine which of the two alternatives has the highest expected monetary value (EMV), where A_1 is to produce Zapest and A_2 is not to produce.

$$\text{EMV}(A_1) = .70(600,000) + .30(-900,000) = \$150,000$$
$$\text{EMV}(A_2) = .70(-200,000) + .30(800,000) = \$100,000$$

On the basis of this prior analysis, A_1 has the greater EMV of $150,000, and we would commence production of Zapest.

We can at this point, also evaluate the situation using the criteria of expected opportunity loss (EOL). The matrix in Table 12-20 is obtained by comparing each payoff with the payoff if the best alternative had been selected. For instance, if we had selected A_2, and S_1 turned out to be true, we could have done better by $600,000 − (−$200,000), or $800,000. The EOL of each action is calculated as follows:

$$\text{EOL}(A_1) = .70(0) + .30(1,700,000) = \$510,000$$
$$\text{EOL}(A_2) = .70(800,000) + .30(0) = \$560,000$$

Since A_2 has the highest EOL, our decision again would be to select A_1. Both methods indicate the decision is to go ahead and produce Zapest.

We have selected probabilities to represent S_1 and S_2 and they are values that could be in error. Since we could be wrong on these values, we might logically ask ourselves what would be the lowest probability value for S_1 which would still indicate A_1 to be the best decision? This would be the value that would result in equal payoffs for A_1 and A_2 and

TABLE 12-20. Opportunity Loss Matrix for Zapest

Product Effectiveness	Alternative Actions	
	A_1: Produce Zapest	A_2: Do Not Produce Zapest
Effective, S_1 (.7)	0	$800,000
Inadequate, S_2 (.3)	$1,700,000	0

is therefore referred to as the *indifference probability* because at that point we are indifferent as to which alternative to select. In that case, the prior analysis would show one alternative to be just as good as the other. The indifference probability (P_i) can be calculated by equating the expected losses of A_1 and A_2:

$$P_i(0) + (1 - P_i)1,700,000 = P_i(800,000) + (1 - P_i)0$$
$$1,700,000 - 1,700,000 P_i = 800,000 P_i$$
$$1,700,000 = 2,500,000 P_i$$
$$P_i = \frac{17}{25} = .68$$

We conclude that if $P(S_1)$ is greater than .68, the indication will be to select A_1, whereas if $P(S_1)$ is less than .68, we should select A_2. The same values could have been calculated from the payoff table, Table 12-19.

Posterior Analysis

We have made a prior analysis which indicated that the best payoff results if we go ahead and produce Zapest. Suppose instead that we decide to obtain additional information through the process of sampling farmers who had experimentally used our product. We note that our probability of .7 is not far from the indifference probability of .68. We shall now review the process developed in Chapter 4 to revise our probabilities based on the new information. To explain Bayes' theorem, we shall utilize another example and then return to further analysis of our Zapest problem.

Bayes' Theorem: A Review

An example dealing with transformer production will be used to illustrate Bayes' theorem.

Assume that we have been stocking transformers from two assembly lines, B_1 and B_2. In the past we assumed that items produced on either line were identical, so boxes of items from either line were marked the same and

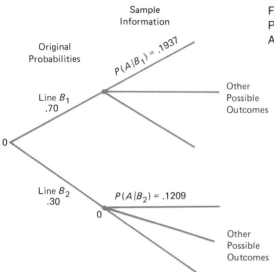

FIGURE 12.3. Tree Showing Probabilities of Events in the Assembly Line Example

stored together. This month, a new type of problem turned up in reports received from customers. In checking products from both assembly lines, it was determined that 10 per cent of the items from line B_1 and 40 per cent of the items from line B_2 had this defect. There were a large number of cartons of transformers in inventory, each containing 2,000 items per box. It is known from past production records that 70 per cent of the cartons came from line B_1 and 30 per cent from B_2, but the cartons cannot be identified individually as to the line on which they were produced.

A box is selected at random and a sample of 10 is taken, resulting in 8 good items and 2 bad items, or 20 per cent defective in the sample. Based on these sample data, what is the probability that the carton came from line B_1? From B_2? Figure 12-3 shows the data in the form of a tree.

We shall call A the event that the sample of 8 good and 2 bad was selected. This is represented by the branches shown. There would also be other branches for all other possible samples such as 7 good out of 10, 6 good out of 10, etc., if we desired to complete the tree. If we took a random sample from production line B_1, we can calculate the probability that A (8 good out of 10) would occur. From the binomial probability table (Appendix C),

$$P(A|B_1) = P(r = 8|n = 10, p = .90) = .1937$$
$$P(A|B_2) = P(r = 8|n = 10, p = .60) = .1209$$

From Bayes' rule developed in Chapter 4,

$$P(B_1|A) = \frac{P(B_1) \times P(A|B_1)}{P(B_1) \times P(A|B_1) + P(B_2) \times P(A|B_2)}$$
$$= \frac{.7 \times .1937}{.7 \times .1937 + .3 \times .1209} = \frac{.136}{.136 + .036}$$
$$= \frac{.136}{.172} = .79$$

We can state at this point that there is a probability of .79 that this particular carton came from assembly line B_1. We can also state that there is a probability of $1.0 - .79$, or .21, that the carton came from line 2, which is $P(B_2|A)$. If we had picked a carton at random, there would be a probability of .7 that it had come from line B_1. Based on the data from the sample (8 good out of 10), we have revised this probability so that there is now a probability of .79 that the carton came from line B_1. It becomes apparent that Bayes' theorem is simply a conditional probability, but to find $P(A|B_1)$ we utilize $P(B_1|A)$. We are calculating the probability that the sample came from B_1, with the condition that a sample of 8 good transformers out of 10 was obtained.

This type of analysis can now be related to the Zapest problem. We started with a probability of .7 that the product was effective, and .3 that it was inadequate. Suppose that an additional year of testing generated a sample from 10 tracts where 8 out of the 10 were effective. We can now ask: What is the probability that this sample came from an effective product? Figure 12-4 is a tree showing the original probabilities of .7 for B_1 (effective) and .3 for B_2 (inadequate). The event A is that we obtained a sample of 10 with 8 being effective. The numbers on the tree are the same as for Figure 12-3. The probability that this result came from an effective product is calculated exactly the same as in the production line problem and is .79. We have essentially revised our probability of effectiveness from .7 to .79, based on the sample.

FIGURE 12-4. Tree of Probabilities of Event in the Zapest Example

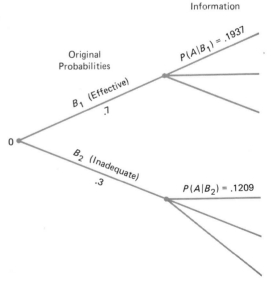

Preposterior Analysis

The greatest value of Bayesian decision theory comes where we want to examine the advisability of collecting additional information. We approach this by evaluating the potential payoffs that can result from the use of additional information. Preposterior analysis deals with the type of decision-making situation where a judgment is to be made as to whether or not it is economically desirable to collect additional information. In the Zapest pesticide example, if we were not satisfied with the adequacy of the information used to render the decision in the prior analysis, a further evaluation could be carried out to determine whether or not to collect additional information. In essence, we are asking ourselves how much it is potentially worth to our enterprise to secure the added data.

Let us consider specifically how this might apply to the Zapest example. As part of a preposterior analysis, it is necessary to estimate the accuracy of the contemplated experiment. On the basis of past experience with a product, a firm's research department can often estimate how accurate an additional series of experiments will be. We recognize, of course, that the experimental results would not provide perfect information. Table 12-21 gives some assumed conditional probabilities of obtaining experimental results that correspond to the states of nature S_1 (effective) and S_2 (inadequate). Remember that we are carrying out this analysis before any of the experiments are actually performed.

From Table 12-21, it can be observed that the research and development department feels that the results of the experiment will be slightly more accurate (.80 versus .70) if the product is actually effective than if the product is not effective. In addition, the possibility exists that the additional experimental sample will provide inconclusive results. For example, if the product is actually inadequate (S_2), there is a .15 probability that the experiment will give inconclusive results.

Our primary objective is to determine probabilities for the two states of nature (effective or inadequate), given that the experimental results are

TABLE 12-21. Conditional Probabilities of Experimental Results Given States of Nature: $P(Z_i|S_i)$

Product (States of Nature)	Experimental Results		
	Effective, Z_1	Inadequate, Z_2	Inconclusive Results, Z_3
Effective, S_1	.80	.15	.05
Inadequate, S_2	.15	.70	.15

one or the other. This brings in another conditional probability. We are attempting to determine $P(S_1|Z_i)$ and $P(S_2|Z_i)$, whereas all we have available is $P(Z_i|S_1)$ and $P(Z_i|S_2)$. Bayes' rule will help us in this evaluation. The rule states that

$$P(S_1|Z_i) = \frac{P(Z_i|S_1) \cdot P(S_1)}{P(Z_i|S_1) \cdot P(S_1) + P(Z_i|S_2) \cdot P(S_2)}$$

However, we must include each $P(Z_i|S_1)$ in the denominator. This can be related to the material covered in the last section of Chapter 4 and the example used for posterior analysis in the previous section. The probability $P(S_1|Z_i)$ is a conditional probability, and a conditional probability is obtained by dividing a joint probability by a marginal probability. The term in the denominator is the summation of all the joint probabilities that contain the particular Z that we are calculating. This summation of joint probabilities is the marginal probability of the Z we are working with. In our case, it could be the Z_1, Z_2 or Z_3. The numerator is the joint probability which consists of all terms that correspond to those for the conditional probability we are seeking. As an example, we can express $P(S_1|Z_1)$ as

$$P(S_1|Z_1) = \frac{P(S_1 \text{ and } Z_1)}{P(Z_1)}$$

We recall that our prior probabilities were

$$P(S_1) = .70$$
$$P(S_2) = .30$$

These can be used to establish a new table of revised probabilities as shown in Table 12-22. In this table, we have calculated the joint probabilities in the body of the table by multiplying the conditional probabilities by the corresponding prior probabilities. The marginal probability of each Z is derived by adding the joint probabilities. The marginal probabilities for S_1 and S_2 are summed across as a verification. Table 12-23 then shows the revised prior probabilities, which are found by dividing each joint probability by the corresponding marginal probability of Z in accordance with Bayes' rule. The following is a sample calculation:

$$P(S_1|Z_1) = \frac{\text{joint probability of } S_1 \text{ and } Z_1}{\text{marginal probability of } Z_1}$$
$$= \frac{P(S_1 \text{ and } Z_1)}{P(Z_1)} = \frac{.56}{.605} = .926$$

TABLE 12-22. Joint and Marginal Probabilities for Zapest Pesticide Example

| | Experimental Results | | | |
Probability	Z_1	Z_2	Z_3	$P(S_i)$
$P(Z_i\|S_1) \cdot P(S_1)$ or $P(S_1 \text{ and } Z_i)$	$.80 \times .70 = .560$	$.15 \times .70 = .105$	$.05 \times .70 = .035$	.70
$P(Z_i\|S_2) \cdot P(S_2)$ or $P(S_2 \text{ and } Z_i)$	$.15 \times .30 = .045$	$.70 \times .30 = .210$	$.15 \times .30 = .045$	.30
$P(Z_i)$	.605	.315	.080	1.00

The probabilities in the body of Table 12-23 are posterior probabilities. Each represents the probability of a particular state of nature S_i, conditional on the occurrence of a specific experimental result Z_i. To state it another way, it indicates the probability that S_i will occur if a particular Z occurs. From the above calculation, we can state, for example, that if the experimental results show Z_1 (effective), then there is a .926 probability that the product is actually effective (S_1). The figures tabulated in Table 12-23 will now permit us to evaluate the value of collecting more information from an experiment. Remember that all of the calculations we have been conducting are done before any sample data are collected. All of our tabulations can be placed in the form of the decision tree in Figure 12-5. Working with this tree will aid in understanding the use of decision trees.

TABLE 12-23. Revised Prior Probabilities for Zapest

| | Experimental Results | | |
Probability	Z_1	Z_2	Z_3
$P(S_1\|Z_i)$	$\dfrac{.56}{.605} = .926$	$\dfrac{.105}{.315} = .333$	$\dfrac{.035}{.08} = .438$
$P(S_2\|Z_i)$	$\dfrac{.045}{.605} = .074$	$\dfrac{.21}{.315} = .667$	$\dfrac{.045}{.08} = .563$

414 Business Statistics

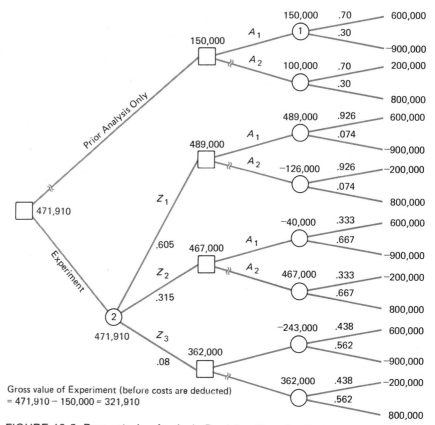

Gross value of Experiment (before costs are deducted)
= 471,910 − 150,000 = 321,910

FIGURE 12-5. Preposterior Analysis Decision Tree for Zapest

As we work with the decision tree, we will compute from right to left, a process called *backward induction*. Also note that there are two types of vertices (nodes) on the decision tree. A vertex or node is a branch point in the tree. A *probabilistic node* is a branch point that is *not* under the control of the decision maker. The branch taken depends upon uncontrollable factors, such as our states of nature or potential experiment outcomes. These probabilistic nodes are represented by circles on the decision tree. *Nodes of decision* are the second type of node. These are represented by squares and are points at which a manager is to make a choice from the available alternatives. The strategy of the manager then is to select the branch that yields the maximum gain or payoff.

The values of Z_1 are not controllable, since they are experimental results. The steps in completing the tree and making the evaluation are as follows:

1. Enter the conditional payoffs from Table 12-19.
2. Enter the probabilities in the last branches from the prior probabilities or from Table 12-23, as appropriate.
3. Calculate values of the circle nodes; for example, at node 1:

$$.7(600,000) + .3(-900,000) = \$150,000$$

4. At the square branching point (decision point), select the branch with the greatest value and enter at the square. Cross off the other branch with two marks.
5. Evaluate the next circle (probability node) as before; for example, at node 2:

$$.605(489,000) + .315(467,000) + .08(362,000)$$
$$= 295,845 + 147,105 + 28,960 = \$471,910$$

6. Select "collect more experimental data" as the choice since the value $471,910 is greater than the value $150,000.

The gross value of the experiment can be determined by taking the difference between the experiment payoff and the prior analysis payoff:

$$\text{gross value} = 471,910 - 150,000 = \$321,910$$

This is called a *gross result* since it does not consider the costs of conducting the experiments. We might reason at this point that the manager could pay as much as $321,910 for the information provided by the experiment. He should really pay less than this amount so as to obtain a *net* benefit from the experiment. We would never pay more than $321,910 because we would be better off to select the prior analysis strategy and anticipate the resulting expected payoff.

Benefits of Bayesian Analysis

Up through this point, we presented several examples illustrating the procedures for Bayesian analysis. It is appropriate at this point to review some of the benefits of this methodology and advantages of Bayesian analysis in general. These are summarized as follows:

1. The prior analysis provides useful information for making initial strategy selections from alternatives.
2. The technique then allows us to consider the desirability of collecting additional information. Collecting more information then becomes an alternative to compare with other alternatives available. Ordinary statistical approaches often neglect the consideration of costs of gathering more information.
3. The procedures allow for consideration of the potential costs arising when a wrong decision is made.
4. The process of analysis forces the user to carefully define the alternatives and associated probabilities in quantitative terms. Even though the values may be subjective, the user is at least forced to consider them carefully.
5. As part of the Bayesian analysis process, the decision maker is forced to consider and define other alternative actions. The systematic process of analysis can often turn up feasible alternative solutions that may have otherwise not been considered.
6. Having defined the alternatives, the effects of the variables (S_i or Z_i) must be considered. The systematic process forces the user to fill in values for each combination of alternatives. Subjective probabilities are used wherever necessary.

Limitations of Bayesian Analysis

In the use of any analysis technique, it is important to recognize the limitations as well as the benefits. These limitations, either in the theory or in the application to real problems, must be recognized so that any effect on the usage of results can be taken into account by the decision maker.

1. In some practical problems, the computations can become very involved.
2. Subjective probabilities are utilized, and many people are not as

willing to accept their validity as compared to objective probabilities.

3. The payoffs used in the calculation are estimates. The estimates of payoffs and prior distributions are utilized throughout the computations and are never questioned as the analysis proceeds. The user does not raise such questions as "What would happen if the payoff were different?" However, in some cases, the user can raise this type of question and carry out added analyses.

There is continual effort to remedy some of these limitations. Use of Bayesian techniques in combination with the traditional statistical approaches is one possibility. Experimental work is in progress in the areas of Bayesian regression analysis, applications to sampling, and use in various models. Although the rigorous statistician might object to use of subjective probabilities, their use is certainly better than no values at all. This expression of concern is certainly worth noting; however, in many cases, there is no valid alternative. In any particular decision-making situation, the estimation of values will require extensive experience. Thus, we see that the "judgment of management" is not left out at all. We have enhanced its value, however, by providing a systematic method of utilizing the judgments.

Relative to the limitation of excessive computation, much of this can be made easier by use of the computer. Even though we can computerize the procedure, however, there is still an extensive amount of effort required to evaluate and set up the problem.

Exercises

12.16 A service station has decided to construct a car wash to be operated in conjunction with the service station. It is recognized that the operation will result in a loss; however, the primary intent is to help gasoline sales. The manager has developed two alternative strategies, A_1 and A_2, involving free washes and fees to charge relative to gas purchases. He determines that if the market reactions is S_1 (probability = .60), then the loss will be $10,000 per year on strategy A_1 and a $4,000 loss on strategy A_2. If the market reaction is S_2, the loss will be $5,000 for strategy A_1 and a loss of $12,000 on strategy A_2. Use prior analysis to recommend a course of action.

12.17 With reference to problem 12-16, the parent firm has offered to furnish a survey team for a cost of $500. Should the station manager purchase the survey?

12.18 Starting with Table 12-21 we selected the best strategy based on use of tables in the analysis. Carry out the analysis using decision trees.

12.19 The Karpuss Insurance Company is considering handling automobile insurance in addition to its present line of life and health insurance. The president, after some extensive consultation with his accounting firm, has preposed a matrix of the expected payoffs for the first 2 years. The payoffs depend on success in sales and upon potential losses.

	Alternative Actions	
Sales and Service	A_1: Add Auto Insurance	A_2: Stay with Present Lines
S_1 (Good)	$14,000,000	0
S_2 (Poor)	−$7,000,000	0

As a result of a preliminary analysis, the expected probabilities are

$$P(S_1) = .70 \qquad P(S_2) = .30$$

There are three possible alternatives:

D_0: Make a decision now.
D_1: Conduct a survey on its own at a cost of $500,000, at a 70 per cent reliability.

$$P(Z_1 S_1) = .70$$
$$P(Z_2 S_2) = .70$$
$$P(Z_1 S_2) = .30$$
$$P(Z_2 S_1) = .30$$

D_2: Purchase a two-step survey at a cost of $250,000 for the first step and an additional $550,000 for the second step. The first step is 65 per cent reliable and the second step is 70 per cent reliable. Karpuss can decide to stop after the first step.

(a) Calculate prior payoffs for decision D_0.
(b) Use decision trees to compute expected payoffs for D_1 and D_2.
(c) Make a recommendation for one of the three alternative strategies, D_0, D_1, or D_2.

Learning Objectives

Index Numbers
  Types of Index Numbers
  Price Indexes

Considerations in Constructing an Index
  Selection of Items
  Weighting
  Base Period

Construction of Price Indexes
  Simple Aggregative Index
  Weighted Aggregative Index
  Simple Average of Price Relatives
  Weighted Average of Price Relatives

Other Indexes
  Quantity Indexes
  Value Indexes

Shifting the Base and Splicing

Deflation of a Series

Review of Index Uses

Time Series

Trends

Components of Time Series
  Seasonal Variations
  Cyclical Variations
  Irregular Variations

Techniques in Time-Series Analysis

Index Numbers and Time Series

Learning Objectives

The primary learning objectives for this chapter are:

1. To understand index numbers and how they are used in business.
2. To be able to derive index numbers from data.
3. To understand time series and their use in business.

Index Numbers

Managers of business enterprises often need information concerning the business environment in which they operate. They need to be aware of changes in economic variables so that they can make appropriate decisions in both the day-to-day operations and in their long-term planning. Some measures of change useful to the manager are expressed in absolute quantities, such as automobiles produced or steel ingot production. These are called *business indicators*. Other factors are more useful when expressed relative to a base period or place. These relative numbers are usually expressed as a percentage and are referred to as *index numbers*. Although they may be a percentage, the per cent sign (%) is not usually utilized. The important characteristic of an index number is that it expresses the relationship between two numbers. One of these numbers represents a particular time, place, or period and is called the *base*. Index numbers are also used by enterprises for internal administrative evaluation.

Types of Index Numbers

Index numbers can be classified either as to what they measure or as to how they are compiled. There are three important categories of index numbers when we consider what is measured. (1) An index that measures a change in prices paid or received by consumers or producers is called a *price index*. The consumer *cost-of-living* index is an example of a widely referenced price index. (2) A *value index* measures changes in the value of a commodity. As an example, many companies report sales relative to a selected base year. Another example is an index showing the value of the dollar as an index of buying power. (3) The third category is the *quantity index*.

TABLE 13-1. Simple Index for Commodity A

Year	Price	Index (1974 = 100)
1974	$38.50	100
1975	39.66	103
1976	41.58	108
1977	43.12	112

Such an index might measure the amount of steel produced or the number of tons of cargo shipped related to a base period.

An index can also be categorized as either a simple index or a composite index. A *simple index* is one constructed from a single item, such as the price of cotton. A simple index is shown in Table 13-1 for a commodity. The indexes that are more important in business, however, combine figures related to several factors. The consumer price index utilizes several different items, and this type of index is called a *composite index*. Our discussion here will deal primarily with the composite index because of its relatively greater importance.

A further classification of index numbers relates to the mathematical method used to construct the index. An index may be weighted or unweighted. A *weighted index* assigns different weights to the items used in the calculation of the index. If all items are weighted equally, it is called an *unweighted*, or *simple, index*. Also, an index can either be aggregative or an average of relatives. Those will be considered in detail later.

Price Indexes

Price indexes are used frequently by business as well as for economic or governmental purposes. As one example, a company's sales may have increased in 5 years from $2,000,000 to $4,000,000. In order to judge whether the volume of goods sold really doubled, it would be necessary to adjust this value to reflect any price change. If prices had also doubled in the 5-year period, the quantity of sales would actually be unchanged.

Consider as another example the effect of prices of consumer goods on the employee's salary. In determining the amount of salary increase to be paid to employees, the executive should certainly consider this factor. For over 2 million workers, an escalator clause in the union contract provides for wage increases based on an index. As another example, a company may be transferring a manager from Dallas to New York. Should his salary be adjusted to reflect the higher cost of living in New

York? To measure this difference in cost of living, some type of index is necessary. There are many other ways in which a price index is important in business and industry.

Considerations in Constructing an Index

When a price index (or other type) is to be constructed, several factors must be decided. We shall continue to work with price indexes, since they are quite widely used and they enable us to better understand the important elements of constructing and using any index. These factors are:

1. What items or commodities are to be included in the index?
2. How are each of these items to be weighted?
3. What will be the base period?
4. What will be the procedure for compiling the data periodically to prepare the index?
5. How will the data be mathematically combined to form the index?

Each of these five factors will be considered in turn.

Selection of Items

Items to make up the index must be carefully selected; however, there are a variety of techniques used in the selection. Take the Consumer Price Index as an example. There are about 2,000 items that moderate-income wage earners normally purchase. The items to be used in the Consumer Price Index (CPI) are selected systematically from these rather than being selected at random. The CPI is based upon about 400 specific items used by the typical family. These could be goods or services, such as meat, other food items, housing, clothing, and medical costs. The Bureau of Labor Statistics (BLS) has selected these 400 items and the same items are used each time an index is prepared. Over the years, however, certain items would be dropped out and others added to reflect typical family purchases. The items would be representative of those currently used by moderate-income families. Since we are selecting the items and probability or chance is not involved, the index is strictly a descriptive statistic.

Weighting

In the construction of an index, some items may be given a greater weight than others. In preparing a grain price index, corn may be given a heavier weight than oats because the amount of corn produced is greater. Those items felt to be of greater importance are given more weight. In the case of the CPI, food may be given greater weight than clothing because the typical family would spend a greater percentage of their income for food during a typical month.

A common choice of weighting is the quantity from the base year. The corresponding weighting for each year should not be used since this will introduce a second variable. The objective is to measure one variable only with any one index. Alternative weighting could be the current-year quantity instead of the base year, or else an average of a period of years.

In constructing an index, sometimes only one or a few food items would be included. Fresh tomatoes might be selected to represent all fresh vegetables; therefore, the price of tomatoes would be collected periodically and given a weight to represent fresh vegetables as a whole. Sometimes it is necessary to change weighting factors to reflect changes in buying patterns over periods of time. As frozen foods have become more popular, they were given greater weight. It is important, however, when changing weights, to remember that the primary purpose of a price index is to measure price changes. Therefore, these other changes should be kept to a minimum so that the index will tend to measure the same items insofar as possible.

Base Period

A price index reflects the price of a particular group of items related to a base period. The index would indicate the percentage change from that base. Many government-prepared indexes have used such base periods as 1947–1949, 1957–1959, or 1967. If a 1977 house-construction price index were expressed as 145 with 1967 as the base, this would be interpreted as saying that house construction prices are 45 per cent higher in 1977 than they were in 1967. The choice of the base period is important, and it should represent a fairly normal period insofar as possible. If the base period were given as 1957–1959, it means that a span of 3 years is averaged out to form the base period.

As an additional consideration in selecting the base period, it should not be too far distant in the past—say, no more than 15 years ago, since base periods too far in the past are less meaningful to people. For purposes of comparison, it is better to reference a period for which the majority of people have some recollection. Agencies thus tend to update their base periods as necessary. A base period may also be changed if a period originally thought of as typical is later viewed as no longer typical.

We also find that many different governmental agencies tend to select the same base period for their indexes. This makes it easier to compare one commodity or item with another.

Construction of Price Indexes

Different equations can be utilized to construct an index. Each of four basic types of equations will be illustrated, after which some of the advantages of each method will be discussed. The four types are

1. Simple aggregative
2. Weighted aggregative
3. Simple average of price relatives
4. Weighted average of price relatives

Simple Aggregative Index

The simple aggregative index shows the cost to purchase one of each of several items in a particular year as compared to the base year. Table 13-2 shows prices of 4 food items. In 1975 it would cost $6.99 to buy one unit of each of these items. We call this the *price aggregate* for 1975, calculated as

$$\Sigma P_{75} = 1.19 + 1.43 + 3.92 + .45 = \$6.99$$

TABLE 13-2. Prices for Certain Food Items

		Year		
Item	Unit	1975	1976	1977
Ground sirloin	Pound	$1.19	$1.29	$1.33
Milk	Gallon	1.43	1.58	1.57
Potatoes	Bushel	3.92	4.28	4.55
Bread	Large loaf	.45	.59	.64
Total		$6.99	$7.74	$8.09
Simple Aggregative Index		100	111	116

The year 1975 is used as the base year, so we can say that 1975 = 100. In the same manner the price aggregate for 1976 is calculated as follows:

$$\Sigma P_{76} = 1.29 + 1.58 + 4.28 + .59 = \$7.74$$

Using 1975 as the base, the simple aggregative index is expressed as

$$\text{index} = \frac{\Sigma P_n \times 100}{\Sigma P \text{ base}} = \frac{\Sigma P_{76} \times 100}{\Sigma P_{75}} = \frac{7.74}{6.99} \times 100 = 110.73$$

Usually it is customary to round off an index so that

$$\text{index} = 111 \text{ (rounded)}$$

We have multiplied by 100 since the index is a per cent of the base year cost; however, the per cent sign is always omitted from the index. The index for the 3 years is shown in the table.

In our example we observe that all the prices do not rise and fall together. The simple aggregative index represents the sum of the items, and, therefore, the item that has the greatest price per unit (in this case potatoes) has the greatest effect on the aggregative index. If we changed the units for potatoes from bushel to pound, the effect of potatoes on the index would be reduced considerably. Why, then, don't we put them all in the same units? This could be done but it is not usually practical. In this case, the price per pound of milk would not be really meaningful. The simple aggregative index does have its advantages, though. It is simple to compute and understand, and it is sufficient for many purposes. Its primary disadvantage is that it does not represent quantities of each item in proportion to their use by a typical family. The next index to be considered will weight the items by their importance.

Weighted Aggregative Index

There were two limitations on the simple (unweighted) aggregative index. The first was the undue influence of the higher-priced items; the second was due to the arbitrary use of units of measurement. We had observed that if the unit for potatoes was pound rather than bushel, the index would have been different. These limitations may be overcome by use of a weighted index. Each item is weighted according to its importance to the index being prepared. We could take the price data in Table 13-2 and assign weights according to the importance of each item to the typical family budget.

Table 13-3 shows the quantities of each of the items (from Table 13-2)

TABLE 13-3. Prices for Certain Food Items and Quantity Weights for a 3-Month Period in 1975

Item	Unit	Quantity, Q_{75}	Unit Price 1975	Unit Price 1976	$P_{75}Q_{75}$	$P_{76}Q_{75}$
Ground sirloin	Pound	40	$1.19	$1.29	$47.60	$51.60
Milk	Gallon	36	1.43	1.58	51.48	56.88
Potatoes	Bushel	1	3.92	4.28	3.92	4.28
Bread	Large loaf	24	.45	.59	10.80	14.16
Total					$113.80	$126.92
Weighted Aggregative Index					100	111.5

which may be used by a typical family of four in a set period. Each of the item prices is then weighted by the quantity used by the typical family of four in a 3-month period of 1975. The $P_n Q_n$ value is obtained by multiplying the price by the quantity.

In this particular case we find that the quantity of items consumed in the stated period has risen from $113.80 in 1975 to $126.92 in 1976. The weighted aggregative price index can then be computed as follows:

$$\text{price index} = \frac{\Sigma P_1 Q_0}{\Sigma P_0 Q_0} \times 100 = \frac{126.92}{113.80} = 111.5$$

In other words, the price of the package of foods has risen 11.5 per cent for 1976 related to the base year.

Simple Average of Price Relatives

The simple average of price relatives places equal weights on all items making up the index. Thus it avoids the problem of emphasis upon the item having the highest price. An index will be formed based on the same items used before, as shown in Table 13-4.

The simple average of price relatives can be expressed as follows:

$$\text{price index} = \frac{\Sigma (P_n/P_0 \times 100)}{\Sigma (P_0/P_0)} = \frac{\Sigma (P_n/P_0 \times 100)}{n}$$

We observe that a price relative is merely a given year price expressed as a percentage of the price in a base year. The price index is then the average of these price relatives.

TABLE 13-4. Simple Average of Price Relatives Index

	Prices			$P_n/P_0 \times 100$		
Item	1975	1976	1977	1975	1976	1977
Ground sirloin	$1.19	$1.29	$1.33	100.0	108.4	111.8
Milk	1.43	1.58	1.57	100.0	110.5	109.8
Potatoes	3.92	4.28	4.55	100.0	109.2	116.1
Bread	.45	.59	.64	100.0	131.1	142.2
Total				400.0	459.2	479.9
Price Index				100.0	114.8	120.0

Weighted Average of Price Relatives

The same quantities that were used to weight the aggregatives can be utilized to weight the price relatives. If this is done, the results of each method should be identical. These values are shown in Table 13-5. The equation used to calculate the weighted average of price relatives index is as follows:

$$\text{price index} = \frac{\Sigma (P_n/P_0)(P_0 Q_0)}{\Sigma (P_0/P_0)(P_0 Q_0)} \times 100$$

Since P_0/P_0 is 1.0, the equation can be simplified to

$$\text{price index} = \frac{\Sigma (P_n/P_0)(P_0 Q_0)}{\Sigma (P_0 Q_0)}$$

$$\text{price index}_{1976} = \frac{12{,}693}{11{,}380} = 111.5$$

$$\text{price index}_{1977} = \frac{12{,}966}{11{,}380} = 113.9$$

These are the same values obtained for the weighted aggregative index. In this case, the computations were more involved than for the weighted aggregative method; however, there are some advantages. Table 13-5 shows the price of each item for each year as a per cent of the base year, so this is available to the user. These individual price relatives can then be compiled into various groups for other indexes. Although the equations express the relationship between the numerical quantities, the computations are performed in the table.

TABLE 13-5. Calculation of the Weighted Average of Price Relatives Index

Item	Weights $p_0 q_0$	Price Relatives			Weighted Price Relatives		
		1975	1976	1977	1975	1976	1977
Ground sirloin	47.60	100.0	108.4	111.8	4,760	5,160	5,322
Milk	51.48	100.0	110.5	109.8	5,148	5,689	5,653
Potatoes	3.92	100.0	109.2	116.1	392	428	455
Bread	10.80	100.0	131.1	142.2	1,080	1,416	1,536
Total					11,380	12,693	12,966
Price Index					100.0	111.5	113.9

Other Indexes

Although price indexes are more widely known, other indexes are frequently used by a business enterprise. Changes in physical quantities are measured by quantity indexes, while value indexes are also frequently used. The methods of computation of these indexes are similar to those used for the price index. These indexes are discussed briefly.

Quantity Indexes

A quantity index measures the change in a physical quantity, such as the number of automobiles manufactured, oil consumption, physical volume of imports or exports, or number of shares sold on a stock exchange.

A simple quantity index can be established by a ratio between the quantity in a specified year to the quantity in a base year:

$$\text{simple quantity index} = \frac{Q_n}{Q_0} \times 100$$

Just as we found the desirability of a weighted price index, the most widely used quantity indexes are weighted also. In the weighted price index, we weighted prices according to a price in a base year. Likewise, in a weighted quantity index, the various period quantities are weighted by a price held constant. The equation would then be

$$\text{weighted quantity index} = \frac{\Sigma P_0 Q_n}{\Sigma P_0 Q_0}$$

where

$P_0 =$ price in base year
$Q_0 =$ quantity in base year
$Q_n =$ quantity in selected year

In the computation of a quantity index it is necessary to have quantities for each year and a price for a base year. Since the price has been held constant, any change would be due to increase or decrease in quantity.

Value Indexes

Value indexes are not widely used as published economic indicators; however, they have important use by companies. Frequently, they are used without really being called indexes. The company financial statement usually contains sales such as in the following example:

	Year			
	1974	1975	1976	1977
Sales	$1,200,000	$1,312,000	$1,418,000	$1,616,000
Per cent	100.0	109.3	118.2	134.7

The 1974 sales have actually been used as a value index with 1974 as a base to show relative increases for the other years. This value index can be expressed as

$$\text{value index} = \frac{\Sigma P_n Q_n}{\Sigma P_0 Q_0} \times 100$$

or

$$\text{value index for 1976} = \frac{1{,}418{,}000}{1{,}200{,}000} \times 100 = 118.2$$

The total sales figure in any year is a summation of prices of each item times the number of each item sold.

Shifting the Base and Splicing

Numerous occasions arise when it becomes necessary to shift the base of a price or other index from one year to another. This can be done without going back to the original raw data and recalculating the data for the entire span of time. Quite often the original data are not available to the user; however, he or she can readily shift the base. Consider the following index:

Year	Price Index (1968 = 100.0)	Price Index (1970 = 100.0)
1968	100.0	88.9
1969	105.3	93.6
1970	112.5	100.0
1971	119.0	105.8
1972	123.3	109.6

Starting with the index using 1968 as a base, a new index has been calculated using 1970 as the base. The following shows the calculations for 2 years at the 1970 base:

$$I_{1971} = \frac{119.0}{112.5} \times 100 = 105.8$$

$$I_{1968} = \frac{100.0}{112.5} \times 100 = 88.9$$

Frequently it is necessary to combine two indexes into one index. Assume that two sets of numbers are available as shown in the first two columns of the following table:

Year	Price Index (1968 = 100)	Price Index (1974 = 100)	Spliced Index (1974 = 100)
1968	100.0		77.9
1969	106.8		83.2
1970	109.2		85.1
1971	113.4		88.3
1972	118.3		92.2
1973	124.6		97.1
1974	128.3	100.0	100.0
1975		108.4	108.4
1976		114.3	114.3
1977		119.6	119.6

A single index has been prepared in the third column using 1974 as the base year, although 1968 or any other year could have been used as the new base. This permitted us to obtain a single series over the period of interest. It is necessary, however, to recognize that the items used in one index may have differed from those used in the other. But, despite such problems, the spliced index may be very helpful in understanding a relationship over the total span of years.

Deflation of a Series

The objective of an index is to show the fluctuation of a price or value over a period of time. If the index is reported in terms of dollars, it will contain fluctuations due to changes in the price level over that period of time. The index in Table 13-6 shows sales over a period of 6 years in terms of dollars. We know that over the period of time, there were changes in prices for the group of products. This is shown in the second column of the table. Our objective is to show sales in terms of volume. The first step is to divide each sales figure by the price index for the year to give the deflated sales. It is observed that the sales expressed in 1972 dollars do not climb at as fast a rate as does the actual dollar sales indicated.

The physical volume index is then obtained by dividing each year's deflated sales into the 1972 sales. This makes possible a comparison between volume of goods sold over the 6-year period.

TABLE 13-6. Company Sales and Deflation of Sales for 1972–1977

Year	Sales	Price Index (1972 = 100)	Deflated Sales (1972 dollars)	Physical Volume Index
1972	$2,600,000	100.0	$2,600,000	100.0
1973	2,840,000	106.5	2,667,000	102.6
1974	3,200,000	111.0	2,882,000	110.8
1975	3,940,000	117.8	3,345,000	128.6
1976	4,460,000	123.3	3,617,000	139.1
1977	5,100,000	129.8	3,929,000	151.1

Review of Index Uses

Index number have many uses in decision making related to a business or other enterprise. Many wage agreements contain an escalator clause such that workers' wages are modified to reflect changes in the Consumer Price Index. The objective sought by unions in negotiating this provision is to avoid a loss in the workers' purchasing power over periods of inflation.

In many other situations, agreements between companies and their suppliers may provide price changes related to the Bureau of Labor Statistics Wholesale Price Index. Contracts between the government and companies also often contain a similar provision.

Exercises

13.1 The following are a representative group of commodities purchased by the Garmuth Company

Item	Units	Quantities Used			Prices (per unit)		
		1974	1975	1976	1974	1975	1976
A	Dozen	100	120	130	$24.00	$26.40	$28.00
B	Pound	400	440	480	32.00	31.20	33.40
C	Liter	200	240	220	60.00	64.80	66.10
D	Yard	800	580	610	12.10	12.80	14.40

(a) Calculate an unweighted simple aggregative index of prices for each year using 1974 as the base year.
(b) Calculate a weighted aggregative index of prices for each year using 1974 as the base year and 1974 quantities as weights.

13.2 Use the data from problem 13-1 to calculate a simple average of price relatives index using 1974 as the base year.

13.3 Use the data from problem 13-2 to determine a weighted average of price relatives using 1974 as the base year and 1974 quantities as weights.

13.4 Compare the results from the previous three problems and discuss any similarities or differences in the values obtained.

13.5 The following represent sales of the Harset Company for a 5-year period.

Year	Sales	Price Index (1965 base)
1973	$32,814,000	120.1
1974	34,215,000	123.2
1975	41,818,000	128.4
1976	54,212,000	132.8
1977	59,814,000	136.2

Prepare an index of dollar value of sales with 1973 as the base year (ignore price index).

13.6 Use the price index data to adjust the sales figures to compensate for price-level variation. Express the sales of each year in terms of 1973 dollars.

13.7 Calculate an index to show a comparison between product volume sold in each year as compared to 1973 as the base year. Compare the results obtained to the sales growth and discuss any differences.

13.8 The following represent the dollar value of the average inventory of each year over a 4-year period:

Year	Average Inventory	Price Index
1974	$786,000	124.8
1975	814,000	131.4
1976	844,000	133.6
1977	911,000	138.3

Prepare an index of deflated inventories to help in the evaluation of changes in the size of the average inventory over the 4-year period, using 1974 as a base.

13.9 The following represents an index of prices charged by the Ingrest Company for its products using a 1967 base. Shift the base of the index so that 1969 is the base year.

Year	Price Index (1967 = 100)
1967	100.0
1968	103.3
1969	107.4
1970	111.8
1971	114.2
1972	119.6
1973	121.8

13.10 The following index was obtained later for the Ingrest Company:

Year	Price Index
1973	100.0
1974	106.5
1975	111.2
1976	117.7
1977	121.3

Splice this index to that in problem 13-9 using a base year of 1967.

13.11 In problems 13-9 and 13-10, splice the index using a base year of 1970.

13.12 A company employee has had his pay increase 34 per cent over the past 6 years. During the same period the purchasing power of a dollar has decreased 31 per cent. What pay adjustment would provide the employee with the same buying power he had 6 years ago?

13.13 The Krestler Company uses three basic raw materials in its products. The quantities consumed and the average prices are shown below for each of the three products.

Product	1975		1976	
	Price	Quantity	Price	Quantity
A	$40	10	$50	15
B	2	50	4	60
C	10	25	12	35

Compute an appropriate weighted aggregative price index for 1976 using 1975 as a base.

13.14 The Lampler Company sells three models of lamps. The average item selling price and quantity sold for each model for 1974 and 1976 are given below

	1974		1976	
Model	Price	Quantity	Price	Quantity
C	$40	12,000	$44	15,000
D	60	4,000	72	4,000
G	70	8,000	80	9,000

Calculate the weighted average of price relatives index using a base year of 1974 and interpret the results.

Time Series

Many vital decisions made by the business manager depend upon knowledge of present economic conditions and perceptions of future conditions. We have seen how index numbers help to relate data into usable indicators. We shall next discuss some techniques for identifying trends and forecasting future conditions. These forecasts will help the business manager in making decisions from among various available alternatives as he sets and tries to meet business objectives.

Time series analysis is an important technique available to the business manager for use in forecasting. A *time series* is a set of observations taken at regular intervals and arranged in chronological order. In an individual business, the quarterly sales figures (or annual sales) over a period of years would form a time series. Other examples of time series would include annual population growth figures, the quarterly Consumer Price Index, or weekly closing stock prices.

In each case, the use of these time series takes on a set pattern. The first step in this pattern involves the understanding of past conditions based on the analysis of historical data, where the objective is to observe trends. These trends are then considered in preparing future forcasts. Based on these future forecasts, business management can then establish plans for the short term, as well as make projections for the next 5 or 10 years.

This section of the chapter will be primarily concerned with the analysis of time-series data and the determination of trends. The traditional, or classical, methods of time-series analysis are essentially descriptive in nature. Other more involved methods involve probabilities and are called *econometric methods;* however, we shall not deal with them in this text. In past usage, time-series analysis has demonstrated its usefulness when correctly interpreted and applied. Just as in the use of other statistical data and analysis, however, the use of trends developed from time series must be considered in conjunction with other subjective data and sound judgment.

Trends

The ability to identify long-term trends and to interpret their impact on a business is an important function of management. *Trends* are defined as long-term movements of a time series that are characterized by steady or slightly varying rates of change. Trends over a long period of time are often referred to as *secular trends*. Figure 13-1 shows such a trend for

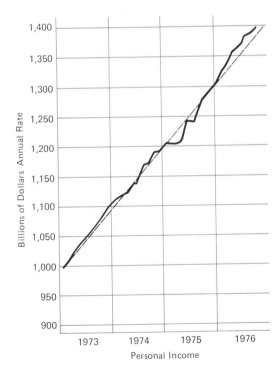

FIGURE 13-1. Personal Income in Billions of Dollars Source: U.S. Dept. of Commerce.

personal income over a 4-year period. This trend could be described by a straight line, as shown in the figure. We could determine the equation for this straight line by use of regression analysis, as described in Chapter 9. If we extend the straight line beyond the point where we have data, we are extrapolating the data. Care must be taken in using data to forecast how this trend may continue.

In the case above, a straight line was used to show the trend. In other cases, a curve may be a better representation of the trend, as shown in Figure 13-2. A curve can also be fitted to data by use of regression analysis. Standard computer programs are available to evaluate data and determine the best-fitting regression line or curve. These programs also determine the standard error and correlation coefficients for both a straight line or a curve. The curve can also be extrapolated to forecast future conditions; however, care must always be taken in their use for this purpose.

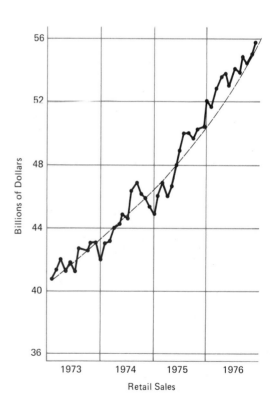

FIGURE 13-2. Retail Sales over a Four-year Period Source: U.S. Dept. of Commerce.

Index Numbers and Time Series

Components of Time Series

When trying to evaluate time-series data and determine trends, we find that there are fluctuations in the data which may make it more difficult to identify the trend line. Fluctuations can be noted in Figure 13-2. In order to better understand the time-series data and to be able to define a trend more accurately, it is necessary to separate fluctuations into groups by cause, as discussed in the following paragraphs.

Seasonal Variations

Frequently, it is possible to identify movements of time series which are similar at the same time each year. Sales of ski equipment would certainly be higher over certain portions of the year. Variations that occur at predictable times of the year are called *seasonal variations*. In order for these types of variations to be evident, the time-series data must be recorded at monthly or weekly intervals. A plot of yearly sales would certainly not show up this type of fluctuation. You can notice in Figure 13-3 that there seem to be consistent peaks at the year-end points and

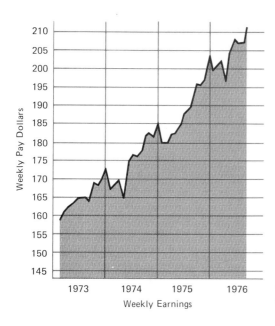

FIGURE 13-3. Average Weekly Pay of Factory Workers

consistent low points around the latter part of the first quarter of each year.

Cyclical Variations

Cyclical variations are movements in a time series that are recurrent in cycles longer than one year. Usually we require data over a period of 15 to 20 years in order to be able to identify these variations. The data could be recorded annually, as in Figure 13-4, showing housing starts over a 17-year period. The cycles may have a variation of anywhere from 2 to 15 years. The duration of the cycles is measured from a peak to the next peak, or from a low point to the next low point in the data. These cycles do not recur at set periods as we saw in the seasonal variations, but rather they are said to be *recurrent*. Often there is no reasonable explanation of why the cycles occur, or they may follow periods of economic expansion or contraction which terminate with a peak or low point, respectively. Sometimes one cyclical variation may be out of phase with another. Frequently, the dollar value of machine-tool orders goes in cycles and leads business-volume cycles. It is thought that when a business is starting to expand, one of the first steps is to order more machine tools. Therefore, the volume of machine tool orders is often considered to be a predictor of future business cycles.

Irregular Variations

Movements in the time-series data that do not fit into any of the other three categories are classified as *irregular variations*. They are not periodic in nature, and historical data are not of value in predicting

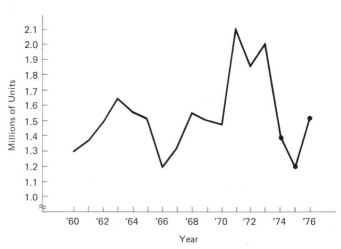

FIGURE 13-4. Housing Starts 1960–1976

trends. They can be caused by such factors as war, strikes, or changes in governmental legislation, weather, or other factors. Sometimes harsh or unusual weather conditions will affect the production of crops. A strike in a vital industry may produce a substantial negative effect on the economy. The energy crisis in 1974 caused a number of unpredicted changes. All these unusual changes are classified as irregular variations. We can usually determine their cause; however, it is not possible to predict their occurrence.

Techniques in Time-Series Analysis

The main objective in using time-series analysis is to determine trends for use in forecasting. In this process it is desirable to identify any cyclical fluctuations or other variations that would be useful in understanding the data. This analysis process is often referred to as the *classical time-series model*, comprising the following components: (1) trend, (2) seasonal variation, (3) cyclical variation, and (4) irregular movement. The objective in the time-series analysis is to compile and plot available data and identify each of these components in the data if they are present. Following this analysis, if a trend is identified, a trend curve can be fitted to the trend component. These steps can be summarized as follows:

STEP 1. DESEASONALIZE DATA

The first step is to isolate patterns of seasonal variation. These are sometimes obvious; however, there are analytical methods to isolate these patterns. Retail sales, for example, are usually reported on a seasonally adjusted basis.

STEP 2. PLOT A TREND LINE

Computer programs are available to evaluate data and plot the data in various forms.

1. *Straight line.* In Chapter 9, we learned how to take data and determine the best-fitting regression line and the associated equation in the form

$$y = a + bx.$$

2. *Curves.* In Figure 13-2, we observed that a curve would fit rather closely. Actually, one computer program can plot the same data in a

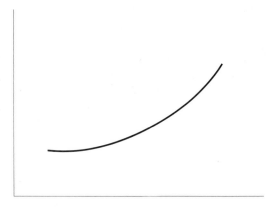

FIGURE 13-5. Curve Which Can Be Represented by Second Degree Equation $y = a + bx + cx^2$

straight line as well as different curves and determine the corresponding errors for each so that the best equation can be selected. A curve like that shown in Figure 13-5 can be fitted by the equation

$$y = a + bx + cx^2$$

Other curves, such as the one in Figure 13-6, are best fitted with higher-degree equations such as

$$y = a + bx + cx^2 + dx^3$$

or

$$y = a + bx + cx^2 + dx^3 + ex^4$$

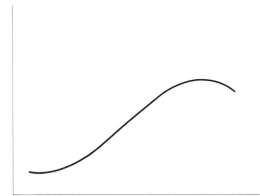

FIGURE 13-6. Curve Which Can Be Represented by Third Degree Equation $y = a + bx + cx^2 + dx^3$

Once the data are collected, the computer program can select the best form for the particular set of data.

Step 3. Identify Cyclical Fluctuations

Usually these fluctuations can be recognized from the plot of the data, as in Figure 13.4. Sometimes these patterns can be separated out into shorter-term periods of growth or decline. In other cases, the plotting of the trend line tends to identify them but still allows the overall trend to be apparent.

Supplemental Readings on Index Numbers and Time Series

Hamburg, Morris, *Statistical Analysis for Decision Making.* New York: Harcourt Brace Jovanovich, Inc., 1970, Chapters 10–12.

Huntsberger, David V., Patrick Billingsley, and **D. James Croft,** *Statistical Inference.* Boston: Allyn and Bacon, Inc., 1975, Chapters 14 and 15.

Lapin, Lawrence L., *Statistics for Modern Business Decisions.* Harcourt Brace Jovanovich, Inc., 1973, Chapters 13–16.

Newton, Byron L., *Statistics for Business.* Chicago: Science Research Associates, Inc., 1973, Chapters 10 and 11.

LIST OF TABLES

Appendix	Title
A	Normal Distribution Areas
B	t Distribution
C	Binomial Distribution: Individual Probabilities
D	Binomial Distribution: Cumulative Probabilities
E	Poisson Distribution: Individual Probabilities
F	Poisson Distribution: Cumulative Probabilities
G	Chi-Square Distribution
H	F Distribution
I	Squares and Square Roots
J	Random Digits
K	Common Logarithms
L	The Greek Alphabet

Appendixes

APPENDIX A. Normal Probability Distribution Areas

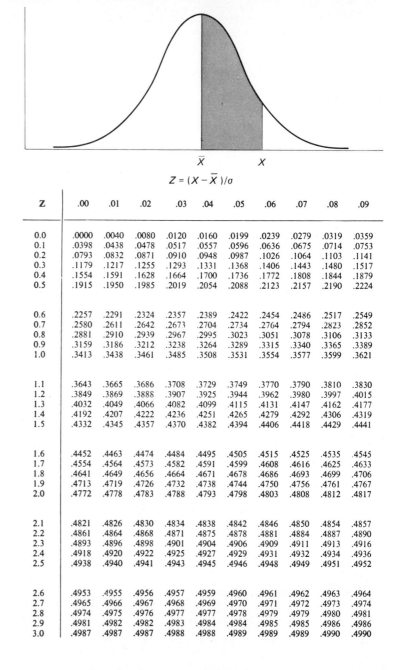

$Z = (X - \bar{X})/\sigma$

Z	.00	.01	.02	.03	.04	.05	.06	.07	.08	.09
0.0	.0000	.0040	.0080	.0120	.0160	.0199	.0239	.0279	.0319	.0359
0.1	.0398	.0438	.0478	.0517	.0557	.0596	.0636	.0675	.0714	.0753
0.2	.0793	.0832	.0871	.0910	.0948	.0987	.1026	.1064	.1103	.1141
0.3	.1179	.1217	.1255	.1293	.1331	.1368	.1406	.1443	.1480	.1517
0.4	.1554	.1591	.1628	.1664	.1700	.1736	.1772	.1808	.1844	.1879
0.5	.1915	.1950	.1985	.2019	.2054	.2088	.2123	.2157	.2190	.2224
0.6	.2257	.2291	.2324	.2357	.2389	.2422	.2454	.2486	.2517	.2549
0.7	.2580	.2611	.2642	.2673	.2704	.2734	.2764	.2794	.2823	.2852
0.8	.2881	.2910	.2939	.2967	.2995	.3023	.3051	.3078	.3106	.3133
0.9	.3159	.3186	.3212	.3238	.3264	.3289	.3315	.3340	.3365	.3389
1.0	.3413	.3438	.3461	.3485	.3508	.3531	.3554	.3577	.3599	.3621
1.1	.3643	.3665	.3686	.3708	.3729	.3749	.3770	.3790	.3810	.3830
1.2	.3849	.3869	.3888	.3907	.3925	.3944	.3962	.3980	.3997	.4015
1.3	.4032	.4049	.4066	.4082	.4099	.4115	.4131	.4147	.4162	.4177
1.4	.4192	.4207	.4222	.4236	.4251	.4265	.4279	.4292	.4306	.4319
1.5	.4332	.4345	.4357	.4370	.4382	.4394	.4406	.4418	.4429	.4441
1.6	.4452	.4463	.4474	.4484	.4495	.4505	.4515	.4525	.4535	.4545
1.7	.4554	.4564	.4573	.4582	.4591	.4599	.4608	.4616	.4625	.4633
1.8	.4641	.4649	.4656	.4664	.4671	.4678	.4686	.4693	.4699	.4706
1.9	.4713	.4719	.4726	.4732	.4738	.4744	.4750	.4756	.4761	.4767
2.0	.4772	.4778	.4783	.4788	.4793	.4798	.4803	.4808	.4812	.4817
2.1	.4821	.4826	.4830	.4834	.4838	.4842	.4846	.4850	.4854	.4857
2.2	.4861	.4864	.4868	.4871	.4875	.4878	.4881	.4884	.4887	.4890
2.3	.4893	.4896	.4898	.4901	.4904	.4906	.4909	.4911	.4913	.4916
2.4	.4918	.4920	.4922	.4925	.4927	.4929	.4931	.4932	.4934	.4936
2.5	.4938	.4940	.4941	.4943	.4945	.4946	.4948	.4949	.4951	.4952
2.6	.4953	.4955	.4956	.4957	.4959	.4960	.4961	.4962	.4963	.4964
2.7	.4965	.4966	.4967	.4968	.4969	.4970	.4971	.4972	.4973	.4974
2.8	.4974	.4975	.4976	.4977	.4977	.4978	.4979	.4979	.4980	.4981
2.9	.4981	.4982	.4982	.4983	.4984	.4984	.4985	.4985	.4986	.4986
3.0	.4987	.4987	.4987	.4988	.4988	.4989	.4989	.4989	.4990	.4990

APPENDIX B. t Distribution

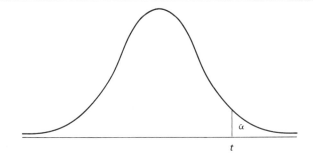

Degrees of Freedom	Confidence Level					
	.8	.9	.95	.975	.99	.995
1	1.38	3.08	6.31	12.7	31.8	63.7
2	1.06	1.89	2.92	4.30	6.96	9.92
3	.978	1.64	2.35	3.18	4.54	5.84
4	.941	1.53	2.13	2.78	3.75	4.60
5	.920	1.48	2.01	2.57	3.36	4.03
6	.906	1.44	1.94	2.45	3.14	3.71
7	.896	1.42	1.90	2.36	3.00	3.50
8	.889	1.40	1.86	2.31	2.90	3.36
9	.883	1.38	1.83	2.26	2.82	3.25
10	.879	1.37	1.81	2.23	2.76	3.17
11	.876	1.36	1.80	2.20	2.72	3.11
12	.873	1.36	1.78	2.18	2.68	3.06
13	.870	1.35	1.77	2.16	2.65	3.01
14	.868	1.34	1.76	2.14	2.62	2.98
15	.866	1.34	1.75	2.13	2.60	2.95
16	.865	1.34	1.75	2.12	2.58	2.92
17	.863	1.33	1.74	2.11	2.57	2.90
18	.862	1.33	1.73	2.10	2.55	2.88
19	.861	1.33	1.73	2.09	2.54	2.86
20	.860	1.32	1.72	2.09	2.53	2.84
21	.859	1.32	1.72	2.08	2.52	2.83
22	.858	1.32	1.72	2.07	2.51	2.82
23	.858	1.32	1.71	2.07	2.50	2.81
24	.857	1.32	1.71	2.06	2.49	2.80
25	.856	1.32	1.71	2.06	2.48	2.79
26	.856	1.32	1.71	2.06	2.48	2.78
27	.855	1.31	1.70	2.05	2.47	2.77
28	.855	1.31	1.70	2.05	2.47	2.76
29	.854	1.31	1.70	2.04	2.46	2.76
30	.854	1.31	1.70	2.04	2.46	2.75
∞	.842	1.28	1.64	1.96	2.33	2.58

NOTE: The table value is for a one-tail test. For example, at a significance level of .05 and 10 degrees of freedom, the confidence level is $1 - .05$ or .95 and t is 1.81. For a two-tail test, the tail area is $\alpha/2$ and the confidence level is $1 - .05/2$ or .975. The two-tail t value for $\alpha = .05$, 10 df is then 2.23.

APPENDIX C. Binomial Distribution — Individual Probabilities

N	R	.05	.10	.15	.20	.25	.30	.35	.40	.45	.50	.55	.60	.65	.70	.75	.80	.85	.90	.95
1	0	.9500	.9000	.8500	.8000	.7500	.7000	.6500	.6000	.5500	.5000	.4500	.4000	.3500	.3000	.2500	.2000	.1500	.1000	.0500
	1	.0500	.1000	.1500	.2000	.2500	.3000	.3500	.4000	.4500	.5000	.5500	.6000	.6500	.7000	.7500	.8000	.8500	.9000	.9500
2	0	.9025	.8100	.7225	.6400	.5625	.4900	.4225	.3600	.3025	.2500	.2025	.1600	.1225	.0900	.0625	.0400	.0225	.0100	.0025
	1	.0950	.1800	.2550	.3200	.3750	.4200	.4550	.4800	.4950	.5000	.4950	.4800	.4550	.4200	.3750	.3200	.2550	.1800	.0950
	2	.0025	.0100	.0225	.0400	.0625	.0900	.1225	.1600	.2025	.2500	.3025	.3600	.4225	.4900	.5625	.6400	.7225	.8100	.9025
3	0	.8574	.7290	.6141	.5120	.4219	.3430	.2746	.2160	.1664	.1250	.0911	.0640	.0429	.0270	.0156	.0080	.0034	.0010	.0001
	1	.1354	.2430	.3251	.3840	.4219	.4410	.4436	.4320	.4084	.3750	.3341	.2880	.2389	.1890	.1406	.0960	.0574	.0270	.0071
	2	.0071	.0270	.0574	.0960	.1406	.1890	.2389	.2880	.3341	.3750	.4084	.4320	.4436	.4410	.4219	.3840	.3251	.2430	.1354
	3	.0001	.0010	.0034	.0080	.0156	.0270	.0429	.0640	.0911	.1250	.1664	.2160	.2746	.3430	.4219	.5120	.6141	.7290	.8574
4	0	.8145	.6561	.5220	.4096	.3164	.2401	.1785	.1296	.0915	.0625	.0410	.0256	.0150	.0081	.0039	.0016	.0005	.0000	.0000
	1	.1715	.2916	.3685	.4096	.4219	.4116	.3845	.3456	.2995	.2500	.2005	.1536	.1115	.0756	.0469	.0256	.0115	.0036	.0005
	2	.0135	.0486	.0975	.1536	.2109	.2646	.3105	.3456	.3675	.3750	.3675	.3456	.3105	.2646	.2109	.1536	.0975	.0486	.0135
	3	.0005	.0036	.0115	.0256	.0469	.0756	.1115	.1536	.2005	.2500	.2995	.3456	.3845	.4116	.4219	.4096	.3685	.2916	.1715
	4	.0000	.0001	.0005	.0016	.0039	.0081	.0150	.0256	.0410	.0625	.0915	.1296	.1785	.2401	.3164	.4096	.5220	.6561	.8145
5	0	.7738	.5905	.4437	.3277	.2373	.1681	.1160	.0778	.0503	.0313	.0185	.0102	.0053	.0024	.0010	.0003	.0001	.0000	.0000
	1	.2036	.3280	.3915	.4096	.3955	.3601	.3124	.2592	.2059	.1563	.1128	.0768	.0488	.0284	.0146	.0064	.0022	.0005	.0000
	2	.0214	.0729	.1382	.2048	.2637	.3087	.3364	.3456	.3369	.3125	.2757	.2304	.1811	.1323	.0879	.0512	.0244	.0081	.0011
	3	.0011	.0081	.0244	.0512	.0879	.1323	.1811	.2304	.2757	.3125	.3369	.3456	.3364	.3087	.2637	.2048	.1382	.0729	.0214
	4	.0000	.0004	.0022	.0064	.0146	.0283	.0488	.0768	.1128	.1562	.2059	.2592	.3124	.3601	.3955	.4096	.3915	.3281	.2036
	5	.0000	.0000	.0001	.0003	.0010	.0024	.0053	.0102	.0185	.0312	.0503	.0778	.1160	.1681	.2373	.3277	.4437	.5905	.7738
6	0	.7351	.5314	.3771	.2621	.1780	.1176	.0754	.0467	.0277	.0156	.0083	.0041	.0018	.0007	.0002	.0001	.0000	.0000	.0000
	1	.2321	.3543	.3993	.3932	.3560	.3025	.2437	.1866	.1359	.0938	.0609	.0369	.0205	.0102	.0044	.0015	.0004	.0001	.0000
	2	.0305	.0984	.1762	.2458	.2966	.3241	.3280	.3110	.2780	.2344	.1861	.1382	.0951	.0595	.0330	.0154	.0055	.0012	.0001
	3	.0021	.0146	.0415	.0819	.1318	.1852	.2355	.2765	.3032	.3125	.3032	.2765	.2355	.1852	.1318	.0819	.0415	.0146	.0021
	4	.0001	.0012	.0055	.0154	.0330	.0595	.0951	.1382	.1861	.2344	.2779	.3110	.3280	.3241	.2966	.2458	.1762	.0984	.0305
	5	.0000	.0001	.0004	.0015	.0044	.0102	.0205	.0369	.0609	.0937	.1359	.1866	.2437	.3025	.3560	.3932	.3993	.3543	.2321
	6	.0000	.0000	.0000	.0001	.0002	.0007	.0018	.0041	.0083	.0156	.0277	.0467	.0754	.1176	.1780	.2621	.3771	.5314	.7351
7	0	.6983	.4783	.3206	.2097	.1335	.0824	.0490	.0280	.0152	.0078	.0037	.0016	.0006	.0002	.0001	.0000	.0000	.0000	.0000
	1	.2573	.3720	.3960	.3670	.3115	.2471	.1848	.1306	.0872	.0547	.0320	.0172	.0084	.0036	.0013	.0004	.0001	.0000	.0000
	2	.0406	.1240	.2097	.2753	.3115	.3177	.2985	.2613	.2140	.1641	.1172	.0774	.0466	.0250	.0115	.0043	.0012	.0002	.0000
	3	.0036	.0230	.0617	.1147	.1730	.2269	.2679	.2903	.2918	.2734	.2388	.1935	.1442	.0972	.0577	.0287	.0109	.0026	.0002
	4	.0002	.0026	.0109	.0287	.0577	.0972	.1442	.1935	.2388	.2734	.2918	.2903	.2679	.2269	.1730	.1147	.0617	.0230	.0036

APPENDIX C. (cont'd)

n	x	.05	.10	.15	.20	.25	.30	.35	.40	.45	.50	.55	.60	.65	.70	.75	.80	.85	.90	.95
7	5	.0000	.0002	.0012	.0043	.0115	.0250	.0466	.0774	.1172	.1641	.2140	.2613	.2985	.3177	.3115	.2753	.2097	.1240	.0406
7	6	.0000	.0000	.0001	.0004	.0013	.0036	.0084	.0172	.0320	.0547	.0872	.1306	.1848	.2471	.3115	.3670	.3960	.3720	.2573
7	7	.0000	.0000	.0000	.0000	.0001	.0002	.0006	.0016	.0037	.0078	.0152	.0280	.0490	.0824	.1335	.2097	.3206	.4783	.6983
8	0	.6634	.4305	.2725	.1678	.1001	.0576	.0319	.0168	.0084	.0039	.0017	.0007	.0002	.0001	.0000	.0000	.0000	.0000	.0000
8	1	.2793	.3826	.3847	.3355	.2670	.1977	.1373	.0896	.0548	.0313	.0164	.0079	.0033	.0012	.0004	.0001	.0000	.0000	.0000
8	2	.0515	.1488	.2376	.2936	.3115	.2965	.2587	.2090	.1569	.1094	.0703	.0413	.0217	.0100	.0038	.0011	.0002	.0000	.0000
8	3	.0054	.0331	.0839	.1468	.2076	.2541	.2786	.2787	.2568	.2188	.1719	.1239	.0808	.0467	.0231	.0092	.0026	.0004	.0000
8	4	.0004	.0046	.0185	.0459	.0865	.1361	.1875	.2322	.2627	.2734	.2627	.2322	.1875	.1361	.0865	.0459	.0185	.0046	.0004
8	5	.0000	.0004	.0026	.0092	.0231	.0467	.0808	.1239	.1719	.2188	.2568	.2787	.2786	.2541	.2076	.1468	.0839	.0331	.0054
8	6	.0000	.0000	.0002	.0011	.0038	.0100	.0217	.0413	.0703	.1094	.1569	.2090	.2587	.2965	.3115	.2936	.2376	.1488	.0515
8	7	.0000	.0000	.0000	.0001	.0004	.0012	.0033	.0079	.0164	.0313	.0548	.0896	.1373	.1977	.2670	.3355	.3847	.3826	.2793
8	8	.0000	.0000	.0000	.0000	.0000	.0001	.0002	.0007	.0017	.0039	.0084	.0168	.0319	.0576	.1001	.1678	.2725	.4305	.6634
9	0	.6302	.3874	.2316	.1342	.0751	.0404	.0207	.0101	.0046	.0020	.0008	.0003	.0001	.0000	.0000	.0000	.0000	.0000	.0000
9	1	.2985	.3874	.3679	.3020	.2253	.1556	.1004	.0605	.0339	.0176	.0083	.0035	.0013	.0004	.0001	.0000	.0000	.0000	.0000
9	2	.0629	.1722	.2597	.3020	.3003	.2668	.2162	.1612	.1110	.0703	.0407	.0212	.0098	.0039	.0012	.0003	.0000	.0000	.0000
9	3	.0077	.0446	.1069	.1762	.2336	.2668	.2716	.2508	.2119	.1641	.1160	.0743	.0424	.0210	.0087	.0028	.0006	.0001	.0000
9	4	.0006	.0074	.0283	.0661	.1168	.1715	.2194	.2508	.2600	.2461	.2128	.1672	.1181	.0735	.0389	.0165	.0050	.0008	.0000
9	5	.0000	.0008	.0050	.0165	.0389	.0735	.1181	.1672	.2128	.2461	.2600	.2508	.2194	.1715	.1168	.0661	.0283	.0074	.0006
9	6	.0000	.0001	.0006	.0028	.0087	.0210	.0424	.0743	.1160	.1641	.2119	.2508	.2716	.2668	.2336	.1762	.1069	.0446	.0077
9	7	.0000	.0000	.0000	.0003	.0012	.0039	.0098	.0212	.0407	.0703	.1110	.1612	.2162	.2668	.3003	.3020	.2597	.1722	.0629
9	8	.0000	.0000	.0000	.0000	.0001	.0004	.0013	.0035	.0083	.0176	.0339	.0605	.1004	.1556	.2253	.3020	.3679	.3874	.2985
9	9	.0000	.0000	.0000	.0000	.0000	.0000	.0001	.0003	.0008	.0020	.0046	.0101	.0207	.0404	.0751	.1342	.2316	.3874	.6302
10	0	.5987	.3487	.1969	.1074	.0563	.0282	.0135	.0060	.0025	.0010	.0003	.0001	.0000	.0000	.0000	.0000	.0000	.0000	.0000
10	1	.3151	.3874	.3474	.2684	.1877	.1211	.0725	.0403	.0207	.0098	.0042	.0016	.0005	.0001	.0000	.0000	.0000	.0000	.0000
10	2	.0746	.1937	.2759	.3020	.2816	.2335	.1757	.1209	.0763	.0439	.0229	.0106	.0043	.0014	.0004	.0001	.0000	.0000	.0000
10	3	.0105	.0574	.1298	.2013	.2503	.2668	.2522	.2150	.1665	.1172	.0746	.0425	.0212	.0090	.0031	.0008	.0001	.0000	.0000
10	4	.0010	.0112	.0401	.0881	.1460	.2001	.2377	.2508	.2384	.2051	.1596	.1115	.0689	.0368	.0162	.0055	.0012	.0001	.0000
10	5	.0001	.0015	.0085	.0264	.0584	.1029	.1536	.2007	.2340	.2461	.2340	.2007	.1536	.1029	.0584	.0264	.0085	.0015	.0001
10	6	.0000	.0001	.0012	.0055	.0162	.0368	.0689	.1115	.1596	.2051	.2384	.2508	.2377	.2001	.1460	.0881	.0401	.0112	.0010
10	7	.0000	.0000	.0001	.0008	.0031	.0090	.0212	.0425	.0746	.1172	.1665	.2150	.2522	.2668	.2503	.2013	.1298	.0574	.0105
10	8	.0000	.0000	.0000	.0001	.0004	.0014	.0043	.0106	.0229	.0439	.0763	.1209	.1757	.2335	.2816	.3020	.2759	.1937	.0746
10	9	.0000	.0000	.0000	.0000	.0000	.0001	.0005	.0016	.0042	.0098	.0207	.0403	.0725	.1211	.1877	.2684	.3474	.3874	.3151
10	10	.0000	.0000	.0000	.0000	.0000	.0000	.0000	.0001	.0003	.0010	.0025	.0060	.0135	.0282	.0563	.1074	.1969	.3487	.5987

APPENDIX C. (cont'd)

APPENDIX C. (cont'd)

n = 14

x	.05	.10	.15	.20	.25	.30	.35	.40	.45	.50	.55	.60	.65	.70	.75	.80	.85	.90	.95
0	.4877	.2288	.1028	.0440	.0178	.0068	.0024	.0008	.0002	.0001	.0000	.0000	.0000	.0000	.0000	.0000	.0000	.0000	.0000
1	.3593	.3559	.2539	.1539	.0832	.0407	.0181	.0073	.0027	.0009	.0002	.0001	.0000	.0000	.0000	.0000	.0000	.0000	.0000
2	.1229	.2570	.2912	.2501	.1802	.1134	.0634	.0317	.0141	.0056	.0019	.0005	.0001	.0000	.0000	.0000	.0000	.0000	.0000
3	.0259	.1142	.2056	.2501	.2402	.1943	.1366	.0845	.0462	.0222	.0093	.0033	.0010	.0002	.0000	.0000	.0000	.0000	.0000
4	.0037	.0349	.0998	.1720	.2202	.2290	.2022	.1549	.1040	.0611	.0312	.0136	.0049	.0014	.0003	.0000	.0000	.0000	.0000
5	.0004	.0078	.0352	.0860	.1468	.1963	.2178	.2066	.1701	.1222	.0762	.0408	.0183	.0066	.0014	.0002	.0000	.0000	.0000
6	.0000	.0013	.0093	.0322	.0734	.1262	.1759	.2066	.2088	.1833	.1398	.0918	.0510	.0232	.0082	.0020	.0003	.0000	.0000
7	.0000	.0002	.0019	.0092	.0280	.0618	.1082	.1574	.1952	.2095	.1952	.1574	.1082	.0618	.0280	.0092	.0019	.0002	.0000
8	.0000	.0000	.0003	.0020	.0082	.0232	.0510	.0918	.1398	.1833	.2088	.2066	.1759	.1262	.0734	.0322	.0093	.0013	.0000
9	.0000	.0000	.0000	.0003	.0018	.0066	.0183	.0408	.0762	.1222	.1701	.2066	.2178	.1963	.1468	.0860	.0352	.0078	.0004
10	.0000	.0000	.0000	.0000	.0003	.0014	.0049	.0136	.0312	.0611	.1040	.1549	.2022	.2290	.2202	.1720	.0998	.0349	.0037
11	.0000	.0000	.0000	.0000	.0000	.0002	.0010	.0033	.0093	.0222	.0462	.0845	.1366	.1943	.2402	.2501	.2056	.1142	.0259
12	.0000	.0000	.0000	.0000	.0000	.0000	.0001	.0005	.0019	.0056	.0141	.0317	.0634	.1134	.1802	.2501	.2912	.2570	.1229
13	.0000	.0000	.0000	.0000	.0000	.0000	.0000	.0001	.0002	.0009	.0027	.0073	.0181	.0407	.0832	.1539	.2539	.3559	.3593
14	.0000	.0000	.0000	.0000	.0000	.0000	.0000	.0000	.0000	.0001	.0002	.0008	.0024	.0068	.0178	.0440	.1028	.2288	.4877

n = 15

x	.05	.10	.15	.20	.25	.30	.35	.40	.45	.50	.55	.60	.65	.70	.75	.80	.85	.90	.95
0	.4633	.2059	.0874	.0352	.0134	.0047	.0016	.0005	.0001	.0000	.0000	.0000	.0000	.0000	.0000	.0000	.0000	.0000	.0000
1	.3658	.3432	.2312	.1319	.0668	.0305	.0126	.0047	.0016	.0005	.0001	.0000	.0000	.0000	.0000	.0000	.0000	.0000	.0000
2	.1348	.2669	.2856	.2309	.1559	.0916	.0476	.0219	.0090	.0032	.0010	.0003	.0001	.0000	.0000	.0000	.0000	.0000	.0000
3	.0307	.1285	.2184	.2501	.2252	.1700	.1110	.0634	.0318	.0139	.0052	.0016	.0004	.0001	.0000	.0000	.0000	.0000	.0000
4	.0049	.0428	.1156	.1876	.2252	.2186	.1792	.1268	.0780	.0417	.0191	.0074	.0024	.0006	.0001	.0000	.0000	.0000	.0000
5	.0006	.0105	.0449	.1032	.1651	.2061	.2123	.1859	.1404	.0916	.0515	.0245	.0096	.0030	.0007	.0001	.0000	.0000	.0000
6	.0000	.0019	.0132	.0430	.0917	.1472	.1906	.2066	.1914	.1527	.1048	.0612	.0298	.0116	.0034	.0007	.0001	.0000	.0000
7	.0000	.0003	.0030	.0138	.0393	.0811	.1319	.1771	.2013	.1964	.1647	.1181	.0710	.0348	.0131	.0035	.0005	.0000	.0000
8	.0000	.0000	.0005	.0035	.0131	.0348	.0710	.1181	.1647	.1964	.2013	.1771	.1319	.0811	.0393	.0138	.0030	.0003	.0000
9	.0000	.0000	.0001	.0007	.0034	.0116	.0298	.0612	.1048	.1527	.1914	.2066	.1906	.1472	.0917	.0430	.0132	.0019	.0000
10	.0000	.0000	.0000	.0001	.0007	.0030	.0096	.0245	.0515	.0916	.1404	.1859	.2123	.2061	.1651	.1032	.0449	.0105	.0006
11	.0000	.0000	.0000	.0000	.0001	.0006	.0024	.0074	.0191	.0417	.0780	.1268	.1792	.2186	.2252	.1876	.1156	.0428	.0049
12	.0000	.0000	.0000	.0000	.0000	.0001	.0004	.0016	.0052	.0139	.0318	.0634	.1110	.1700	.2252	.2501	.2184	.1285	.0307
13	.0000	.0000	.0000	.0000	.0000	.0000	.0001	.0003	.0010	.0032	.0090	.0219	.0476	.0916	.1559	.2309	.2856	.2669	.1348
14	.0000	.0000	.0000	.0000	.0000	.0000	.0000	.0000	.0001	.0005	.0016	.0047	.0126	.0305	.0668	.1319	.2312	.3432	.3658
15	.0000	.0000	.0000	.0000	.0000	.0000	.0000	.0000	.0000	.0000	.0001	.0005	.0016	.0047	.0134	.0352	.0874	.2059	.4633

n = 16

x	.05	.10	.15	.20	.25	.30	.35	.40	.45	.50	.55	.60	.65	.70	.75	.80	.85	.90	.95
0	.4401	.1853	.0743	.0281	.0100	.0033	.0010	.0003	.0001	.0000	.0000	.0000	.0000	.0000	.0000	.0000	.0000	.0000	.0000
1	.3706	.3294	.2097	.1126	.0535	.0228	.0087	.0030	.0009	.0002	.0001	.0000	.0000	.0000	.0000	.0000	.0000	.0000	.0000
2	.1463	.2745	.2775	.2111	.1336	.0732	.0353	.0150	.0056	.0018	.0005	.0001	.0000	.0000	.0000	.0000	.0000	.0000	.0000
3	.0359	.1423	.2285	.2463	.2079	.1465	.0888	.0468	.0215	.0085	.0029	.0008	.0002	.0000	.0000	.0000	.0000	.0000	.0000
4	.0061	.0514	.1311	.2001	.2252	.2040	.1553	.1014	.0572	.0278	.0115	.0040	.0011	.0002	.0000	.0000	.0000	.0000	.0000

APPENDIX C. (cont'd)

	5	.0008	.0137	.0555	.1201	.1802	.2099	.2008	.1623	.1123	.0667	.0337	.0142	.0049	.0013	.0002	.0000	.0000
	6	.0001	.0028	.0180	.0550	.1101	.1649	.1982	.1983	.1684	.1222	.0755	.0392	.0167	.0056	.0014	.0000	.0000
	7	.0000	.0004	.0045	.0197	.0524	.1010	.1524	.1889	.1969	.1746	.1318	.0840	.0442	.0185	.0058	.0012	.0001
	8	.0000	.0000	.0009	.0055	.0197	.0487	.0923	.1417	.1812	.1964	.1812	.1417	.0923	.0487	.0197	.0055	.0009
	9	.0000	.0000	.0001	.0012	.0058	.0185	.0442	.0840	.1318	.1746	.1969	.1889	.1524	.1010	.0524	.0197	.0045
	10	.0000	.0000	.0000	.0002	.0014	.0056	.0167	.0392	.0755	.1222	.1684	.1982	.2008	.1649	.1101	.0550	.0180
	11	.0000	.0000	.0000	.0000	.0002	.0013	.0049	.0142	.0337	.0666	.1123	.1623	.1983	.2099	.1802	.1201	.0555
	12	.0000	.0000	.0000	.0000	.0000	.0002	.0011	.0040	.0115	.0278	.0572	.1014	.1553	.2040	.2252	.2001	.1311
	13	.0000	.0000	.0000	.0000	.0000	.0000	.0002	.0008	.0029	.0085	.0215	.0468	.0888	.1465	.2079	.2463	.2285
	14	.0000	.0000	.0000	.0000	.0000	.0000	.0000	.0001	.0005	.0018	.0056	.0150	.0353	.0732	.1336	.2111	.2775
	15	.0000	.0000	.0000	.0000	.0000	.0000	.0000	.0000	.0001	.0002	.0009	.0030	.0087	.0228	.0535	.1126	.2096
	16	.0000	.0000	.0000	.0000	.0000	.0000	.0000	.0000	.0000	.0000	.0001	.0003	.0010	.0033	.0100	.0281	.0742
17	0	.4181	.1668	.0631	.0225	.0075	.0023	.0007	.0002	.0000	.0000	.0000	.0000	.0000	.0000	.0000	.0000	.0000
	1	.3741	.3150	.1893	.0957	.0426	.0169	.0060	.0019	.0005	.0001	.0000	.0000	.0000	.0000	.0000	.0000	.0000
	2	.1575	.2800	.2673	.1914	.1136	.0581	.0260	.0102	.0035	.0010	.0003	.0000	.0000	.0000	.0000	.0000	.0000
	3	.0415	.1556	.2359	.2393	.1893	.1245	.0701	.0341	.0144	.0052	.0016	.0004	.0001	.0000	.0000	.0000	.0000
	4	.0076	.0605	.1457	.2093	.2209	.1868	.1320	.0796	.0411	.0182	.0068	.0021	.0005	.0001	.0000	.0000	.0000
	5	.0010	.0175	.0668	.1361	.1914	.2081	.1849	.1379	.0875	.0472	.0215	.0081	.0024	.0006	.0001	.0000	.0000
	6	.0001	.0039	.0236	.0680	.1276	.1784	.1991	.1839	.1432	.0944	.0525	.0242	.0090	.0026	.0005	.0001	.0000
	7	.0000	.0007	.0065	.0267	.0668	.1201	.1685	.1927	.1841	.1484	.1008	.0571	.0263	.0095	.0025	.0004	.0000
	8	.0000	.0001	.0014	.0084	.0279	.0644	.1134	.1606	.1883	.1855	.1540	.1070	.0611	.0276	.0093	.0021	.0003
	9	.0000	.0000	.0003	.0021	.0093	.0276	.0611	.1070	.1540	.1883	.1855	.1606	.1134	.0644	.0279	.0084	.0014
	10	.0000	.0000	.0000	.0004	.0025	.0095	.0263	.0571	.1008	.1484	.1841	.1927	.1685	.1201	.0668	.0267	.0065
	11	.0000	.0000	.0000	.0001	.0005	.0026	.0090	.0242	.0525	.0944	.1432	.1839	.1991	.1784	.1276	.0680	.0236
	12	.0000	.0000	.0000	.0000	.0001	.0006	.0024	.0081	.0215	.0472	.0875	.1379	.1849	.2081	.1914	.1361	.0668
	13	.0000	.0000	.0000	.0000	.0000	.0001	.0005	.0021	.0068	.0182	.0411	.0796	.1320	.1868	.2209	.2093	.1457
	14	.0000	.0000	.0000	.0000	.0000	.0000	.0001	.0004	.0016	.0052	.0144	.0341	.0701	.1245	.1893	.2359	.2393
	15	.0000	.0000	.0000	.0000	.0000	.0000	.0000	.0000	.0003	.0010	.0035	.0102	.0260	.0581	.1136	.1914	.2673
	16	.0000	.0000	.0000	.0000	.0000	.0000	.0000	.0000	.0000	.0001	.0005	.0019	.0060	.0169	.0426	.0957	.1893
	17	.0000	.0000	.0000	.0000	.0000	.0000	.0000	.0000	.0000	.0000	.0000	.0002	.0007	.0023	.0075	.0225	.0631
18	0	.3972	.1501	.0536	.0180	.0056	.0016	.0004	.0001	.0000	.0000	.0000	.0000	.0000	.0000	.0000	.0000	.0000
	1	.3763	.3002	.1704	.0811	.0338	.0126	.0042	.0012	.0003	.0001	.0000	.0000	.0000	.0000	.0000	.0000	.0000
	2	.1683	.2835	.2556	.1723	.0958	.0458	.0190	.0069	.0022	.0006	.0001	.0000	.0000	.0000	.0000	.0000	.0000
	3	.0473	.1680	.2406	.2297	.1704	.1046	.0547	.0246	.0095	.0031	.0009	.0002	.0000	.0000	.0000	.0000	.0000
	4	.0093	.0700	.1592	.2153	.2130	.1681	.1104	.0614	.0291	.0117	.0039	.0011	.0002	.0000	.0000	.0000	.0000

APPENDIX C. (cont'd)

n	x	.05	.10	.15	.20	.25	.30	.35	.40	.45	.50	.55	.60	.65	.70	.75	.80	.85	.90	.95
	5	.0014	.0218	.0787	.1507	.1988	.2017	.1664	.1146	.0666	.0327	.0134	.0045	.0012	.0002	.0000	.0000	.0000	.0000	.0000
	6	.0002	.0052	.0301	.0816	.1436	.1873	.1941	.1655	.1181	.0708	.0354	.0145	.0047	.0012	.0002	.0000	.0000	.0000	.0000
	7	.0000	.0010	.0091	.0350	.0820	.1376	.1792	.1892	.1657	.1214	.0742	.0374	.0151	.0047	.0010	.0001	.0000	.0000	.0000
	8	.0000	.0002	.0022	.0120	.0376	.0811	.1327	.1734	.1864	.1669	.1248	.0771	.0385	.0149	.0042	.0008	.0001	.0000	.0000
	9	.0000	.0000	.0004	.0033	.0139	.0386	.0794	.1284	.1694	.1855	.1694	.1284	.0794	.0386	.0139	.0033	.0004	.0000	.0000
	10	.0000	.0000	.0000	.0008	.0042	.0149	.0385	.0771	.1248	.1669	.1864	.1734	.1327	.0811	.0376	.0120	.0022	.0002	.0000
	11	.0000	.0000	.0000	.0001	.0010	.0046	.0151	.0374	.0742	.1214	.1657	.1892	.1792	.1376	.0820	.0350	.0091	.0010	.0000
	12	.0000	.0000	.0000	.0000	.0002	.0012	.0047	.0145	.0354	.0708	.1181	.1655	.1941	.1873	.1436	.0816	.0301	.0052	.0002
	13	.0000	.0000	.0000	.0000	.0000	.0002	.0012	.0045	.0134	.0327	.0666	.1146	.1664	.2017	.1988	.1507	.0787	.0218	.0014
	14	.0000	.0000	.0000	.0000	.0000	.0000	.0002	.0011	.0039	.0117	.0291	.0614	.1104	.1681	.2130	.2153	.1592	.0700	.0093
	15	.0000	.0000	.0000	.0000	.0000	.0000	.0000	.0002	.0009	.0031	.0095	.0246	.0547	.1046	.1704	.2297	.2406	.1680	.0473
	16	.0000	.0000	.0000	.0000	.0000	.0000	.0000	.0000	.0001	.0006	.0022	.0069	.0190	.0458	.0958	.1723	.2556	.2835	.1684
	17	.0000	.0000	.0000	.0000	.0000	.0000	.0000	.0000	.0000	.0001	.0003	.0012	.0042	.0126	.0338	.0811	.1704	.3002	.3763
	18	.0000	.0000	.0000	.0000	.0000	.0000	.0000	.0000	.0000	.0000	.0000	.0001	.0004	.0016	.0056	.0180	.0536	.1501	.3972
19	0	.3774	.1351	.0456	.0144	.0042	.0011	.0003	.0001	.0000	.0000	.0000	.0000	.0000	.0000	.0000	.0000	.0000	.0000	.0000
	1	.3774	.2852	.1529	.0685	.0268	.0093	.0029	.0008	.0002	.0000	.0000	.0000	.0000	.0000	.0000	.0000	.0000	.0000	.0000
	2	.1787	.2852	.2428	.1540	.0803	.0358	.0138	.0046	.0013	.0003	.0001	.0000	.0000	.0000	.0000	.0000	.0000	.0000	.0000
	3	.0533	.1796	.2428	.2182	.1517	.0869	.0422	.0175	.0062	.0018	.0005	.0001	.0000	.0000	.0000	.0000	.0000	.0000	.0000
	4	.0112	.0798	.1714	.2182	.2023	.1491	.0909	.0467	.0203	.0074	.0022	.0005	.0001	.0000	.0000	.0000	.0000	.0000	.0000
	5	.0018	.0266	.0907	.1636	.2023	.1916	.1468	.0933	.0497	.0222	.0082	.0024	.0006	.0001	.0000	.0000	.0000	.0000	.0000
	6	.0002	.0069	.0374	.0955	.1574	.1916	.1844	.1451	.0949	.0518	.0233	.0085	.0024	.0005	.0001	.0000	.0000	.0000	.0000
	7	.0000	.0014	.0122	.0443	.0974	.1525	.1844	.1797	.1443	.0961	.0529	.0237	.0083	.0022	.0004	.0000	.0000	.0000	.0000
	8	.0000	.0002	.0032	.0166	.0487	.0981	.1489	.1797	.1771	.1442	.0970	.0532	.0233	.0077	.0018	.0003	.0000	.0000	.0000
	9	.0000	.0000	.0007	.0051	.0198	.0514	.0980	.1464	.1771	.1762	.1449	.0976	.0528	.0220	.0066	.0013	.0001	.0000	.0000
	10	.0000	.0000	.0001	.0013	.0066	.0220	.0528	.0976	.1449	.1762	.1771	.1464	.0980	.0514	.0198	.0051	.0007	.0000	.0000
	11	.0000	.0000	.0000	.0003	.0018	.0077	.0233	.0532	.0970	.1442	.1771	.1797	.1489	.0981	.0487	.0166	.0032	.0002	.0000
	12	.0000	.0000	.0000	.0000	.0004	.0022	.0083	.0237	.0529	.0961	.1443	.1797	.1844	.1525	.0974	.0443	.0122	.0014	.0000
	13	.0000	.0000	.0000	.0000	.0001	.0005	.0024	.0085	.0233	.0518	.0949	.1451	.1844	.1916	.1574	.0955	.0374	.0069	.0002
	14	.0000	.0000	.0000	.0000	.0000	.0001	.0006	.0024	.0082	.0222	.0497	.0933	.1468	.1916	.2023	.1636	.0907	.0266	.0018
	15	.0000	.0000	.0000	.0000	.0000	.0000	.0001	.0005	.0022	.0074	.0203	.0467	.0909	.1491	.2023	.2182	.1714	.0798	.0112
	16	.0000	.0000	.0000	.0000	.0000	.0000	.0000	.0001	.0005	.0018	.0062	.0175	.0422	.0869	.1517	.2182	.2428	.1796	.0533
	17	.0000	.0000	.0000	.0000	.0000	.0000	.0000	.0000	.0001	.0003	.0013	.0046	.0138	.0358	.0803	.1540	.2428	.2852	.1787
	18	.0000	.0000	.0000	.0000	.0000	.0000	.0000	.0000	.0000	.0000	.0002	.0008	.0029	.0093	.0268	.0685	.1529	.2852	.3774
	19	.0000	.0000	.0000	.0000	.0000	.0000	.0000	.0000	.0000	.0000	.0000	.0001	.0003	.0011	.0042	.0144	.0456	.1351	.3774
20	0	.3585	.1216	.0388	.0115	.0032	.0008	.0002	.0000	.0000	.0000	.0000	.0000	.0000	.0000	.0000	.0000	.0000	.0000	.0000
	1	.3774	.2702	.1368	.0576	.0211	.0068	.0020	.0005	.0001	.0000	.0000	.0000	.0000	.0000	.0000	.0000	.0000	.0000	.0000
	2	.1887	.2852	.2293	.1369	.0669	.0278	.0100	.0031	.0008	.0002	.0000	.0000	.0000	.0000	.0000	.0000	.0000	.0000	.0000
	3	.0596	.1901	.2428	.2054	.1339	.0716	.0323	.0123	.0040	.0011	.0002	.0000	.0000	.0000	.0000	.0000	.0000	.0000	.0000
	4	.0133	.0898	.1821	.2182	.1897	.1304	.0738	.0350	.0139	.0046	.0013	.0003	.0000	.0000	.0000	.0000	.0000	.0000	.0000

APPENDIX C. (cont'd)

5	.0022	.0319	.1028	.1746	.2023	.1789	.1272	.0746	.0365	.0148	.0049	.0013	.0003	.0000	.0000	.0000	.0000	.0000	.0000	.0000	.0000
6	.0003	.0089	.0454	.1091	.1686	.1916	.1712	.1244	.0746	.0370	.0150	.0049	.0012	.0002	.0000	.0000	.0000	.0000	.0000	.0000	.0000
7	.0000	.0020	.0160	.0545	.1124	.1643	.1844	.1659	.1221	.0739	.0366	.0146	.0045	.0010	.0002	.0000	.0000	.0000	.0000	.0000	.0000
8	.0000	.0004	.0046	.0222	.0609	.1144	.1614	.1797	.1623	.1201	.0727	.0355	.0136	.0039	.0008	.0001	.0000	.0000	.0000	.0000	.0000
9	.0000	.0001	.0011	.0074	.0271	.0654	.1158	.1597	.1771	.1602	.1185	.0710	.0336	.0120	.0030	.0005	.0001	.0000	.0000	.0000	.0000
10	.0000	.0000	.0002	.0020	.0099	.0308	.0686	.1171	.1593	.1762	.1593	.1171	.0686	.0308	.0099	.0020	.0002	.0000	.0000	.0000	.0000
11	.0000	.0000	.0000	.0005	.0030	.0120	.0336	.0710	.1185	.1602	.1771	.1597	.1158	.0654	.0271	.0074	.0011	.0001	.0000	.0000	.0000
12	.0000	.0000	.0000	.0001	.0008	.0039	.0136	.0355	.0727	.1201	.1623	.1797	.1614	.1144	.0609	.0222	.0046	.0004	.0000	.0000	.0000
13	.0000	.0000	.0000	.0000	.0002	.0010	.0045	.0146	.0366	.0739	.1221	.1659	.1844	.1643	.1124	.0545	.0160	.0020	.0000	.0000	.0000
14	.0000	.0000	.0000	.0000	.0000	.0002	.0012	.0049	.0150	.0370	.0746	.1244	.1712	.1916	.1686	.1091	.0454	.0089	.0003	.0000	.0000
15	.0000	.0000	.0000	.0000	.0000	.0000	.0003	.0013	.0049	.0148	.0365	.0746	.1272	.1789	.2023	.1746	.1028	.0319	.0022	.0000	.0000
16	.0000	.0000	.0000	.0000	.0000	.0000	.0000	.0003	.0013	.0046	.0139	.0350	.0738	.1304	.1897	.2182	.1821	.0898	.0133	.0000	.0000
17	.0000	.0000	.0000	.0000	.0000	.0000	.0000	.0000	.0002	.0011	.0040	.0123	.0323	.0716	.1339	.2054	.2428	.1901	.0596	.0000	.0000
18	.0000	.0000	.0000	.0000	.0000	.0000	.0000	.0000	.0000	.0002	.0008	.0031	.0100	.0278	.0669	.1369	.2293	.2852	.1887	.0320	.0000
19	.0000	.0000	.0000	.0000	.0000	.0000	.0000	.0000	.0000	.0000	.0001	.0005	.0020	.0068	.0211	.0576	.1368	.2702	.3774	.1887	.0003
20	.0000	.0000	.0000	.0000	.0000	.0000	.0000	.0000	.0000	.0000	.0000	.0000	.0002	.0008	.0032	.0115	.0388	.1216	.3585		

21

0	.3406	.1094	.0329	.0092	.0024	.0006	.0001	.0000	.0000	.0000	.0000	.0000	.0000	.0000	.0000	.0000	.0000	.0000	.0000	.0000	.0000
1	.3764	.2553	.1221	.0484	.0166	.0050	.0013	.0003	.0000	.0000	.0000	.0000	.0000	.0000	.0000	.0000	.0000	.0000	.0000	.0000	.0000
2	.1981	.2837	.2155	.1211	.0555	.0215	.0072	.0020	.0005	.0001	.0000	.0000	.0000	.0000	.0000	.0000	.0000	.0000	.0000	.0000	.0000
3	.0660	.1996	.2408	.1917	.1172	.0585	.0245	.0086	.0026	.0006	.0001	.0000	.0000	.0000	.0000	.0000	.0000	.0000	.0000	.0000	.0000
4	.0156	.0998	.1912	.2156	.1757	.1128	.0593	.0259	.0095	.0029	.0007	.0001	.0000	.0000	.0000	.0000	.0000	.0000	.0000	.0000	.0000
5	.0028	.0377	.1147	.1833	.1992	.1643	.1085	.0588	.0263	.0097	.0029	.0007	.0001	.0000	.0000	.0000	.0000	.0000	.0000	.0000	.0000
6	.0004	.0112	.0540	.1222	.1770	.1878	.1558	.1045	.0574	.0259	.0094	.0027	.0006	.0001	.0000	.0000	.0000	.0000	.0000	.0000	.0000
7	.0000	.0027	.0204	.0655	.1265	.1725	.1798	.1493	.1007	.0554	.0247	.0087	.0024	.0005	.0001	.0000	.0000	.0000	.0000	.0000	.0000
8	.0000	.0005	.0063	.0286	.0738	.1294	.1694	.1742	.1442	.0970	.0529	.0229	.0077	.0019	.0003	.0000	.0000	.0000	.0000	.0000	.0000
9	.0000	.0001	.0016	.0103	.0355	.0801	.1318	.1677	.1704	.1402	.0933	.0497	.0206	.0063	.0013	.0002	.0000	.0000	.0000	.0000	.0000
10	.0000	.0000	.0003	.0031	.0142	.0412	.0851	.1342	.1673	.1682	.1369	.0895	.0458	.0176	.0047	.0008	.0001	.0000	.0000	.0000	.0000
11	.0000	.0000	.0001	.0008	.0047	.0176	.0458	.0895	.1369	.1682	.1673	.1342	.0851	.0412	.0142	.0031	.0003	.0000	.0000	.0000	.0000
12	.0000	.0000	.0000	.0002	.0013	.0063	.0206	.0497	.0933	.1402	.1704	.1677	.1318	.0801	.0355	.0103	.0016	.0001	.0000	.0000	.0000
13	.0000	.0000	.0000	.0000	.0003	.0019	.0077	.0229	.0529	.0970	.1442	.1742	.1694	.1294	.0738	.0286	.0063	.0005	.0000	.0000	.0000
14	.0000	.0000	.0000	.0000	.0001	.0005	.0024	.0087	.0247	.0554	.1007	.1493	.1798	.1725	.1265	.0655	.0204	.0027	.0000	.0000	.0000
15	.0000	.0000	.0000	.0000	.0000	.0001	.0006	.0027	.0094	.0259	.0574	.1045	.1558	.1878	.1770	.1222	.0540	.0112	.0004	.0000	.0000
16	.0000	.0000	.0000	.0000	.0000	.0000	.0001	.0007	.0029	.0097	.0263	.0588	.1085	.1643	.1992	.1833	.1147	.0377	.0028	.0000	.0000
17	.0000	.0000	.0000	.0000	.0000	.0000	.0000	.0001	.0007	.0029	.0095	.0259	.0593	.1128	.1757	.2156	.1912	.0998	.0156	.0000	.0000
18	.0000	.0000	.0000	.0000	.0000	.0000	.0000	.0000	.0001	.0006	.0026	.0086	.0245	.0585	.1172	.1917	.2408	.1996	.0660	.0000	.0000
19	.0000	.0000	.0000	.0000	.0000	.0000	.0000	.0000	.0000	.0001	.0005	.0020	.0072	.0215	.0555	.1211	.2155	.2837	.1981	.0000	.0000
20	.0000	.0000	.0000	.0000	.0000	.0000	.0000	.0000	.0000	.0000	.0001	.0003	.0013	.0050	.0166	.0484	.1221	.2553	.3764	.0000	.0000
21	.0000	.0000	.0000	.0000	.0000	.0000	.0000	.0000	.0000	.0000	.0000	.0000	.0001	.0006	.0024	.0092	.0329	.1094	.3406	.0000	.0000

APPENDIX C. (cont'd)

n = 22

r	.05	.10	.15	.20	.25	.30	.35	.40	.45	.50	.55	.60	.65	.70	.75	.80	.85	.90	.95
0	.3235	.0985	.0280	.0074	.0018	.0004	.0001	.0000	.0000	.0000	.0000	.0000	.0000	.0000	.0000	.0000	.0000	.0000	.0000
1	.3746	.2407	.1087	.0406	.0131	.0037	.0009	.0002	.0000	.0000	.0000	.0000	.0000	.0000	.0000	.0000	.0000	.0000	.0000
2	.2070	.2808	.2015	.1065	.0458	.0166	.0051	.0014	.0003	.0001	.0000	.0000	.0000	.0000	.0000	.0000	.0000	.0000	.0000
3	.0726	.2080	.2370	.1775	.1017	.0474	.0184	.0060	.0016	.0003	.0001	.0000	.0000	.0000	.0000	.0000	.0000	.0000	.0000
4	.0182	.1098	.1987	.2108	.1611	.0965	.0471	.0190	.0064	.0017	.0004	.0001	.0000	.0000	.0000	.0000	.0000	.0000	.0000
5	.0034	.0439	.1262	.1898	.1933	.1489	.0913	.0456	.0187	.0063	.0017	.0004	.0001	.0000	.0000	.0000	.0000	.0000	.0000
6	.0005	.0138	.0631	.1344	.1826	.1808	.1393	.0862	.0434	.0178	.0058	.0015	.0003	.0000	.0000	.0000	.0000	.0000	.0000
7	.0001	.0035	.0255	.0768	.1391	.1771	.1714	.1314	.0812	.0407	.0163	.0051	.0012	.0002	.0000	.0000	.0000	.0000	.0000
8	.0000	.0007	.0084	.0360	.0869	.1423	.1730	.1642	.1246	.0762	.0374	.0144	.0042	.0009	.0001	.0000	.0000	.0000	.0000
9	.0000	.0001	.0023	.0140	.0451	.0949	.1449	.1703	.1586	.1186	.0711	.0336	.0122	.0032	.0006	.0001	.0000	.0000	.0000
10	.0000	.0000	.0005	.0046	.0195	.0529	.1015	.1476	.1687	.1542	.1129	.0656	.0294	.0097	.0022	.0003	.0000	.0000	.0000
11	.0000	.0000	.0001	.0012	.0071	.0247	.0596	.1073	.1506	.1682	.1506	.1073	.0596	.0247	.0071	.0012	.0001	.0000	.0000
12	.0000	.0000	.0000	.0003	.0022	.0097	.0294	.0656	.1129	.1542	.1687	.1476	.1015	.0529	.0195	.0046	.0005	.0000	.0000
13	.0000	.0000	.0000	.0001	.0006	.0032	.0122	.0336	.0711	.1186	.1586	.1703	.1449	.0949	.0451	.0140	.0023	.0001	.0000
14	.0000	.0000	.0000	.0000	.0001	.0009	.0042	.0144	.0374	.0762	.1246	.1642	.1730	.1423	.0869	.0360	.0084	.0007	.0000
15	.0000	.0000	.0000	.0000	.0000	.0002	.0012	.0051	.0163	.0407	.0812	.1314	.1714	.1771	.1391	.0768	.0255	.0035	.0001
16	.0000	.0000	.0000	.0000	.0000	.0000	.0003	.0015	.0058	.0178	.0434	.0862	.1393	.1808	.1826	.1344	.0631	.0138	.0005
17	.0000	.0000	.0000	.0000	.0000	.0000	.0001	.0004	.0017	.0063	.0187	.0456	.0913	.1489	.1933	.1898	.1262	.0439	.0034
18	.0000	.0000	.0000	.0000	.0000	.0000	.0000	.0001	.0004	.0017	.0064	.0190	.0471	.0965	.1611	.2108	.1987	.1098	.0182
19	.0000	.0000	.0000	.0000	.0000	.0000	.0000	.0000	.0001	.0004	.0016	.0060	.0184	.0474	.1017	.1775	.2370	.2080	.0726
20	.0000	.0000	.0000	.0000	.0000	.0000	.0000	.0000	.0000	.0001	.0003	.0014	.0051	.0166	.0458	.1065	.2015	.2808	.2070
21	.0000	.0000	.0000	.0000	.0000	.0000	.0000	.0000	.0000	.0000	.0000	.0002	.0009	.0037	.0131	.0406	.1087	.2407	.3746
22	.0000	.0000	.0000	.0000	.0000	.0000	.0000	.0000	.0000	.0000	.0000	.0000	.0001	.0004	.0018	.0074	.0280	.0985	.3235

n = 23

r	.05	.10	.15	.20	.25	.30	.35	.40	.45	.50	.55	.60	.65	.70	.75	.80	.85	.90	.95
0	.3074	.0886	.0238	.0059	.0013	.0003	.0000	.0000	.0000	.0000	.0000	.0000	.0000	.0000	.0000	.0000	.0000	.0000	.0000
1	.3721	.2265	.0966	.0339	.0103	.0027	.0006	.0001	.0000	.0000	.0000	.0000	.0000	.0000	.0000	.0000	.0000	.0000	.0000
2	.2154	.2768	.1875	.0933	.0376	.0127	.0037	.0009	.0002	.0000	.0000	.0000	.0000	.0000	.0000	.0000	.0000	.0000	.0000
3	.0794	.2153	.2317	.1633	.0878	.0382	.0138	.0041	.0010	.0002	.0000	.0000	.0000	.0000	.0000	.0000	.0000	.0000	.0000
4	.0209	.1196	.2044	.2042	.1463	.0818	.0371	.0138	.0042	.0011	.0002	.0000	.0000	.0000	.0000	.0000	.0000	.0000	.0000
5	.0042	.0505	.1371	.1940	.1853	.1332	.0758	.0350	.0132	.0040	.0010	.0002	.0000	.0000	.0000	.0000	.0000	.0000	.0000
6	.0007	.0168	.0726	.1455	.1853	.1712	.1225	.0700	.0323	.0120	.0036	.0008	.0001	.0000	.0000	.0000	.0000	.0000	.0000
7	.0001	.0045	.0311	.0883	.1500	.1782	.1602	.1133	.0642	.0292	.0106	.0029	.0006	.0001	.0000	.0000	.0000	.0000	.0000
8	.0000	.0010	.0110	.0442	.1000	.1527	.1725	.1511	.1051	.0585	.0258	.0088	.0023	.0004	.0000	.0000	.0000	.0000	.0000
9	.0000	.0002	.0032	.0184	.0555	.1091	.1548	.1679	.1433	.0974	.0525	.0221	.0070	.0016	.0002	.0000	.0000	.0000	.0000
10	.0000	.0000	.0008	.0064	.0259	.0655	.1167	.1567	.1642	.1364	.0899	.0464	.0182	.0052	.0010	.0001	.0000	.0000	.0000
11	.0000	.0000	.0002	.0019	.0102	.0332	.0743	.1234	.1587	.1612	.1299	.0823	.0400	.0142	.0034	.0005	.0000	.0000	.0000
12	.0000	.0000	.0000	.0005	.0034	.0142	.0400	.0823	.1299	.1612	.1587	.1234	.0743	.0332	.0102	.0019	.0002	.0000	.0000
13	.0000	.0000	.0000	.0001	.0010	.0052	.0182	.0464	.0899	.1364	.1642	.1567	.1167	.0655	.0259	.0064	.0008	.0000	.0000
14	.0000	.0000	.0000	.0000	.0002	.0016	.0070	.0221	.0525	.0974	.1433	.1679	.1548	.1091	.0555	.0184	.0032	.0002	.0000

455

APPENDIX C. (cont'd)

	15	.0000	.0000	.0000	.0000	.0000	.0000	.0004	.0023	.0088	.0258	.0584	.1051	.1511	.1725	.1527	.1000	.0442	.0110	.0010	.0000				
	16	.0000	.0000	.0000	.0000	.0000	.0000	.0001	.0006	.0029	.0106	.0292	.0642	.1133	.1602	.1782	.1500	.0883	.0311	.0045	.0001				
	17	.0000	.0000	.0000	.0000	.0000	.0000	.0000	.0001	.0008	.0036	.0120	.0323	.0700	.1225	.1712	.1853	.1455	.0726	.0168	.0007				
	18	.0000	.0000	.0000	.0000	.0000	.0000	.0000	.0000	.0000	.0008	.0040	.0132	.0350	.0758	.1331	.1853	.1940	.1371	.0505	.0042				
	19	.0000	.0000	.0000	.0000	.0000	.0000	.0000	.0000	.0000	.0002	.0011	.0042	.0138	.0371	.0818	.1463	.2042	.2044	.1196	.0209				
	20	.0000	.0000	.0000	.0000	.0000	.0000	.0000	.0000	.0000	.0000	.0002	.0010	.0041	.0138	.0382	.0878	.1633	.2317	.2153	.0794				
	21	.0000	.0000	.0000	.0000	.0000	.0000	.0000	.0000	.0000	.0000	.0000	.0002	.0009	.0037	.0127	.0376	.0933	.1875	.2768	.2154				
	22	.0000	.0000	.0000	.0000	.0000	.0000	.0000	.0000	.0000	.0000	.0000	.0000	.0001	.0006	.0027	.0103	.0339	.0966	.2265	.3721				
	23	.0000	.0000	.0000	.0000	.0000	.0000	.0000	.0000	.0000	.0000	.0000	.0000	.0000	.0000	.0003	.0013	.0059	.0238	.0886	.3074				
24	0	.2920	.0798	.0202	.0047	.0010	.0002	.0000	.0000	.0000	.0000	.0000	.0000	.0000	.0000	.0000	.0000	.0000	.0000	.0000	.0000				
	1	.3688	.2127	.0857	.0283	.0080	.0020	.0004	.0001	.0000	.0000	.0000	.0000	.0000	.0000	.0000	.0000	.0000	.0000	.0000	.0000				
	2	.2232	.2718	.1739	.0815	.0308	.0097	.0026	.0006	.0001	.0000	.0000	.0000	.0000	.0000	.0000	.0000	.0000	.0000	.0000	.0000				
	3	.0862	.2215	.2251	.1493	.0752	.0305	.0102	.0028	.0006	.0001	.0000	.0000	.0000	.0000	.0000	.0000	.0000	.0000	.0000	.0000				
	4	.0238	.1292	.2085	.1960	.1316	.0687	.0289	.0099	.0028	.0007	.0001	.0000	.0000	.0000	.0000	.0000	.0000	.0000	.0000	.0000				
	5	.0050	.0574	.1472	.1960	.1755	.1177	.0622	.0265	.0091	.0025	.0006	.0001	.0000	.0000	.0000	.0000	.0000	.0000	.0000	.0000				
	6	.0008	.0202	.0822	.1552	.1853	.1598	.1061	.0560	.0237	.0080	.0021	.0004	.0001	.0000	.0000	.0000	.0000	.0000	.0000	.0000				
	7	.0001	.0058	.0373	.0998	.1588	.1761	.1470	.0960	.0499	.0206	.0067	.0017	.0003	.0000	.0000	.0000	.0000	.0000	.0000	.0000				
	8	.0000	.0014	.0140	.0530	.1125	.1604	.1682	.1360	.0867	.0438	.0174	.0053	.0012	.0002	.0000	.0000	.0000	.0000	.0000	.0000				
	9	.0000	.0003	.0044	.0236	.0667	.1222	.1610	.1612	.1261	.0779	.0378	.0141	.0039	.0008	.0001	.0000	.0000	.0000	.0000	.0000				
	10	.0000	.0000	.0012	.0088	.0333	.0785	.1300	.1612	.1548	.1169	.0694	.0318	.0109	.0026	.0004	.0000	.0000	.0000	.0000	.0000				
	11	.0000	.0000	.0003	.0028	.0141	.0428	.0891	.1367	.1612	.1488	.1079	.0608	.0258	.0079	.0016	.0002	.0000	.0000	.0000	.0000				
	12	.0000	.0000	.0000	.0008	.0051	.0199	.0520	.0988	.1429	.1612	.1429	.0988	.0520	.0199	.0051	.0008	.0000	.0000	.0000	.0000				
	13	.0000	.0000	.0000	.0002	.0016	.0079	.0258	.0608	.1079	.1488	.1612	.1367	.0891	.0428	.0141	.0028	.0003	.0000	.0000	.0000				
	14	.0000	.0000	.0000	.0000	.0004	.0026	.0109	.0318	.0694	.1169	.1548	.1612	.1300	.0785	.0333	.0088	.0012	.0000	.0000	.0000				
	15	.0000	.0000	.0000	.0000	.0001	.0008	.0039	.0141	.0378	.0779	.1261	.1612	.1610	.1222	.0667	.0236	.0044	.0003	.0000	.0000				
	16	.0000	.0000	.0000	.0000	.0000	.0002	.0012	.0053	.0174	.0438	.0867	.1360	.1682	.1604	.1125	.0530	.0140	.0014	.0000	.0000				
	17	.0000	.0000	.0000	.0000	.0000	.0000	.0003	.0016	.0067	.0206	.0499	.0960	.1470	.1761	.1588	.0998	.0373	.0058	.0001	.0000				
	18	.0000	.0000	.0000	.0000	.0000	.0000	.0001	.0004	.0021	.0080	.0237	.0560	.1061	.1598	.1853	.1588	.0815	.0202	.0008	.0000				
	19	.0000	.0000	.0000	.0000	.0000	.0000	.0000	.0001	.0006	.0025	.0091	.0265	.0622	.1177	.1755	.1960	.1493	.0574	.0050	.0000				
	20	.0000	.0000	.0000	.0000	.0000	.0000	.0000	.0000	.0001	.0006	.0028	.0099	.0289	.0687	.1316	.1960	.2085	.1292	.0238	.0001				
	21	.0000	.0000	.0000	.0000	.0000	.0000	.0000	.0000	.0000	.0001	.0007	.0028	.0102	.0305	.0752	.1493	.2251	.2215	.0862	.0014				
	22	.0000	.0000	.0000	.0000	.0000	.0000	.0000	.0000	.0000	.0000	.0001	.0006	.0026	.0097	.0308	.0815	.1739	.2718	.2232	.0232				
	23	.0000	.0000	.0000	.0000	.0000	.0000	.0000	.0000	.0000	.0000	.0000	.0001	.0004	.0020	.0080	.0283	.0857	.2127	.3688	.2232				
	24	.0000	.0000	.0000	.0000	.0000	.0000	.0000	.0000	.0000	.0000	.0000	.0000	.0000	.0002	.0010	.0047	.0202	.0798	.2920					
25	0	.2774	.0718	.0172	.0038	.0008	.0001	.0000	.0000	.0000	.0000	.0000	.0000	.0000	.0000	.0000	.0000	.0000	.0000	.0000	.0000				
	1	.3650	.1994	.0759	.0236	.0063	.0014	.0003	.0000	.0000	.0000	.0000	.0000	.0000	.0000	.0000	.0000	.0000	.0000	.0000	.0000				
	2	.2305	.2659	.1607	.0708	.0251	.0074	.0018	.0004	.0001	.0000	.0000	.0000	.0000	.0000	.0000	.0000	.0000	.0000	.0000	.0000				
	3	.0930	.2265	.2174	.1358	.0641	.0243	.0076	.0019	.0004	.0001	.0000	.0000	.0000	.0000	.0000	.0000	.0000	.0000	.0000	.0000				
	4	.0269	.1384	.2110	.1867	.1175	.0572	.0224	.0071	.0018	.0004	.0001	.0000	.0000	.0000	.0000	.0000	.0000	.0000	.0000	.0000				

APPENDIX C. (cont'd)

x	.05	.10	.15	.20	.25	.30	.35	.40	.45	.50	.55	.60	.65	.70	.75	.80	.85	.90	.95
5	.0060	.0646	.1564	.1960	.1645	.1030	.0506	.0199	.0063	.0016	.0003	.0000	.0000	.0000	.0000	.0000	.0000	.0000	.0000
6	.0010	.0239	.0920	.1633	.1828	.1472	.0908	.0442	.0172	.0053	.0013	.0002	.0000	.0000	.0000	.0000	.0000	.0000	.0000
7	.0001	.0072	.0441	.1108	.1654	.1712	.1327	.0800	.0381	.0143	.0042	.0009	.0001	.0000	.0000	.0000	.0000	.0000	.0000
8	.0000	.0018	.0175	.0623	.1241	.1651	.1607	.1200	.0701	.0322	.0115	.0031	.0006	.0001	.0000	.0000	.0000	.0000	.0000
9	.0000	.0004	.0058	.0294	.0781	.1336	.1635	.1511	.1084	.0609	.0266	.0088	.0021	.0004	.0000	.0000	.0000	.0000	.0000
10	.0000	.0001	.0016	.0118	.0417	.0916	.1409	.1612	.1419	.0974	.0520	.0212	.0064	.0013	.0002	.0000	.0000	.0000	.0000
11	.0000	.0000	.0004	.0040	.0189	.0536	.1034	.1465	.1583	.1328	.0867	.0434	.0161	.0042	.0007	.0001	.0000	.0000	.0000
12	.0000	.0000	.0001	.0012	.0074	.0268	.0650	.1140	.1511	.1550	.1236	.0760	.0350	.0115	.0025	.0003	.0000	.0000	.0000
13	.0000	.0000	.0000	.0003	.0025	.0115	.0350	.0760	.1236	.1550	.1511	.1140	.0650	.0268	.0074	.0012	.0001	.0000	.0000
14	.0000	.0000	.0000	.0001	.0007	.0042	.0161	.0434	.0867	.1328	.1583	.1465	.1034	.0536	.0189	.0040	.0004	.0000	.0000
15	.0000	.0000	.0000	.0000	.0002	.0013	.0064	.0212	.0520	.0974	.1419	.1612	.1409	.0916	.0417	.0118	.0016	.0001	.0000
16	.0000	.0000	.0000	.0000	.0000	.0004	.0021	.0088	.0266	.0609	.1084	.1511	.1635	.1336	.0781	.0294	.0058	.0004	.0000
17	.0000	.0000	.0000	.0000	.0000	.0001	.0006	.0031	.0115	.0322	.0701	.1200	.1607	.1651	.1241	.0623	.0175	.0018	.0000
18	.0000	.0000	.0000	.0000	.0000	.0000	.0001	.0009	.0042	.0143	.0381	.0800	.1327	.1712	.1654	.1108	.0441	.0072	.0001
19	.0000	.0000	.0000	.0000	.0000	.0000	.0000	.0002	.0013	.0053	.0172	.0442	.0908	.1472	.1828	.1633	.0920	.0239	.0010
20	.0000	.0000	.0000	.0000	.0000	.0000	.0000	.0000	.0003	.0016	.0063	.0199	.0506	.1030	.1645	.1960	.1564	.0646	.0060
21	.0000	.0000	.0000	.0000	.0000	.0000	.0000	.0000	.0000	.0004	.0018	.0071	.0224	.0572	.1175	.1867	.2110	.1384	.0269
22	.0000	.0000	.0000	.0000	.0000	.0000	.0000	.0000	.0000	.0001	.0004	.0019	.0076	.0243	.0641	.1358	.2174	.2265	.0930
23	.0000	.0000	.0000	.0000	.0000	.0000	.0000	.0000	.0000	.0000	.0001	.0004	.0018	.0074	.0251	.0708	.1607	.2659	.2305
24	.0000	.0000	.0000	.0000	.0000	.0000	.0000	.0000	.0000	.0000	.0000	.0001	.0003	.0014	.0063	.0236	.0759	.1994	.3650
25	.0000	.0000	.0000	.0000	.0000	.0000	.0000	.0000	.0000	.0000	.0000	.0000	.0000	.0001	.0008	.0038	.0172	.0718	.2774

APPENDIX D. Binomial Distribution — Cumulative Probabilities

n	r	.05	.10	.15	.20	.25	.30	.35	.40	.45	.50
1	0	1.0000	1.0000	1.0000	1.0000	1.0000	1.0000	1.0000	1.0000	1.0000	1.0000
	1	0.0500	0.1000	0.1500	0.2000	0.2500	0.3000	0.3500	0.4000	0.4500	0.5000
2	0	1.0000	1.0000	1.0000	1.0000	1.0000	1.0000	1.0000	1.0000	1.0000	1.0000
	1	0.0975	0.1900	0.2775	0.3600	0.4375	0.5100	0.5775	0.6400	0.6975	0.7500
	2	0.0025	0.0100	0.0225	0.0400	0.0625	0.0900	0.1225	0.1600	0.2025	0.2500
3	0	1.0000	1.0000	1.0000	1.0000	1.0000	1.0000	1.0000	1.0000	1.0000	1.0000
	1	0.1426	0.2710	0.3859	0.4880	0.5781	0.6570	0.7254	0.7840	0.8336	0.8750
	2	0.0072	0.0280	0.0607	0.1040	0.1562	0.2160	0.2817	0.3520	0.4252	0.5000
	3	0.0001	0.0010	0.0034	0.0080	0.0156	0.0270	0.0429	0.0640	0.0911	0.1250
4	0	1.0000	1.0000	1.0000	1.0000	1.0000	1.0000	1.0000	1.0000	1.0000	1.0000
	1	0.1855	0.3439	0.4780	0.5904	0.6836	0.7599	0.8215	0.8704	0.9085	0.9375
	2	0.0140	0.0523	0.1095	0.1808	0.2617	0.3483	0.4370	0.5248	0.6090	0.6875
	3	0.0005	0.0037	0.0120	0.0272	0.0508	0.0837	0.1265	0.1792	0.2415	0.3125
	4	0.0000	0.0001	0.0005	0.0016	0.0039	0.0081	0.0150	0.0256	0.0410	0.0625
5	0	1.0000	1.0000	1.0000	1.0000	1.0000	1.0000	1.0000	1.0000	1.0000	1.0000
	1	0.2262	0.4095	0.5563	0.6723	0.7627	0.8319	0.8840	0.9222	0.9497	0.9687
	2	0.0226	0.0815	0.1648	0.2627	0.3672	0.4718	0.5716	0.6630	0.7438	0.8125
	3	0.0012	0.0086	0.0266	0.0579	0.1035	0.1631	0.2352	0.3174	0.4069	0.5000
	4	0.0000	0.0005	0.0022	0.0067	0.0156	0.0308	0.0540	0.0870	0.1312	0.1875
	5	0.0000	0.0000	0.0001	0.0003	0.0010	0.0024	0.0053	0.0102	0.0185	0.0312
6	0	1.0000	1.0000	1.0000	1.0000	1.0000	1.0000	1.0000	1.0000	1.0000	1.0000
	1	0.2649	0.4686	0.6229	0.7379	0.8220	0.8824	0.9246	0.9533	0.9723	0.9844
	2	0.0328	0.1143	0.2235	0.3446	0.4661	0.5798	0.6809	0.7667	0.8364	0.8906
	3	0.0022	0.0158	0.0473	0.0989	0.1694	0.2557	0.3529	0.4557	0.5585	0.6562
	4	0.0001	0.0013	0.0059	0.0170	0.0376	0.0705	0.1174	0.1792	0.2553	0.3437
	5	0.0000	0.0001	0.0004	0.0016	0.0046	0.0109	0.0223	0.0410	0.0692	0.1094
	6	0.0000	0.0000	0.0000	0.0001	0.0002	0.0007	0.0018	0.0041	0.0083	0.0156

APPENDIX D. (cont'd)

				p				
.55	.60	.65	.70	.75	.80	.85	.90	.95
1.0000	1.0000	1.0000	1.0000	1.0000	1.0000	1.0000	1.0000	1.0000
0.5500	0.6000	0.6500	0.7000	0.7500	0.8000	0.8500	0.9000	0.9500
1.0000	1.0000	1.0000	1.0000	1.0000	1.0000	1.0000	1.0000	1.0000
0.7975	0.8400	0.8775	0.9100	0.9375	0.9600	0.9775	0.9900	0.9975
0.3025	0.3600	0.4225	0.4900	0.5625	0.6400	0.7225	0.8100	0.9025
1.0000	1.0000	1.0000	1.0000	1.0000	1.0000	1.0000	1.0000	1.0000
0.9089	0.9360	0.9571	0.9730	0.9844	0.9920	0.9966	0.9990	0.9999
0.5747	0.6480	0.7182	0.7840	0.8437	0.8960	0.9392	0.9720	0.9927
0.1664	0.2160	0.2746	0.3430	0.4219	0.5120	0.6141	0.7290	0.8574
1.0000	1.0000	1.0000	1.0000	1.0000	1.0000	1.0000	1.0000	1.0000
0.9590	0.9744	0.9850	0.9919	0.9961	0.9984	0.9995	0.9999	1.0000
0.7585	0.8208	0.8735	0.9163	0.9492	0.9728	0.9880	0.9963	0.9995
0.3910	0.4752	0.5630	0.6517	0.7383	0.8192	0.8905	0.9477	0.9860
0.0915	0.1296	0.1785	0.2401	0.3164	0.4096	0.5220	0.6561	0.8145
1.0000	1.0000	1.0000	1.0000	1.0000	1.0000	1.0000	1.0000	1.0000
0.9815	0.9898	0.9947	0.9976	0.9990	0.9997	0.9999	1.0000	1.0000
0.8688	0.9130	0.9460	0.9692	0.9844	0.9933	0.9978	0.9995	1.0000
0.5931	0.6826	0.7648	0.8369	0.8965	0.9421	0.9734	0.9914	0.9988
0.2562	0.3370	0.4284	0.5282	0.6328	0.7373	0.8352	0.9185	0.9774
0.0503	0.0778	0.1160	0.1681	0.2373	0.3277	0.4437	0.5905	0.7738
1.0000	1.0000	1.0000	1.0000	1.0000	1.0000	1.0000	1.0000	1.0000
0.9917	0.9959	0.9982	0.9993	0.9998	0.9999	1.0000	1.0000	1.0000
0.9308	0.9590	0.9777	0.9891	0.9954	0.9984	0.9996	0.9999	1.0000
0.7447	0.8208	0.8826	0.9295	0.9624	0.9830	0.9941	0.9987	0.9999
0.4415	0.5443	0.6471	0.7443	0.8306	0.9011	0.9527	0.9841	0.9978
0.1636	0.2333	0.3191	0.4202	0.5339	0.6554	0.7765	0.8857	0.9672
0.0277	0.0467	0.0754	0.1176	0.1780	0.2621	0.3771	0.5314	0.7351

APPENDIX D. (cont'd)

7	0	1.0000	1.0000	1.0000	1.0000	1.0000	1.0000	1.0000	1.0000	1.0000	1.0000
	1	0.3017	0.5217	0.6794	0.7903	0.8665	0.9176	0.9510	0.9720	0.9848	0.9922
	2	0.0444	0.1497	0.2834	0.4233	0.5551	0.6706	0.7662	0.8414	0.8976	0.9375
	3	0.0038	0.0257	0.0738	0.1480	0.2436	0.3529	0.4677	0.5801	0.6836	0.7734
	4	0.0002	0.0027	0.0121	0.0333	0.0706	0.1260	0.1998	0.2898	0.3917	0.5000
	5	0.0000	0.0002	0.0012	0.0047	0.0129	0.0288	0.0556	0.0963	0.1529	0.2266
	6	0.0000	0.0000	0.0001	0.0004	0.0013	0.0038	0.0090	0.0188	0.0357	0.0625
	7	0.0000	0.0000	0.0000	0.0000	0.0001	0.0002	0.0006	0.0016	0.0037	0.0078
8	0	1.0000	1.0000	1.0000	1.0000	1.0000	1.0000	1.0000	1.0000	1.0000	1.0000
	1	0.3366	0.5695	0.7275	0.8322	0.8999	0.9424	0.9681	0.9832	0.9916	0.9961
	2	0.0572	0.1869	0.3428	0.4967	0.6329	0.7447	0.8309	0.8936	0.9368	0.9648
	3	0.0058	0.0381	0.1052	0.2031	0.3215	0.4482	0.5722	0.6846	0.7799	0.8555
	4	0.0004	0.0050	0.0214	0.0563	0.1138	0.1941	0.2936	0.4059	0.5230	0.6367
	5	0.0000	0.0004	0.0029	0.0104	0.0273	0.0580	0.1061	0.1737	0.2604	0.3633
	6	0.0000	0.0000	0.0002	0.0012	0.0042	0.0113	0.0253	0.0498	0.0885	0.1445
	7	0.0000	0.0000	0.0000	0.0001	0.0004	0.0013	0.0036	0.0085	0.0181	0.0352
	8	0.0000	0.0000	0.0000	0.0000	0.0000	0.0001	0.0002	0.0007	0.0017	0.0039
9	0	1.0000	1.0000	1.0000	1.0000	1.0000	1.0000	1.0000	1.0000	1.0000	1.0000
	1	0.3698	0.6126	0.7684	0.8658	0.9249	0.9596	0.9793	0.9899	0.9954	0.9980
	2	0.0712	0.2252	0.4005	0.5638	0.6997	0.8040	0.8789	0.9295	0.9615	0.9805
	3	0.0084	0.0530	0.1409	0.2618	0.3993	0.5372	0.6627	0.7682	0.8505	0.9102
	4	0.0006	0.0083	0.0339	0.0856	0.1657	0.2703	0.3911	0.5174	0.6386	0.7461
	5	0.0000	0.0009	0.0056	0.0196	0.0489	0.0988	0.1717	0.2666	0.3786	0.5000
	6	0.0000	0.0001	0.0006	0.0031	0.0100	0.0253	0.0536	0.0994	0.1658	0.2539
	7	0.0000	0.0000	0.0000	0.0003	0.0013	0.0043	0.0112	0.0250	0.0498	0.0898
	8	0.0000	0.0000	0.0000	0.0000	0.0001	0.0004	0.0014	0.0038	0.0091	0.0195
	9	0.0000	0.0000	0.0000	0.0000	0.0000	0.0000	0.0001	0.0003	0.0008	0.0020
10	0	1.0000	1.0000	1.0000	1.0000	1.0000	1.0000	1.0000	1.0000	1.0000	1.0000
	1	0.4013	0.6513	0.8031	0.8926	0.9437	0.9718	0.9865	0.9940	0.9975	0.9990
	2	0.0861	0.2639	0.4557	0.6242	0.7560	0.8507	0.9140	0.9536	0.9767	0.9893
	3	0.0115	0.0702	0.1798	0.3222	0.4744	0.6172	0.7384	0.8327	0.9004	0.9453
	4	0.0010	0.0128	0.0500	0.1209	0.2241	0.3504	0.4862	0.6177	0.7340	0.8281
	5	0.0001	0.0016	0.0099	0.0328	0.0781	0.1503	0.2485	0.3669	0.4956	0.6230
	6	0.0000	0.0001	0.0014	0.0064	0.0197	0.0473	0.0949	0.1662	0.2616	0.3770
	7	0.0000	0.0000	0.0001	0.0009	0.0035	0.0106	0.0260	0.0548	0.1020	0.1719
	8	0.0000	0.0000	0.0000	0.0001	0.0004	0.0016	0.0048	0.0123	0.0274	0.0547
	9	0.0000	0.0000	0.0000	0.0000	0.0000	0.0001	0.0005	0.0017	0.0045	0.0107
	10	0.0000	0.0000	0.0000	0.0000	0.0000	0.0000	0.0000	0.0001	0.0003	0.0010
11	0	1.0000	1.0000	1.0000	1.0000	1.0000	1.0000	1.0000	1.0000	1.0000	1.0000
	1	0.4312	0.6862	0.8327	0.9141	0.9578	0.9802	0.9912	0.9964	0.9986	0.9995
	2	0.1019	0.3026	0.5078	0.6779	0.8029	0.8870	0.9394	0.9698	0.9861	0.9941
	3	0.0152	0.0896	0.2212	0.3826	0.5448	0.6873	0.7999	0.8811	0.9348	0.9673
	4	0.0016	0.0185	0.0694	0.1611	0.2867	0.4304	0.5744	0.7037	0.8089	0.8867
	5	0.0001	0.0028	0.0159	0.0504	0.1146	0.2103	0.3317	0.4672	0.6029	0.7256
	6	0.0000	0.0003	0.0027	0.0117	0.0343	0.0782	0.1487	0.2465	0.3669	0.5000
	7	0.0000	0.0000	0.0003	0.0020	0.0076	0.0216	0.0501	0.0994	0.1738	0.2744
	8	0.0000	0.0000	0.0000	0.0002	0.0012	0.0043	0.0122	0.0293	0.0610	0.1133
	9	0.0000	0.0000	0.0000	0.0000	0.0001	0.0006	0.0020	0.0059	0.0148	0.0327
	10	0.0000	0.0000	0.0000	0.0000	0.0000	0.0000	0.0002	0.0007	0.0022	0.0059
	11	0.0000	0.0000	0.0000	0.0000	0.0000	0.0000	0.0000	0.0000	0.0002	0.0005

APPENDIX D. (cont'd)

```
1.0000  1.0000  1.0000  1.0000  1.0000  1.0000  1.0000  1.0000  1.0000
0.9963  0.9984  0.9994  0.9998  0.9999  1.0000  1.0000  1.0000  1.0000
0.9643  0.9812  0.9910  0.9962  0.9987  0.9996  0.9999  1.0000  1.0000
0.8471  0.9037  0.9444  0.9712  0.9871  0.9953  0.9988  0.9998  1.0000
0.6083  0.7102  0.8002  0.8740  0.9294  0.9667  0.9879  0.9973  0.9998

0.3164  0.4199  0.5323  0.6471  0.7564  0.8520  0.9262  0.9743  0.9962
0.1024  0.1586  0.2338  0.3294  0.4449  0.5767  0.7166  0.8503  0.9556
0.0152  0.0280  0.0490  0.0824  0.1335  0.2097  0.3206  0.4783  0.6983


1.0000  1.0000  1.0000  1.0000  1.0000  1.0000  1.0000  1.0000  1.0000
0.9983  0.9993  0.9998  0.9999  1.0000  1.0000  1.0000  1.0000  1.0000
0.9819  0.9915  0.9964  0.9987  0.9996  0.9999  1.0000  1.0000  1.0000
0.9115  0.9502  0.9747  0.9887  0.9958  0.9988  0.9998  1.0000  1.0000
0.7396  0.8263  0.8939  0.9420  0.9727  0.9896  0.9971  0.9996  1.0000

0.4770  0.5941  0.7064  0.8059  0.8862  0.9437  0.9786  0.9950  0.9996
0.2201  0.3154  0.4278  0.5518  0.6785  0.7969  0.8948  0.9619  0.9942
0.0632  0.1064  0.1691  0.2553  0.3671  0.5033  0.6572  0.8131  0.9428
0.0084  0.0168  0.0319  0.0576  0.1001  0.1678  0.2725  0.4305  0.6634


1.0000  1.0000  1.0000  1.0000  1.0000  1.0000  1.0000  1.0000  1.0000
0.9992  0.9997  0.9999  1.0000  1.0000  1.0000  1.0000  1.0000  1.0000
0.9909  0.9962  0.9986  0.9996  0.9999  1.0000  1.0000  1.0000  1.0000
0.9502  0.9750  0.9888  0.9957  0.9987  0.9997  1.0000  1.0000  1.0000
0.8342  0.9006  0.9464  0.9747  0.9900  0.9969  0.9994  0.9999  1.0000

0.6214  0.7334  0.8283  0.9012  0.9511  0.9804  0.9944  0.9991  1.0000
0.3614  0.4826  0.6089  0.7297  0.8343  0.9144  0.9661  0.9917  0.9994
0.1495  0.2318  0.3373  0.4628  0.6007  0.7382  0.8591  0.9470  0.9916
0.0385  0.0705  0.1211  0.1960  0.3003  0.4362  0.5995  0.7748  0.9288
0.0046  0.0101  0.0207  0.0404  0.0751  0.1342  0.2316  0.3874  0.6302



1.0000  1.0000  1.0000  1.0000  1.0000  1.0000  1.0000  1.0000  1.0000
0.9997  0.9999  1.0000  1.0000  1.0000  1.0000  1.0000  1.0000  1.0000
0.9955  0.9983  0.9995  0.9999  1.0000  1.0000  1.0000  1.0000  1.0000
0.9726  0.9877  0.9952  0.9984  0.9996  0.9999  1.0000  1.0000  1.0000
0.8980  0.9452  0.9740  0.9894  0.9965  0.9991  0.9999  1.0000  1.0000

0.7384  0.8338  0.9051  0.9526  0.9803  0.9936  0.9986  0.9999  1.0000
0.5044  0.6331  0.7515  0.8497  0.9219  0.9672  0.9901  0.9984  0.9999
0.2660  0.3823  0.5138  0.6496  0.7759  0.8791  0.9500  0.9872  0.9990
0.0996  0.1673  0.2616  0.3828  0.5256  0.6778  0.8202  0.9298  0.9885
0.0233  0.0464  0.0860  0.1493  0.2440  0.3758  0.5443  0.7361  0.9139

0.0025  0.0060  0.0135  0.0282  0.0563  0.1074  0.1969  0.3487  0.5987


1.0000  1.0000  1.0000  1.0000  1.0000  1.0000  1.0000  1.0000  1.0000
0.9998  1.0000  1.0000  1.0000  1.0000  1.0000  1.0000  1.0000  1.0000
0.9978  0.9993  0.9998  1.0000  1.0000  1.0000  1.0000  1.0000  1.0000
0.9852  0.9941  0.9980  0.9994  0.9999  1.0000  1.0000  1.0000  1.0000
0.9390  0.9707  0.9878  0.9957  0.9988  0.9998  1.0000  1.0000  1.0000

0.8262  0.9006  0.9499  0.9784  0.9924  0.9980  0.9997  1.0000  1.0000
0.6331  0.7535  0.8513  0.9218  0.9657  0.9883  0.9973  0.9997  1.0000
0.3971  0.5328  0.6683  0.7897  0.8854  0.9496  0.9841  0.9972  0.9999
0.1911  0.2963  0.4255  0.5696  0.7133  0.8389  0.9306  0.9815  0.9984
0.0652  0.1189  0.2001  0.3127  0.4552  0.6174  0.7788  0.9104  0.9848

0.0139  0.0302  0.0606  0.1130  0.1971  0.3221  0.4922  0.6974  0.8981
0.0014  0.0036  0.0088  0.0198  0.0422  0.0859  0.1673  0.3138  0.5688
```

APPENDIX D. (cont'd)

12	0	1.0000	1.0000	1.0000	1.0000	1.0000	1.0000	1.0000	1.0000	1.0000	1.0000
	1	0.4596	0.7176	0.8578	0.9313	0.9683	0.9862	0.9943	0.9978	0.9992	0.9998
	2	0.1184	0.3410	0.5565	0.7251	0.8416	0.9150	0.9576	0.9804	0.9917	0.9968
	3	0.0196	0.1109	0.2642	0.4417	0.6093	0.7472	0.8487	0.9166	0.9579	0.9807
	4	0.0022	0.0256	0.0922	0.2054	0.3512	0.5075	0.6533	0.7747	0.8655	0.9270
	5	0.0002	0.0043	0.0239	0.0726	0.1576	0.2763	0.4167	0.5618	0.6956	0.8062
	6	0.0000	0.0005	0.0046	0.0194	0.0544	0.1178	0.2127	0.3348	0.4731	0.6128
	7	0.0000	0.0001	0.0007	0.0039	0.0143	0.0386	0.0846	0.1582	0.2607	0.3872
	8	0.0000	0.0000	0.0001	0.0006	0.0028	0.0095	0.0255	0.0573	0.1117	0.1938
	9	0.0000	0.0000	0.0000	0.0001	0.0004	0.0017	0.0056	0.0153	0.0356	0.0730
	10	0.0000	0.0000	0.0000	0.0000	0.0000	0.0002	0.0008	0.0028	0.0079	0.0193
	11	0.0000	0.0000	0.0000	0.0000	0.0000	0.0000	0.0001	0.0003	0.0011	0.0032
	12	0.0000	0.0000	0.0000	0.0000	0.0000	0.0000	0.0000	0.0000	0.0001	0.0002
13	0	1.0000	1.0000	1.0000	1.0000	1.0000	1.0000	1.0000	1.0000	1.0000	1.0000
	1	0.4867	0.7458	0.8791	0.9450	0.9762	0.9903	0.9963	0.9987	0.9996	0.9999
	2	0.1354	0.3787	0.6017	0.7664	0.8733	0.9363	0.9704	0.9874	0.9951	0.9983
	3	0.0245	0.1339	0.3080	0.4983	0.6674	0.7975	0.8868	0.9421	0.9731	0.9888
	4	0.0031	0.0342	0.1180	0.2527	0.4157	0.5794	0.7217	0.8314	0.9071	0.9539
	5	0.0003	0.0065	0.0342	0.0991	0.2060	0.3457	0.4995	0.6470	0.7720	0.8666
	6	0.0000	0.0009	0.0075	0.0300	0.0802	0.1654	0.2841	0.4256	0.5732	0.7095
	7	0.0000	0.0001	0.0013	0.0070	0.0243	0.0624	0.1295	0.2288	0.3563	0.5000
	8	0.0000	0.0000	0.0002	0.0012	0.0056	0.0182	0.0462	0.0977	0.1788	0.2905
	9	0.0000	0.0000	0.0000	0.0002	0.0010	0.0040	0.0126	0.0321	0.0698	0.1334
	10	0.0000	0.0000	0.0000	0.0000	0.0001	0.0007	0.0025	0.0078	0.0203	0.0461
	11	0.0000	0.0000	0.0000	0.0000	0.0000	0.0001	0.0003	0.0013	0.0041	0.0112
	12	0.0000	0.0000	0.0000	0.0000	0.0000	0.0000	0.0000	0.0001	0.0005	0.0017
	13	0.0000	0.0000	0.0000	0.0000	0.0000	0.0000	0.0000	0.0000	0.0000	0.0001
14	0	1.0000	1.0000	1.0000	1.0000	1.0000	1.0000	1.0000	1.0000	1.0000	1.0000
	1	0.5123	0.7712	0.8972	0.9560	0.9822	0.9932	0.9976	0.9992	0.9998	0.9999
	2	0.1530	0.4154	0.6433	0.8021	0.8990	0.9525	0.9795	0.9919	0.9971	0.9991
	3	0.0301	0.1584	0.3521	0.5519	0.7189	0.8392	0.9161	0.9602	0.9830	0.9935
	4	0.0042	0.0441	0.1465	0.3018	0.4787	0.6448	0.7795	0.8757	0.9368	0.9713
	5	0.0004	0.0092	0.0467	0.1298	0.2585	0.4158	0.5773	0.7207	0.8328	0.9102
	6	0.0000	0.0015	0.0115	0.0439	0.1117	0.2195	0.3595	0.5141	0.6627	0.7880
	7	0.0000	0.0002	0.0022	0.0116	0.0383	0.0933	0.1836	0.3075	0.4539	0.6047
	8	0.0000	0.0000	0.0003	0.0024	0.0103	0.0315	0.0753	0.1501	0.2586	0.3953
	9	0.0000	0.0000	0.0000	0.0004	0.0022	0.0083	0.0243	0.0583	0.1189	0.2120
	10	0.0000	0.0000	0.0000	0.0000	0.0003	0.0017	0.0060	0.0175	0.0426	0.0898
	11	0.0000	0.0000	0.0000	0.0000	0.0000	0.0002	0.0011	0.0039	0.0114	0.0287
	12	0.0000	0.0000	0.0000	0.0000	0.0000	0.0000	0.0001	0.0006	0.0022	0.0065
	13	0.0000	0.0000	0.0000	0.0000	0.0000	0.0000	0.0000	0.0001	0.0003	0.0009
	14	0.0000	0.0000	0.0000	0.0000	0.0000	0.0000	0.0000	0.0000	0.0000	0.0001
15	0	1.0000	1.0000	1.0000	1.0000	1.0000	1.0000	1.0000	1.0000	1.0000	1.0000
	1	0.5367	0.7941	0.9126	0.9648	0.9866	0.9953	0.9984	0.9995	0.9999	1.0000
	2	0.1710	0.4510	0.6814	0.8329	0.9198	0.9647	0.9858	0.9948	0.9983	0.9995
	3	0.0362	0.1841	0.3958	0.6020	0.7639	0.8732	0.9383	0.9729	0.9893	0.9963
	4	0.0055	0.0556	0.1773	0.3518	0.5387	0.7031	0.8273	0.9095	0.9576	0.9824
	5	0.0006	0.0127	0.0617	0.1642	0.3135	0.4845	0.6481	0.7827	0.8796	0.9408
	6	0.0001	0.0022	0.0168	0.0611	0.1484	0.2784	0.4357	0.5968	0.7392	0.8491
	7	0.0000	0.0003	0.0036	0.0181	0.0566	0.1311	0.2452	0.3902	0.5478	0.6964
	8	0.0000	0.0000	0.0006	0.0042	0.0173	0.0500	0.1132	0.2131	0.3465	0.5000
	9	0.0000	0.0000	0.0001	0.0008	0.0042	0.0152	0.0422	0.0950	0.1818	0.3036

APPENDIX D. (cont'd)

```
1.0000  1.0000  1.0000  1.0000  1.0000  1.0000  1.0000  1.0000  1.0000
0.9999  1.0000  1.0000  1.0000  1.0000  1.0000  1.0000  1.0000  1.0000
0.9989  0.9997  0.9999  1.0000  1.0000  1.0000  1.0000  1.0000  1.0000
0.9921  0.9972  0.9992  0.9998  1.0000  1.0000  1.0000  1.0000  1.0000
0.9644  0.9847  0.9944  0.9983  0.9996  0.9999  1.0000  1.0000  1.0000

0.8883  0.9427  0.9745  0.9905  0.9972  0.9994  0.9999  1.0000  1.0000
0.7393  0.8418  0.9154  0.9614  0.9857  0.9961  0.9993  0.9999  1.0000
0.5269  0.6652  0.7873  0.8821  0.9456  0.9806  0.9954  0.9995  1.0000
0.3044  0.4382  0.5833  0.7237  0.8424  0.9274  0.9761  0.9957  0.9998
0.1345  0.2253  0.3467  0.4925  0.6488  0.7946  0.9078  0.9744  0.9978

0.0421  0.0834  0.1513  0.2528  0.3907  0.5583  0.7358  0.8891  0.9804
0.0083  0.0196  0.0424  0.0850  0.1584  0.2749  0.4435  0.6590  0.8816
0.0008  0.0022  0.0057  0.0138  0.0317  0.0687  0.1422  0.2824  0.5404


1.0000  1.0000  1.0000  1.0000  1.0000  1.0000  1.0000  1.0000  1.0000
1.0000  1.0000  1.0000  1.0000  1.0000  1.0000  1.0000  1.0000  1.0000
0.9995  0.9999  1.0000  1.0000  1.0000  1.0000  1.0000  1.0000  1.0000
0.9959  0.9987  0.9997  0.9999  1.0000  1.0000  1.0000  1.0000  1.0000
0.9797  0.9922  0.9975  0.9993  0.9999  1.0000  1.0000  1.0000  1.0000

0.9301  0.9679  0.9874  0.9960  0.9990  0.9998  1.0000  1.0000  1.0000
0.8212  0.9023  0.9538  0.9818  0.9943  0.9988  0.9998  1.0000  1.0000
0.6437  0.7712  0.8705  0.9376  0.9757  0.9930  0.9987  0.9999  1.0000
0.4268  0.5744  0.7159  0.8346  0.9198  0.9700  0.9925  0.9991  1.0000
0.2279  0.3530  0.5005  0.6543  0.7940  0.9009  0.9658  0.9935  0.9997

0.0929  0.1686  0.2783  0.4206  0.5842  0.7473  0.8820  0.9658  0.9969
0.0269  0.0579  0.1132  0.2025  0.3326  0.5016  0.6920  0.8661  0.9755
0.0049  0.0126  0.0296  0.0637  0.1267  0.2336  0.3983  0.6213  0.8646
0.0004  0.0013  0.0037  0.0097  0.0238  0.0550  0.1209  0.2542  0.5133


1.0000  1.0000  1.0000  1.0000  1.0000  1.0000  1.0000  1.0000  1.0000
1.0000  1.0000  1.0000  1.0000  1.0000  1.0000  1.0000  1.0000  1.0000
0.9997  0.9999  1.0000  1.0000  1.0000  1.0000  1.0000  1.0000  1.0000
0.9978  0.9994  0.9999  1.0000  1.0000  1.0000  1.0000  1.0000  1.0000
0.9886  0.9961  0.9989  0.9998  1.0000  1.0000  1.0000  1.0000  1.0000

0.9574  0.9825  0.9940  0.9983  0.9997  1.0000  1.0000  1.0000  1.0000
0.8811  0.9417  0.9757  0.9917  0.9978  0.9996  0.9999  1.0000  1.0000
0.7414  0.8499  0.9247  0.9685  0.9897  0.9976  0.9997  1.0000  1.0000
0.5461  0.6924  0.8164  0.9067  0.9617  0.9884  0.9978  0.9998  1.0000
0.3373  0.4859  0.6405  0.7805  0.8883  0.9561  0.9885  0.9985  1.0000

0.1672  0.2793  0.4227  0.5842  0.7415  0.8702  0.9533  0.9908  0.9996
0.0632  0.1243  0.2205  0.3552  0.5213  0.6982  0.8535  0.9559  0.9958
0.0170  0.0398  0.0839  0.1608  0.2811  0.4480  0.6479  0.8416  0.9699
0.0029  0.0081  0.0205  0.0475  0.1010  0.1979  0.3567  0.5846  0.8470
0.0002  0.0008  0.0024  0.0068  0.0178  0.0440  0.1028  0.2288  0.4877


1.0000  1.0000  1.0000  1.0000  1.0000  1.0000  1.0000  1.0000  1.0000
1.0000  1.0000  1.0000  1.0000  1.0000  1.0000  1.0000  1.0000  1.0000
0.9999  1.0000  1.0000  1.0000  1.0000  1.0000  1.0000  1.0000  1.0000
0.9989  0.9997  0.9999  1.0000  1.0000  1.0000  1.0000  1.0000  1.0000
0.9937  0.9981  0.9995  0.9999  1.0000  1.0000  1.0000  1.0000  1.0000

0.9745  0.9906  0.9972  0.9993  0.9999  1.0000  1.0000  1.0000  1.0000
0.9231  0.9662  0.9876  0.9963  0.9992  0.9999  1.0000  1.0000  1.0000
0.8182  0.9049  0.9578  0.9848  0.9958  0.9992  0.9999  1.0000  1.0000
0.6535  0.7869  0.8868  0.9500  0.9827  0.9958  0.9994  1.0000  1.0000
0.4522  0.6098  0.7548  0.8689  0.9434  0.9819  0.9964  0.9997  1.0000
```

APPENDIX D. (cont'd)

	n	k									
			0.05	0.10	0.15	0.20	0.25	0.30	0.35	0.40	0.45	0.50
		10	0.0000	0.0000	0.0000	0.0001	0.0008	0.0037	0.0124	0.0338	0.0769	0.1509
		11	0.0000	0.0000	0.0000	0.0000	0.0001	0.0007	0.0028	0.0093	0.0255	0.0592
		12	0.0000	0.0000	0.0000	0.0000	0.0000	0.0001	0.0005	0.0019	0.0063	0.0176
		13	0.0000	0.0000	0.0000	0.0000	0.0000	0.0000	0.0001	0.0003	0.0011	0.0037
		14	0.0000	0.0000	0.0000	0.0000	0.0000	0.0000	0.0000	0.0000	0.0001	0.0005
		15	0.0000	0.0000	0.0000	0.0000	0.0000	0.0000	0.0000	0.0000	0.0000	0.0000
	16	0	1.0000	1.0000	1.0000	1.0000	1.0000	1.0000	1.0000	1.0000	1.0000	1.0000
		1	0.5599	0.8147	0.9257	0.9719	0.9900	0.9967	0.9990	0.9997	0.9999	1.0000
		2	0.1892	0.4853	0.7161	0.8593	0.9365	0.9739	0.9902	0.9967	0.9990	0.9997
		3	0.0429	0.2108	0.4386	0.6482	0.8029	0.9006	0.9549	0.9817	0.9934	0.9979
		4	0.0070	0.0684	0.2101	0.4019	0.5950	0.7541	0.8661	0.9349	0.9719	0.9894
		5	0.0009	0.0170	0.0791	0.2018	0.3698	0.5501	0.7108	0.8334	0.9147	0.9616
		6	0.0001	0.0033	0.0235	0.0817	0.1897	0.3402	0.5100	0.6712	0.8024	0.8949
		7	0.0000	0.0005	0.0056	0.0267	0.0796	0.1753	0.3119	0.4728	0.6340	0.7727
		8	0.0000	0.0001	0.0011	0.0070	0.0271	0.0744	0.1594	0.2839	0.4371	0.5982
		9	0.0000	0.0000	0.0002	0.0015	0.0075	0.0257	0.0671	0.1423	0.2559	0.4018
		10	0.0000	0.0000	0.0000	0.0002	0.0016	0.0071	0.0229	0.0583	0.1241	0.2272
		11	0.0000	0.0000	0.0000	0.0000	0.0003	0.0016	0.0062	0.0191	0.0486	0.1051
		12	0.0000	0.0000	0.0000	0.0000	0.0000	0.0003	0.0013	0.0049	0.0149	0.0384
		13	0.0000	0.0000	0.0000	0.0000	0.0000	0.0000	0.0002	0.0009	0.0035	0.0106
		14	0.0000	0.0000	0.0000	0.0000	0.0000	0.0000	0.0000	0.0001	0.0006	0.0021
		15	0.0000	0.0000	0.0000	0.0000	0.0000	0.0000	0.0000	0.0000	0.0001	0.0003
		16	0.0000	0.0000	0.0000	0.0000	0.0000	0.0000	0.0000	0.0000	0.0000	0.0000
	17	0	1.0000	1.0000	1.0000	1.0000	1.0000	1.0000	1.0000	1.0000	1.0000	1.0000
		1	0.5819	0.8332	0.9369	0.9775	0.9925	0.9977	0.9993	0.9998	1.0000	1.0000
		2	0.2078	0.5182	0.7475	0.8818	0.9499	0.9807	0.9933	0.9979	0.9994	0.9999
		3	0.0503	0.2382	0.4802	0.6904	0.8363	0.9226	0.9673	0.9877	0.9959	0.9988
		4	0.0088	0.0826	0.2444	0.4511	0.6470	0.7981	0.8972	0.9536	0.9815	0.9936
		5	0.0012	0.0221	0.0987	0.2418	0.4261	0.6113	0.7652	0.8740	0.9404	0.9755
		6	0.0001	0.0047	0.0319	0.1057	0.2347	0.4032	0.5803	0.7361	0.8529	0.9283
		7	0.0000	0.0008	0.0083	0.0377	0.1071	0.2248	0.3812	0.5522	0.7098	0.8338
		8	0.0000	0.0001	0.0017	0.0109	0.0402	0.1046	0.2128	0.3595	0.5257	0.6855
		9	0.0000	0.0000	0.0003	0.0026	0.0124	0.0403	0.0994	0.1989	0.3374	0.5000
		10	0.0000	0.0000	0.0000	0.0005	0.0031	0.0127	0.0383	0.0919	0.1834	0.3145
		11	0.0000	0.0000	0.0000	0.0001	0.0006	0.0032	0.0120	0.0348	0.0826	0.1662
		12	0.0000	0.0000	0.0000	0.0000	0.0001	0.0007	0.0030	0.0106	0.0301	0.0717
		13	0.0000	0.0000	0.0000	0.0000	0.0000	0.0001	0.0006	0.0025	0.0086	0.0245
		14	0.0000	0.0000	0.0000	0.0000	0.0000	0.0000	0.0001	0.0005	0.0019	0.0064
		15	0.0000	0.0000	0.0000	0.0000	0.0000	0.0000	0.0000	0.0001	0.0003	0.0012
		16	0.0000	0.0000	0.0000	0.0000	0.0000	0.0000	0.0000	0.0000	0.0000	0.0001
		17	0.0000	0.0000	0.0000	0.0000	0.0000	0.0000	0.0000	0.0000	0.0000	0.0000
	18	0	1.0000	1.0000	1.0000	1.0000	1.0000	1.0000	1.0000	1.0000	1.0000	1.0000
		1	0.6028	0.8499	0.9464	0.9820	0.9944	0.9984	0.9996	0.9999	1.0000	1.0000
		2	0.2265	0.5497	0.7759	0.9009	0.9605	0.9858	0.9954	0.9987	0.9997	0.9999
		3	0.0581	0.2662	0.5203	0.7287	0.8647	0.9400	0.9764	0.9918	0.9975	0.9993
		4	0.0109	0.0982	0.2798	0.4990	0.6943	0.8354	0.9217	0.9672	0.9880	0.9962
		5	0.0015	0.0282	0.1206	0.2836	0.4813	0.6673	0.8114	0.9058	0.9589	0.9846
		6	0.0002	0.0064	0.0419	0.1329	0.2825	0.4656	0.6450	0.7912	0.8923	0.9519
		7	0.0000	0.0012	0.0118	0.0513	0.1390	0.2783	0.4509	0.6257	0.7742	0.8811
		8	0.0000	0.0002	0.0027	0.0163	0.0569	0.1407	0.2717	0.4366	0.6085	0.7597
		9	0.0000	0.0000	0.0005	0.0043	0.0193	0.0596	0.1391	0.2632	0.4221	0.5927
		10	0.0000	0.0000	0.0001	0.0009	0.0054	0.0210	0.0597	0.1347	0.2527	0.4073
		11	0.0000	0.0000	0.0000	0.0002	0.0012	0.0061	0.0212	0.0576	0.1280	0.2403
		12	0.0000	0.0000	0.0000	0.0000	0.0002	0.0014	0.0062	0.0203	0.0537	0.1189
		13	0.0000	0.0000	0.0000	0.0000	0.0000	0.0003	0.0014	0.0058	0.0183	0.0481
		14	0.0000	0.0000	0.0000	0.0000	0.0000	0.0000	0.0003	0.0013	0.0049	0.0154

APPENDIX D. (cont'd)

```
0.2608  0.4032  0.5643  0.7216  0.8516  0.9389  0.9832  0.9977  0.9999
0.1204  0.2173  0.3519  0.5155  0.6865  0.8358  0.9383  0.9873  0.9994
0.0424  0.0905  0.1727  0.2969  0.4613  0.6482  0.8227  0.9444  0.9945
0.0107  0.0271  0.0617  0.1268  0.2361  0.3980  0.6042  0.8159  0.9638
0.0017  0.0052  0.0142  0.0353  0.0802  0.1671  0.3186  0.5490  0.8290

0.0001  0.0005  0.0016  0.0047  0.0134  0.0352  0.0874  0.2059  0.4633


1.0000  1.0000  1.0000  1.0000  1.0000  1.0000  1.0000  1.0000  1.0000
1.0000  1.0000  1.0000  1.0000  1.0000  1.0000  1.0000  1.0000  1.0000
0.9999  1.0000  1.0000  1.0000  1.0000  1.0000  1.0000  1.0000  1.0000
0.9994  0.9999  1.0000  1.0000  1.0000  1.0000  1.0000  1.0000  1.0000
0.9965  0.9991  0.9998  1.0000  1.0000  1.0000  1.0000  1.0000  1.0000

0.9851  0.9951  0.9987  0.9997  1.0000  1.0000  1.0000  1.0000  1.0000
0.9514  0.9809  0.9938  0.9984  0.9997  1.0000  1.0000  1.0000  1.0000
0.8759  0.9417  0.9771  0.9929  0.9984  0.9998  1.0000  1.0000  1.0000
0.7441  0.8577  0.9329  0.9743  0.9925  0.9985  0.9998  1.0000  1.0000
0.5629  0.7161  0.8406  0.9256  0.9729  0.9930  0.9989  0.9999  1.0000

0.3660  0.5272  0.6881  0.8247  0.9204  0.9733  0.9944  0.9995  1.0000
0.1976  0.3288  0.4900  0.6598  0.8103  0.9183  0.9765  0.9967  0.9999
0.0853  0.1666  0.2892  0.4499  0.6302  0.7982  0.9209  0.9830  0.9991
0.0281  0.0651  0.1339  0.2459  0.4050  0.5981  0.7899  0.9316  0.9930
0.0066  0.0183  0.0451  0.0994  0.1971  0.3518  0.5614  0.7892  0.9571

0.0010  0.0033  0.0098  0.0261  0.0635  0.1407  0.2839  0.5147  0.8108
0.0001  0.0003  0.0010  0.0033  0.0100  0.0281  0.0742  0.1853  0.4401


1.0000  1.0000  1.0000  1.0000  1.0000  1.0000  1.0000  1.0000  1.0000
1.0000  1.0000  1.0000  1.0000  1.0000  1.0000  1.0000  1.0000  1.0000
1.0000  1.0000  1.0000  1.0000  1.0000  1.0000  1.0000  1.0000  1.0000
0.9997  0.9999  1.0000  1.0000  1.0000  1.0000  1.0000  1.0000  1.0000
0.9981  0.9995  0.9999  1.0000  1.0000  1.0000  1.0000  1.0000  1.0000

0.9914  0.9975  0.9994  0.9999  1.0000  1.0000  1.0000  1.0000  1.0000
0.9699  0.9894  0.9970  0.9993  0.9999  1.0000  1.0000  1.0000  1.0000
0.9174  0.9652  0.9880  0.9968  0.9994  0.9999  1.0000  1.0000  1.0000
0.8166  0.9081  0.9617  0.9873  0.9969  0.9995  1.0000  1.0000  1.0000
0.6626  0.8011  0.9006  0.9597  0.9876  0.9974  0.9997  1.0000  1.0000

0.4743  0.6405  0.7872  0.8954  0.9598  0.9891  0.9983  0.9999  1.0000
0.2902  0.4478  0.6188  0.7752  0.8929  0.9623  0.9917  0.9992  1.0000
0.1471  0.2639  0.4197  0.5968  0.7653  0.8943  0.9681  0.9953  0.9999
0.0596  0.1260  0.2348  0.3887  0.5739  0.7582  0.9013  0.9779  0.9988
0.0184  0.0464  0.1028  0.2019  0.3530  0.5489  0.7556  0.9174  0.9912

0.0041  0.0123  0.0327  0.0774  0.1637  0.3096  0.5198  0.7618  0.9497
0.0006  0.0021  0.0067  0.0193  0.0501  0.1182  0.2525  0.4818  0.7922
0.0000  0.0002  0.0007  0.0023  0.0075  0.0225  0.0631  0.1668  0.4181


1.0000  1.0000  1.0000  1.0000  1.0000  1.0000  1.0000  1.0000  1.0000
1.0000  1.0000  1.0000  1.0000  1.0000  1.0000  1.0000  1.0000  1.0000
1.0000  1.0000  1.0000  1.0000  1.0000  1.0000  1.0000  1.0000  1.0000
0.9999  1.0000  1.0000  1.0000  1.0000  1.0000  1.0000  1.0000  1.0000
0.9990  0.9998  1.0000  1.0000  1.0000  1.0000  1.0000  1.0000  1.0000

0.9951  0.9987  0.9997  1.0000  1.0000  1.0000  1.0000  1.0000  1.0000
0.9817  0.9942  0.9986  0.9997  1.0000  1.0000  1.0000  1.0000  1.0000
0.9463  0.9797  0.9938  0.9986  0.9998  1.0000  1.0000  1.0000  1.0000
0.8720  0.9423  0.9788  0.9939  0.9988  0.9998  1.0000  1.0000  1.0000
0.7473  0.8653  0.9403  0.9790  0.9946  0.9991  0.9999  1.0000  1.0000

0.5778  0.7368  0.8609  0.9404  0.9806  0.9957  0.9995  1.0000  1.0000
0.3915  0.5634  0.7283  0.8593  0.9430  0.9837  0.9973  0.9998  1.0000
0.2258  0.3743  0.5491  0.7217  0.8610  0.9487  0.9882  0.9988  1.0000
0.1077  0.2088  0.3550  0.5344  0.7174  0.8671  0.9581  0.9936  0.9998
0.0411  0.0942  0.1886  0.3327  0.5187  0.7164  0.8794  0.9718  0.9985
```

APPENDIX D. (cont'd)

	15	0.0000	0.0000	0.0000	0.0000	0.0000	0.0000	0.0000	0.0002	0.0010	0.0038
	16	0.0000	0.0000	0.0000	0.0000	0.0000	0.0000	0.0000	0.0000	0.0001	0.0007
	17	0.0000	0.0000	0.0000	0.0000	0.0000	0.0000	0.0000	0.0000	0.0000	0.0001
	18	0.0000	0.0000	0.0000	0.0000	0.0000	0.0000	0.0000	0.0000	0.0000	0.0000
19	0	1.0000	1.0000	1.0000	1.0000	1.0000	1.0000	1.0000	1.0000	1.0000	1.0000
	1	0.6226	0.8649	0.9544	0.9856	0.9958	0.9989	0.9997	0.9999	1.0000	1.0000
	2	0.2453	0.5797	0.8015	0.9171	0.9690	0.9896	0.9969	0.9992	0.9998	1.0000
	3	0.0665	0.2946	0.5587	0.7631	0.8887	0.9538	0.9830	0.9945	0.9985	0.9996
	4	0.0132	0.1150	0.3158	0.5449	0.7369	0.8668	0.9409	0.9770	0.9923	0.9978
	5	0.0020	0.0352	0.1444	0.3267	0.5346	0.7178	0.8500	0.9304	0.9720	0.9904
	6	0.0002	0.0086	0.0537	0.1631	0.3322	0.5261	0.7032	0.8371	0.9223	0.9682
	7	0.0000	0.0017	0.0163	0.0676	0.1749	0.3345	0.5188	0.6919	0.8273	0.9165
	8	0.0000	0.0003	0.0041	0.0233	0.0775	0.1820	0.3344	0.5122	0.6831	0.8204
	9	0.0000	0.0000	0.0008	0.0067	0.0287	0.0839	0.1855	0.3325	0.5060	0.6762
	10	0.0000	0.0000	0.0001	0.0016	0.0089	0.0326	0.0875	0.1861	0.3290	0.5000
	11	0.0000	0.0000	0.0000	0.0003	0.0023	0.0105	0.0347	0.0885	0.1841	0.3238
	12	0.0000	0.0000	0.0000	0.0000	0.0005	0.0028	0.0114	0.0352	0.0871	0.1796
	13	0.0000	0.0000	0.0000	0.0000	0.0001	0.0006	0.0031	0.0116	0.0342	0.0835
	14	0.0000	0.0000	0.0000	0.0000	0.0000	0.0001	0.0007	0.0031	0.0109	0.0318
	15	0.0000	0.0000	0.0000	0.0000	0.0000	0.0000	0.0001	0.0006	0.0028	0.0096
	16	0.0000	0.0000	0.0000	0.0000	0.0000	0.0000	0.0000	0.0001	0.0005	0.0022
	17	0.0000	0.0000	0.0000	0.0000	0.0000	0.0000	0.0000	0.0000	0.0001	0.0004
	18	0.0000	0.0000	0.0000	0.0000	0.0000	0.0000	0.0000	0.0000	0.0000	0.0000
	19	0.0000	0.0000	0.0000	0.0000	0.0000	0.0000	0.0000	0.0000	0.0000	0.0000
20	0	1.0000	1.0000	1.0000	1.0000	1.0000	1.0000	1.0000	1.0000	1.0000	1.0000
	1	0.6415	0.8784	0.9612	0.9885	0.9968	0.9992	0.9998	1.0000	1.0000	1.0000
	2	0.2642	0.6083	0.8244	0.9308	0.9757	0.9924	0.9979	0.9995	0.9999	1.0000
	3	0.0755	0.3231	0.5951	0.7939	0.9087	0.9645	0.9879	0.9964	0.9991	0.9998
	4	0.0159	0.1330	0.3523	0.5885	0.7748	0.8929	0.9556	0.9840	0.9951	0.9987
	5	0.0026	0.0432	0.1702	0.3704	0.5852	0.7625	0.8818	0.9490	0.9811	0.9941
	6	0.0003	0.0113	0.0673	0.1958	0.3828	0.5836	0.7546	0.8744	0.9447	0.9793
	7	0.0000	0.0024	0.0219	0.0867	0.2142	0.3920	0.5834	0.7500	0.8701	0.9423
	8	0.0000	0.0004	0.0059	0.0321	0.1018	0.2277	0.3990	0.5841	0.7480	0.8684
	9	0.0000	0.0001	0.0013	0.0100	0.0409	0.1133	0.2376	0.4044	0.5857	0.7483
	10	0.0000	0.0000	0.0002	0.0026	0.0139	0.0480	0.1218	0.2447	0.4086	0.5881
	11	0.0000	0.0000	0.0000	0.0006	0.0039	0.0171	0.0532	0.1275	0.2493	0.4119
	12	0.0000	0.0000	0.0000	0.0001	0.0009	0.0051	0.0196	0.0565	0.1308	0.2517
	13	0.0000	0.0000	0.0000	0.0000	0.0002	0.0013	0.0060	0.0210	0.0580	0.1316
	14	0.0000	0.0000	0.0000	0.0000	0.0000	0.0003	0.0015	0.0065	0.0214	0.0577
	15	0.0000	0.0000	0.0000	0.0000	0.0000	0.0000	0.0003	0.0016	0.0064	0.0207
	16	0.0000	0.0000	0.0000	0.0000	0.0000	0.0000	0.0000	0.0003	0.0015	0.0059
	17	0.0000	0.0000	0.0000	0.0000	0.0000	0.0000	0.0000	0.0000	0.0003	0.0013
	18	0.0000	0.0000	0.0000	0.0000	0.0000	0.0000	0.0000	0.0000	0.0000	0.0002
	19	0.0000	0.0000	0.0000	0.0000	0.0000	0.0000	0.0000	0.0000	0.0000	0.0000
	20	0.0000	0.0000	0.0000	0.0000	0.0000	0.0000	0.0000	0.0000	0.0000	0.0000
21	0	1.0000	1.0000	1.0000	1.0000	1.0000	1.0000	1.0000	1.0000	1.0000	1.0000
	1	0.6594	0.8906	0.9671	0.9908	0.9976	0.9994	0.9999	1.0000	1.0000	1.0000
	2	0.2830	0.6353	0.8450	0.9424	0.9810	0.9944	0.9985	0.9997	0.9999	1.0000
	3	0.0849	0.3516	0.6295	0.8213	0.9255	0.9729	0.9914	0.9976	0.9994	0.9999
	4	0.0189	0.1520	0.3887	0.6296	0.8083	0.9144	0.9669	0.9890	0.9969	0.9993
	5	0.0032	0.0522	0.1975	0.4140	0.6326	0.8016	0.9076	0.9630	0.9874	0.9964
	6	0.0004	0.0144	0.0827	0.2307	0.4334	0.6373	0.7991	0.9043	0.9611	0.9867
	7	0.0000	0.0033	0.0287	0.1085	0.2564	0.4495	0.6433	0.7998	0.9036	0.9608
	8	0.0000	0.0006	0.0083	0.0431	0.1299	0.2770	0.4635	0.6505	0.8029	0.9054
	9	0.0000	0.0001	0.0020	0.0144	0.0561	0.1476	0.2941	0.4763	0.6587	0.8083

APPENDIX D. (cont'd)

```
0.0120 0.0328 0.0783 0.1645 0.3057 0.5010 0.7202 0.9018 0.9891
0.0025 0.0082 0.0236 0.0600 0.1353 0.2713 0.4797 0.7338 0.9419
0.0003 0.0013 0.0046 0.0142 0.0395 0.0991 0.2240 0.4503 0.7735
0.0000 0.0001 0.0004 0.0016 0.0056 0.0180 0.0536 0.1501 0.3972

1.0000 1.0000 1.0000 1.0000 1.0000 1.0000 1.0000 1.0000 1.0000
1.0000 1.0000 1.0000 1.0000 1.0000 1.0000 1.0000 1.0000 1.0000
1.0000 1.0000 1.0000 1.0000 1.0000 1.0000 1.0000 1.0000 1.0000
0.9999 1.0000 1.0000 1.0000 1.0000 1.0000 1.0000 1.0000 1.0000
0.9995 0.9999 1.0000 1.0000 1.0000 1.0000 1.0000 1.0000 1.0000

0.9972 0.9994 0.9999 1.0000 1.0000 1.0000 1.0000 1.0000 1.0000
0.9891 0.9969 0.9993 0.9999 1.0000 1.0000 1.0000 1.0000 1.0000
0.9658 0.9884 0.9969 0.9994 0.9999 1.0000 1.0000 1.0000 1.0000
0.9129 0.9648 0.9886 0.9972 0.9995 0.9999 1.0000 1.0000 1.0000
0.8159 0.9115 0.9653 0.9895 0.9977 0.9997 1.0000 1.0000 1.0000

0.6710 0.8139 0.9125 0.9674 0.9911 0.9984 0.9999 1.0000 1.0000
0.4940 0.6675 0.8145 0.9161 0.9712 0.9933 0.9992 1.0000 1.0000
0.3169 0.4878 0.6656 0.8180 0.9225 0.9767 0.9959 0.9997 1.0000
0.1727 0.3081 0.4812 0.6655 0.8251 0.9324 0.9837 0.9983 1.0000
0.0777 0.1629 0.2968 0.4739 0.6678 0.8369 0.9463 0.9914 0.9998

0.0280 0.0696 0.1500 0.2822 0.4654 0.6733 0.8556 0.9648 0.9980
0.0077 0.0230 0.0591 0.1332 0.2631 0.4551 0.6841 0.8850 0.9868
0.0015 0.0055 0.0170 0.0462 0.1113 0.2369 0.4413 0.7054 0.9335
0.0002 0.0008 0.0031 0.0104 0.0310 0.0829 0.1985 0.4203 0.7547
0.0000 0.0001 0.0003 0.0011 0.0042 0.0144 0.0456 0.1351 0.3773


1.0000 1.0000 1.0000 1.0000 1.0000 1.0000 1.0000 1.0000 1.0000
1.0000 1.0000 1.0000 1.0000 1.0000 1.0000 1.0000 1.0000 1.0000
1.0000 1.0000 1.0000 1.0000 1.0000 1.0000 1.0000 1.0000 1.0000
1.0000 1.0000 1.0000 1.0000 1.0000 1.0000 1.0000 1.0000 1.0000
0.9997 0.9999 1.0000 1.0000 1.0000 1.0000 1.0000 1.0000 1.0000

0.9985 0.9997 0.9999 1.0000 1.0000 1.0000 1.0000 1.0000 1.0000
0.9936 0.9984 0.9997 1.0000 1.0000 1.0000 1.0000 1.0000 1.0000
0.9786 0.9935 0.9985 0.9997 1.0000 1.0000 1.0000 1.0000 1.0000
0.9420 0.9790 0.9940 0.9987 0.9998 1.0000 1.0000 1.0000 1.0000
0.8692 0.9435 0.9804 0.9949 0.9991 0.9999 1.0000 1.0000 1.0000

0.7507 0.8725 0.9468 0.9829 0.9961 0.9994 1.0000 1.0000 1.0000
0.5914 0.7553 0.8782 0.9520 0.9861 0.9974 0.9997 1.0000 1.0000
0.4143 0.5956 0.7624 0.8867 0.9591 0.9900 0.9987 0.9999 1.0000
0.2520 0.4159 0.6010 0.7723 0.8982 0.9679 0.9941 0.9996 1.0000
0.1299 0.2500 0.4166 0.6080 0.7858 0.9133 0.9781 0.9976 1.0000

0.0553 0.1256 0.2454 0.4164 0.6172 0.8042 0.9327 0.9887 0.9997
0.0189 0.0510 0.1182 0.2375 0.4148 0.6296 0.8298 0.9568 0.9974
0.0049 0.0160 0.0444 0.1071 0.2252 0.4114 0.6477 0.8670 0.9841
0.0009 0.0036 0.0121 0.0355 0.0913 0.2061 0.4049 0.6769 0.9245
0.0001 0.0005 0.0021 0.0076 0.0243 0.0692 0.1756 0.3917 0.7358

0.0000 0.0000 0.0002 0.0008 0.0032 0.0115 0.0388 0.1216 0.3585


1.0000 1.0000 1.0000 1.0000 1.0000 1.0000 1.0000 1.0000 1.0000
1.0000 1.0000 1.0000 1.0000 1.0000 1.0000 1.0000 1.0000 1.0000
1.0000 1.0000 1.0000 1.0000 1.0000 1.0000 1.0000 1.0000 1.0000
1.0000 1.0000 1.0000 1.0000 1.0000 1.0000 1.0000 1.0000 1.0000
0.9999 1.0000 1.0000 1.0000 1.0000 1.0000 1.0000 1.0000 1.0000

0.9992 0.9998 1.0000 1.0000 1.0000 1.0000 1.0000 1.0000 1.0000
0.9963 0.9992 0.9999 1.0000 1.0000 1.0000 1.0000 1.0000 1.0000
0.9868 0.9964 0.9993 0.9999 1.0000 1.0000 1.0000 1.0000 1.0000
0.9621 0.9877 0.9969 0.9994 0.9999 1.0000 1.0000 1.0000 1.0000
0.9092 0.9648 0.9892 0.9976 0.9996 1.0000 1.0000 1.0000 1.0000
```

APPENDIX D. (cont'd)

	10	0.0000	0.0000	0.0004	0.0041	0.0206	0.0676	0.1623	0.3086	0.4883	0.6682
	11	0.0000	0.0000	0.0001	0.0010	0.0064	0.0264	0.0772	0.1744	0.3210	0.5000
	12	0.0000	0.0000	0.0000	0.0002	0.0017	0.0087	0.0313	0.0849	0.1841	0.3318
	13	0.0000	0.0000	0.0000	0.0000	0.0004	0.0024	0.0108	0.0352	0.0908	0.1917
	14	0.0000	0.0000	0.0000	0.0000	0.0001	0.0006	0.0031	0.0123	0.0379	0.0946
	15	0.0000	0.0000	0.0000	0.0000	0.0000	0.0001	0.0007	0.0036	0.0132	0.0392
	16	0.0000	0.0000	0.0000	0.0000	0.0000	0.0000	0.0001	0.0008	0.0037	0.0133
	17	0.0000	0.0000	0.0000	0.0000	0.0000	0.0000	0.0000	0.0002	0.0008	0.0036
	18	0.0000	0.0000	0.0000	0.0000	0.0000	0.0000	0.0000	0.0000	0.0001	0.0007
	19	0.0000	0.0000	0.0000	0.0000	0.0000	0.0000	0.0000	0.0000	0.0000	0.0001
	20	0.0000	0.0000	0.0000	0.0000	0.0000	0.0000	0.0000	0.0000	0.0000	0.0000
	21	0.0000	0.0000	0.0000	0.0000	0.0000	0.0000	0.0000	0.0000	0.0000	0.0000
22	0	1.0000	1.0000	1.0000	1.0000	1.0000	1.0000	1.0000	1.0000	1.0000	1.0000
	1	0.6765	0.9015	0.9720	0.9926	0.9982	0.9996	0.9999	1.0000	1.0000	1.0000
	2	0.3018	0.6608	0.8633	0.9520	0.9851	0.9959	0.9990	0.9998	1.0000	1.0000
	3	0.0948	0.3800	0.6618	0.8455	0.9393	0.9793	0.9939	0.9984	0.9997	0.9999
	4	0.0222	0.1719	0.4248	0.6680	0.8376	0.9319	0.9755	0.9924	0.9980	0.9996
	5	0.0040	0.0621	0.2262	0.4571	0.6765	0.8354	0.9284	0.9734	0.9917	0.9978
	6	0.0006	0.0182	0.0999	0.2674	0.4832	0.6866	0.8371	0.9278	0.9729	0.9915
	7	0.0001	0.0044	0.0368	0.1330	0.3006	0.5058	0.6978	0.8416	0.9295	0.9738
	8	0.0000	0.0009	0.0114	0.0561	0.1615	0.3287	0.5264	0.7102	0.8482	0.9331
	9	0.0000	0.0001	0.0030	0.0201	0.0746	0.1865	0.3534	0.5460	0.7236	0.8569
	10	0.0000	0.0000	0.0007	0.0061	0.0295	0.0916	0.2084	0.3756	0.5650	0.7383
	11	0.0000	0.0000	0.0001	0.0016	0.0100	0.0387	0.1070	0.2280	0.3963	0.5841
	12	0.0000	0.0000	0.0000	0.0003	0.0029	0.0140	0.0474	0.1207	0.2457	0.4159
	13	0.0000	0.0000	0.0000	0.0001	0.0007	0.0043	0.0180	0.0551	0.1328	0.2617
	14	0.0000	0.0000	0.0000	0.0000	0.0001	0.0011	0.0058	0.0215	0.0617	0.1431
	15	0.0000	0.0000	0.0000	0.0000	0.0000	0.0002	0.0016	0.0070	0.0243	0.0669
	16	0.0000	0.0000	0.0000	0.0000	0.0000	0.0000	0.0003	0.0019	0.0080	0.0262
	17	0.0000	0.0000	0.0000	0.0000	0.0000	0.0000	0.0001	0.0004	0.0021	0.0085
	18	0.0000	0.0000	0.0000	0.0000	0.0000	0.0000	0.0000	0.0001	0.0005	0.0022
	19	0.0000	0.0000	0.0000	0.0000	0.0000	0.0000	0.0000	0.0000	0.0001	0.0004
	20	0.0000	0.0000	0.0000	0.0000	0.0000	0.0000	0.0000	0.0000	0.0000	0.0001
	21	0.0000	0.0000	0.0000	0.0000	0.0000	0.0000	0.0000	0.0000	0.0000	0.0000
	22	0.0000	0.0000	0.0000	0.0000	0.0000	0.0000	0.0000	0.0000	0.0000	0.0000
23	0	1.0000	1.0000	1.0000	1.0000	1.0000	1.0000	1.0000	1.0000	1.0000	1.0000
	1	0.6926	0.9114	0.9762	0.9941	0.9987	0.9997	0.9999	1.0000	1.0000	1.0000
	2	0.3206	0.6849	0.8796	0.9602	0.9884	0.9970	0.9993	0.9999	1.0000	1.0000
	3	0.1052	0.4080	0.6920	0.8668	0.9508	0.9843	0.9957	0.9990	0.9998	1.0000
	4	0.0258	0.1927	0.4604	0.7035	0.8630	0.9462	0.9819	0.9948	0.9988	0.9998
	5	0.0049	0.0731	0.2560	0.4993	0.7168	0.8644	0.9449	0.9810	0.9945	0.9987
	6	0.0008	0.0226	0.1189	0.3053	0.5315	0.7312	0.8691	0.9460	0.9814	0.9947
	7	0.0001	0.0058	0.0463	0.1598	0.3463	0.5601	0.7466	0.8760	0.9490	0.9827
	8	0.0000	0.0012	0.0152	0.0715	0.1963	0.3819	0.5864	0.7627	0.8848	0.9534
	9	0.0000	0.0002	0.0042	0.0273	0.0963	0.2291	0.4140	0.6116	0.7797	0.8950
	10	0.0000	0.0000	0.0010	0.0089	0.0408	0.1201	0.2592	0.4438	0.6364	0.7976
	11	0.0000	0.0000	0.0002	0.0025	0.0149	0.0546	0.1425	0.2871	0.4722	0.6612
	12	0.0000	0.0000	0.0000	0.0006	0.0046	0.0214	0.0682	0.1636	0.3135	0.5000
	13	0.0000	0.0000	0.0000	0.0001	0.0012	0.0072	0.0283	0.0813	0.1836	0.3388
	14	0.0000	0.0000	0.0000	0.0000	0.0003	0.0021	0.0100	0.0349	0.0937	0.2024
	15	0.0000	0.0000	0.0000	0.0000	0.0001	0.0005	0.0030	0.0128	0.0411	0.1050
	16	0.0000	0.0000	0.0000	0.0000	0.0000	0.0001	0.0008	0.0040	0.0153	0.0466
	17	0.0000	0.0000	0.0000	0.0000	0.0000	0.0000	0.0002	0.0010	0.0048	0.0173
	18	0.0000	0.0000	0.0000	0.0000	0.0000	0.0000	0.0000	0.0002	0.0012	0.0053
	19	0.0000	0.0000	0.0000	0.0000	0.0000	0.0000	0.0000	0.0000	0.0002	0.0013

APPENDIX D. (cont'd)

```
0.8159  0.9151  0.9687  0.9913  0.9983  0.9998  1.0000  1.0000  1.0000
0.6790  0.8256  0.9228  0.9736  0.9936  0.9990  0.9999  1.0000  1.0000
0.5117  0.6914  0.8377  0.9324  0.9794  0.9959  0.9996  1.0000  1.0000
0.3413  0.5237  0.7059  0.8523  0.9438  0.9856  0.9980  0.9999  1.0000
0.1971  0.3495  0.5365  0.7230  0.8701  0.9569  0.9917  0.9994  1.0000

0.0964  0.2002  0.3567  0.5505  0.7436  0.8915  0.9713  0.9967  1.0000
0.0389  0.0957  0.2009  0.3627  0.5666  0.7693  0.9173  0.9856  0.9996
0.0126  0.0370  0.0924  0.1984  0.3674  0.5860  0.8025  0.9478  0.9968
0.0031  0.0110  0.0331  0.0856  0.1917  0.3704  0.6113  0.8480  0.9811
0.0006  0.0024  0.0086  0.0271  0.0745  0.1787  0.3705  0.6484  0.9151

0.0001  0.0003  0.0014  0.0056  0.0190  0.0576  0.1550  0.3647  0.7170
0.0000  0.0000  0.0001  0.0006  0.0024  0.0092  0.0329  0.1094  0.3406


1.0000  1.0000  1.0000  1.0000  1.0000  1.0000  1.0000  1.0000  1.0000
1.0000  1.0000  1.0000  1.0000  1.0000  1.0000  1.0000  1.0000  1.0000
1.0000  1.0000  1.0000  1.0000  1.0000  1.0000  1.0000  1.0000  1.0000
1.0000  1.0000  1.0000  1.0000  1.0000  1.0000  1.0000  1.0000  1.0000
0.9999  1.0000  1.0000  1.0000  1.0000  1.0000  1.0000  1.0000  1.0000

0.9995  0.9999  1.0000  1.0000  1.0000  1.0000  1.0000  1.0000  1.0000
0.9979  0.9996  0.9999  1.0000  1.0000  1.0000  1.0000  1.0000  1.0000
0.9920  0.9981  0.9996  1.0000  1.0000  1.0000  1.0000  1.0000  1.0000
0.9757  0.9929  0.9984  0.9998  1.0000  1.0000  1.0000  1.0000  1.0000
0.9383  0.9785  0.9942  0.9989  0.9999  1.0000  1.0000  1.0000  1.0000

0.8672  0.9449  0.9820  0.9957  0.9993  0.9999  1.0000  1.0000  1.0000
0.7543  0.8793  0.9526  0.9860  0.9971  0.9996  1.0000  1.0000  1.0000
0.6037  0.7719  0.8930  0.9613  0.9900  0.9984  0.9999  1.0000  1.0000
0.4350  0.6243  0.7916  0.9084  0.9705  0.9939  0.9993  1.0000  1.0000
0.2764  0.4540  0.6466  0.8135  0.9254  0.9799  0.9970  0.9999  1.0000

0.1518  0.2898  0.4736  0.6712  0.8385  0.9439  0.9886  0.9991  1.0000
0.0705  0.1584  0.3022  0.4942  0.6994  0.8670  0.9632  0.9956  0.9999
0.0271  0.0722  0.1629  0.3134  0.5168  0.7326  0.9001  0.9818  0.9994
0.0083  0.0266  0.0716  0.1645  0.3235  0.5429  0.7738  0.9379  0.9960
0.0020  0.0076  0.0245  0.0681  0.1624  0.3320  0.5752  0.8281  0.9778

0.0003  0.0016  0.0061  0.0207  0.0606  0.1545  0.3382  0.6200  0.9052
0.0000  0.0002  0.0010  0.0041  0.0149  0.0480  0.1367  0.3392  0.6981
0.0000  0.0000  0.0001  0.0004  0.0018  0.0074  0.0280  0.0985  0.3235


1.0000  1.0000  1.0000  1.0000  1.0000  1.0000  1.0000  1.0000  1.0000
1.0000  1.0000  1.0000  1.0000  1.0000  1.0000  1.0000  1.0000  1.0000
1.0000  1.0000  1.0000  1.0000  1.0000  1.0000  1.0000  1.0000  1.0000
1.0000  1.0000  1.0000  1.0000  1.0000  1.0000  1.0000  1.0000  1.0000
1.0000  1.0000  1.0000  1.0000  1.0000  1.0000  1.0000  1.0000  1.0000

0.9997  1.0000  1.0000  1.0000  1.0000  1.0000  1.0000  1.0000  1.0000
0.9988  0.9998  1.0000  1.0000  1.0000  1.0000  1.0000  1.0000  1.0000
0.9952  0.9990  0.9998  1.0000  1.0000  1.0000  1.0000  1.0000  1.0000
0.9847  0.9960  0.9992  0.9999  1.0000  1.0000  1.0000  1.0000  1.0000
0.9589  0.9872  0.9970  0.9995  0.9999  1.0000  1.0000  1.0000  1.0000

0.9063  0.9651  0.9900  0.9979  0.9997  1.0000  1.0000  1.0000  1.0000
0.8164  0.9186  0.9717  0.9928  0.9988  0.9999  1.0000  1.0000  1.0000
0.6865  0.8364  0.9318  0.9785  0.9953  0.9994  1.0000  1.0000  1.0000
0.5278  0.7129  0.8575  0.9454  0.9851  0.9975  0.9998  1.0000  1.0000
0.3636  0.5562  0.7408  0.8799  0.9592  0.9911  0.9990  1.0000  1.0000

0.2203  0.3884  0.5860  0.7709  0.9037  0.9727  0.9958  0.9998  1.0000
0.1152  0.2373  0.4136  0.6181  0.8037  0.9285  0.9848  0.9988  1.0000
0.0510  0.1240  0.2534  0.4399  0.6537  0.8402  0.9537  0.9942  0.9999
0.0186  0.0540  0.1309  0.2688  0.4685  0.6947  0.8811  0.9774  0.9992
0.0055  0.0190  0.0551  0.1356  0.2832  0.5007  0.7440  0.9269  0.9951
```

APPENDIX D. (cont'd)

	20	0.0000	0.0000	0.0000	0.0000	0.0000	0.0000	0.0000	0.0000	0.0000	0.0002
	21	0.0000	0.0000	0.0000	0.0000	0.0000	0.0000	0.0000	0.0000	0.0000	0.0000
	22	0.0000	0.0000	0.0000	0.0000	0.0000	0.0000	0.0000	0.0000	0.0000	0.0000
	23	0.0000	0.0000	0.0000	0.0000	0.0000	0.0000	0.0000	0.0000	0.0000	0.0000
24	0	1.0000	1.0000	1.0000	1.0000	1.0000	1.0000	1.0000	1.0000	1.0000	1.0000
	1	0.7080	0.9202	0.9798	0.9953	0.9990	0.9998	1.0000	1.0000	1.0000	1.0000
	2	0.3392	0.7075	0.8941	0.9669	0.9910	0.9978	0.9995	0.9999	1.0000	1.0000
	3	0.1159	0.4357	0.7202	0.8855	0.9602	0.9881	0.9970	0.9993	0.9999	1.0000
	4	0.0298	0.2143	0.4951	0.7361	0.8850	0.9576	0.9867	0.9965	0.9992	0.9999
	5	0.0060	0.0851	0.2866	0.5401	0.7533	0.8889	0.9578	0.9865	0.9964	0.9992
	6	0.0010	0.0277	0.1394	0.3441	0.5778	0.7712	0.8956	0.9600	0.9873	0.9967
	7	0.0001	0.0075	0.0572	0.1889	0.3926	0.6114	0.7894	0.9040	0.9636	0.9887
	8	0.0000	0.0017	0.0199	0.0892	0.2338	0.4353	0.6425	0.8081	0.9137	0.9680
	9	0.0000	0.0003	0.0059	0.0362	0.1213	0.2750	0.4743	0.6721	0.8270	0.9242
	10	0.0000	0.0001	0.0015	0.0126	0.0547	0.1528	0.3133	0.5109	0.7009	0.8463
	11	0.0000	0.0000	0.0003	0.0038	0.0213	0.0742	0.1833	0.3498	0.5461	0.7294
	12	0.0000	0.0000	0.0001	0.0010	0.0072	0.0314	0.0942	0.2130	0.3849	0.5806
	13	0.0000	0.0000	0.0000	0.0002	0.0021	0.0115	0.0423	0.1143	0.2420	0.4194
	14	0.0000	0.0000	0.0000	0.0000	0.0005	0.0036	0.0164	0.0535	0.1341	0.2706
	15	0.0000	0.0000	0.0000	0.0000	0.0001	0.0010	0.0055	0.0217	0.0648	0.1537
	16	0.0000	0.0000	0.0000	0.0000	0.0000	0.0002	0.0016	0.0075	0.0269	0.0758
	17	0.0000	0.0000	0.0000	0.0000	0.0000	0.0000	0.0004	0.0022	0.0095	0.0320
	18	0.0000	0.0000	0.0000	0.0000	0.0000	0.0000	0.0001	0.0005	0.0028	0.0113
	19	0.0000	0.0000	0.0000	0.0000	0.0000	0.0000	0.0000	0.0001	0.0007	0.0033
	20	0.0000	0.0000	0.0000	0.0000	0.0000	0.0000	0.0000	0.0000	0.0001	0.0008
	21	0.0000	0.0000	0.0000	0.0000	0.0000	0.0000	0.0000	0.0000	0.0000	0.0001
	22	0.0000	0.0000	0.0000	0.0000	0.0000	0.0000	0.0000	0.0000	0.0000	0.0000
	23	0.0000	0.0000	0.0000	0.0000	0.0000	0.0000	0.0000	0.0000	0.0000	0.0000
	24	0.0000	0.0000	0.0000	0.0000	0.0000	0.0000	0.0000	0.0000	0.0000	0.0000
25	0	1.0000	1.0000	1.0000	1.0000	1.0000	1.0000	1.0000	1.0000	1.0000	1.0000
	1	0.7226	0.9282	0.9828	0.9962	0.9992	0.9999	1.0000	1.0000	1.0000	1.0000
	2	0.3576	0.7288	0.9069	0.9726	0.9930	0.9984	0.9997	0.9999	1.0000	1.0000
	3	0.1271	0.4629	0.7463	0.9018	0.9679	0.9910	0.9979	0.9996	0.9999	1.0000
	4	0.0341	0.2364	0.5289	0.7660	0.9038	0.9668	0.9903	0.9976	0.9995	0.9999
	5	0.0072	0.0980	0.3179	0.5793	0.7863	0.9095	0.9679	0.9905	0.9977	0.9995
	6	0.0012	0.0334	0.1615	0.3833	0.6217	0.8065	0.9174	0.9706	0.9914	0.9980
	7	0.0002	0.0095	0.0695	0.2200	0.4389	0.6593	0.8266	0.9264	0.9742	0.9927
	8	0.0000	0.0023	0.0255	0.1091	0.2735	0.4881	0.6939	0.8464	0.9361	0.9784
	9	0.0000	0.0005	0.0080	0.0468	0.1494	0.3231	0.5332	0.7265	0.8660	0.9461
	10	0.0000	0.0001	0.0021	0.0173	0.0713	0.1894	0.3697	0.5754	0.7576	0.8852
	11	0.0000	0.0000	0.0005	0.0056	0.0297	0.0978	0.2288	0.4142	0.6157	0.7878
	12	0.0000	0.0000	0.0001	0.0015	0.0107	0.0442	0.1254	0.2677	0.4574	0.6550
	13	0.0000	0.0000	0.0000	0.0004	0.0034	0.0175	0.0604	0.1538	0.3063	0.5000
	14	0.0000	0.0000	0.0000	0.0001	0.0009	0.0060	0.0255	0.0778	0.1827	0.3450
	15	0.0000	0.0000	0.0000	0.0000	0.0002	0.0018	0.0093	0.0344	0.0960	0.2122
	16	0.0000	0.0000	0.0000	0.0000	0.0000	0.0005	0.0029	0.0132	0.0440	0.1148
	17	0.0000	0.0000	0.0000	0.0000	0.0000	0.0001	0.0008	0.0043	0.0174	0.0539
	18	0.0000	0.0000	0.0000	0.0000	0.0000	0.0000	0.0002	0.0012	0.0058	0.0216
	19	0.0000	0.0000	0.0000	0.0000	0.0000	0.0000	0.0000	0.0003	0.0016	0.0073
	20	0.0000	0.0000	0.0000	0.0000	0.0000	0.0000	0.0000	0.0001	0.0004	0.0020
	21	0.0000	0.0000	0.0000	0.0000	0.0000	0.0000	0.0000	0.0000	0.0001	0.0005
	22	0.0000	0.0000	0.0000	0.0000	0.0000	0.0000	0.0000	0.0000	0.0000	0.0001
	23	0.0000	0.0000	0.0000	0.0000	0.0000	0.0000	0.0000	0.0000	0.0000	0.0000
	24	0.0000	0.0000	0.0000	0.0000	0.0000	0.0000	0.0000	0.0000	0.0000	0.0000
	25	0.0000	0.0000	0.0000	0.0000	0.0000	0.0000	0.0000	0.0000	0.0000	0.0000

APPENDIX D. (cont'd)

```
0.0012 0.0052 0.0181 0.0538 0.1370 0.2965 0.5396 0.8073 0.9742
0.0002 0.0010 0.0043 0.0157 0.0492 0.1332 0.3080 0.5920 0.8948
0.0000 0.0001 0.0007 0.0030 0.0116 0.0398 0.1204 0.3151 0.6794
0.0000 0.0000 0.0000 0.0003 0.0013 0.0059 0.0238 0.0886 0.3074


1.0000 1.0000 1.0000 1.0000 1.0000 1.0000 1.0000 1.0000 1.0000
1.0000 1.0000 1.0000 1.0000 1.0000 1.0000 1.0000 1.0000 1.0000
1.0000 1.0000 1.0000 1.0000 1.0000 1.0000 1.0000 1.0000 1.0000
1.0000 1.0000 1.0000 1.0000 1.0000 1.0000 1.0000 1.0000 1.0000
1.0000 1.0000 1.0000 1.0000 1.0000 1.0000 1.0000 1.0000 1.0000

0.9999 1.0000 1.0000 1.0000 1.0000 1.0000 1.0000 1.0000 1.0000
0.9993 0.9999 1.0000 1.0000 1.0000 1.0000 1.0000 1.0000 1.0000
0.9972 0.9995 0.9999 1.0000 1.0000 1.0000 1.0000 1.0000 1.0000
0.9905 0.9978 0.9996 1.0000 1.0000 1.0000 1.0000 1.0000 1.0000
0.9731 0.9925 0.9984 0.9998 1.0000 1.0000 1.0000 1.0000 1.0000

0.9352 0.9783 0.9945 0.9990 0.9999 1.0000 1.0000 1.0000 1.0000
0.8659 0.9465 0.9836 0.9964 0.9995 1.0000 1.0000 1.0000 1.0000
0.7580 0.8857 0.9577 0.9885 0.9979 0.9998 1.0000 1.0000 1.0000
0.6151 0.7870 0.9058 0.9686 0.9928 0.9990 0.9999 1.0000 1.0000
0.4539 0.6502 0.8167 0.9258 0.9787 0.9962 0.9997 1.0000 1.0000

0.2991 0.4891 0.6866 0.8472 0.9453 0.9874 0.9985 0.9999 1.0000
0.1730 0.3279 0.5257 0.7250 0.8787 0.9638 0.9941 0.9997 1.0000
0.0863 0.1919 0.3575 0.5647 0.7662 0.9108 0.9801 0.9983 1.0000
0.0364 0.0960 0.2105 0.3886 0.6074 0.8111 0.9428 0.9925 0.9999
0.0127 0.0400 0.1044 0.2288 0.4222 0.6559 0.8606 0.9723 0.9990

0.0036 0.0134 0.0422 0.1111 0.2466 0.4599 0.7134 0.9149 0.9940
0.0008 0.0035 0.0133 0.0424 0.1150 0.2639 0.5049 0.7857 0.9702
0.0001 0.0007 0.0030 0.0119 0.0398 0.1145 0.2798 0.5643 0.8841
0.0000 0.0001 0.0005 0.0022 0.0090 0.0331 0.1059 0.2925 0.6608
0.0000 0.0000 0.0000 0.0002 0.0010 0.0047 0.0202 0.0798 0.2920


1.0000 1.0000 1.0000 1.0000 1.0000 1.0000 1.0000 1.0000 1.0000
1.0000 1.0000 1.0000 1.0000 1.0000 1.0000 1.0000 1.0000 1.0000
1.0000 1.0000 1.0000 1.0000 1.0000 1.0000 1.0000 1.0000 1.0000
1.0000 1.0000 1.0000 1.0000 1.0000 1.0000 1.0000 1.0000 1.0000
1.0000 1.0000 1.0000 1.0000 1.0000 1.0000 1.0000 1.0000 1.0000

0.9999 1.0000 1.0000 1.0000 1.0000 1.0000 1.0000 1.0000 1.0000
0.9996 0.9999 1.0000 1.0000 1.0000 1.0000 1.0000 1.0000 1.0000
0.9984 0.9997 1.0000 1.0000 1.0000 1.0000 1.0000 1.0000 1.0000
0.9942 0.9988 0.9998 1.0000 1.0000 1.0000 1.0000 1.0000 1.0000
0.9826 0.9957 0.9992 0.9999 1.0000 1.0000 1.0000 1.0000 1.0000

0.9560 0.9868 0.9971 0.9995 1.0000 1.0000 1.0000 1.0000 1.0000
0.9040 0.9656 0.9907 0.9982 0.9998 1.0000 1.0000 1.0000 1.0000
0.8173 0.9222 0.9745 0.9940 0.9991 0.9999 1.0000 1.0000 1.0000
0.6937 0.8462 0.9396 0.9825 0.9966 0.9996 1.0000 1.0000 1.0000
0.5426 0.7323 0.8746 0.9558 0.9893 0.9985 0.9999 1.0000 1.0000

0.3843 0.5858 0.7712 0.9022 0.9703 0.9944 0.9995 1.0000 1.0000
0.2424 0.4246 0.6303 0.8106 0.9287 0.9827 0.9979 0.9999 1.0000
0.1340 0.2735 0.4668 0.6769 0.8506 0.9532 0.9920 0.9995 1.0000
0.0638 0.1535 0.3061 0.5118 0.7265 0.8909 0.9745 0.9977 1.0000
0.0258 0.0736 0.1734 0.3407 0.5611 0.7800 0.9305 0.9905 0.9998

0.0086 0.0294 0.0826 0.1935 0.3783 0.6167 0.8385 0.9666 0.9988
0.0023 0.0095 0.0320 0.0905 0.2137 0.4207 0.6821 0.9020 0.9928
0.0005 0.0024 0.0097 0.0332 0.0962 0.2340 0.4711 0.7636 0.9659
0.0001 0.0004 0.0021 0.0090 0.0321 0.0982 0.2537 0.5371 0.8729
0.0000 0.0001 0.0003 0.0016 0.0070 0.0274 0.0931 0.2712 0.6424

0.0000 0.0000 0.0000 0.0001 0.0008 0.0038 0.0172 0.0718 0.2774
```

APPENDIX E. Poisson Distribution—Individual Probabilities

m

X	.001	.002	.003	.004	.005	.006	.007	.008	.009	.010	.011	.012	.013	.014	.015	.016	.017	.018	.019
0	.9990	.9980	.9970	.9960	.9950	.9940	.9930	.9920	.9910	.9901	.9891	.9881	.9871	.9861	.9851	.9841	.9831	.9822	.9812
1	.0010	.0020	.0030	.0040	.0050	.0060	.0070	.0079	.0089	.0099	.0109	.0119	.0128	.0138	.0148	.0157	.0167	.0177	.0186
2	.0000	.0000	.0000	.0000	.0000	.0000	.0000	.0000	.0000	.0000	.0001	.0001	.0001	.0001	.0001	.0001	.0001	.0002	.0002

x	.020	.030	.040	.050	.060	.070	.080	.090	.100	.110	.120	.130	.140	.150	.160	.170	.180	.190	.200
0	.9802	.9704	.9608	.9512	.9418	.9324	.9231	.9139	.9048	.8958	.8869	.8781	.8694	.8607	.8521	.8437	.8353	.8270	.8187
1	.0196	.0291	.0384	.0476	.0565	.0653	.0738	.0823	.0905	.0985	.1064	.1142	.1217	.1291	.1363	.1434	.1503	.1571	.1637
2	.0002	.0004	.0008	.0012	.0017	.0023	.0030	.0037	.0045	.0054	.0064	.0074	.0085	.0097	.0109	.0122	.0135	.0149	.0164
3	.0000	.0000	.0000	.0000	.0000	.0001	.0001	.0001	.0002	.0002	.0003	.0003	.0004	.0005	.0006	.0007	.0008	.0009	.0011
4	.0000	.0000	.0000	.0000	.0000	.0000	.0000	.0000	.0000	.0000	.0000	.0000	.0000	.0000	.0000	.0000	.0000	.0000	.0001

x	.210	.220	.230	.240	.250	.260	.270	.280	.290	.300	.310	.320	.330	.340	.350	.360	.370	.380	.390
0	.8106	.8025	.7945	.7866	.7788	.7711	.7634	.7558	.7483	.7408	.7334	.7261	.7189	.7118	.7047	.6977	.6907	.6839	.6771
1	.1702	.1766	.1827	.1888	.1947	.2005	.2061	.2116	.2170	.2222	.2274	.2324	.2372	.2420	.2466	.2512	.2556	.2599	.2641
2	.0179	.0194	.0210	.0227	.0243	.0261	.0278	.0296	.0315	.0333	.0352	.0372	.0391	.0411	.0432	.0452	.0473	.0494	.0515
3	.0013	.0014	.0016	.0018	.0020	.0023	.0025	.0028	.0030	.0033	.0036	.0040	.0043	.0047	.0050	.0054	.0058	.0063	.0067
4	.0001	.0001	.0001	.0001	.0001	.0001	.0002	.0002	.0002	.0003	.0003	.0003	.0004	.0004	.0005	.0005	.0005	.0006	.0007
5	.0000	.0000	.0000	.0000	.0000	.0000	.0000	.0000	.0000	.0000	.0000	.0000	.0000	.0000	.0000	.0000	.0000	.0000	.0001

x	0.40	0.45	0.50	0.55	0.60	0.65	0.70	0.75	0.80	0.85	0.90	0.95	1.00	1.05	1.10	1.15	1.20	1.25	1.30
0	.6703	.6376	.6065	.5769	.5488	.5220	.4966	.4724	.4493	.4274	.4066	.3867	.3679	.3499	.3329	.3166	.3012	.2865	.2725
1	.2681	.2869	.3033	.3173	.3293	.3393	.3476	.3543	.3595	.3633	.3659	.3674	.3679	.3674	.3662	.3641	.3614	.3581	.3543
2	.0536	.0646	.0758	.0873	.0988	.1103	.1217	.1329	.1438	.1544	.1647	.1745	.1839	.1929	.2014	.2094	.2169	.2238	.2303
3	.0072	.0097	.0126	.0160	.0198	.0239	.0284	.0332	.0383	.0437	.0494	.0553	.0613	.0675	.0738	.0803	.0867	.0933	.0998
4	.0007	.0011	.0016	.0022	.0030	.0039	.0050	.0062	.0077	.0093	.0111	.0131	.0153	.0177	.0203	.0231	.0260	.0291	.0324
5	.0001	.0001	.0002	.0002	.0004	.0005	.0007	.0009	.0012	.0016	.0020	.0025	.0031	.0037	.0045	.0053	.0062	.0073	.0084
6	.0000	.0000	.0000	.0000	.0000	.0001	.0001	.0001	.0002	.0002	.0003	.0004	.0005	.0007	.0008	.0010	.0012	.0015	.0018
7	.0000	.0000	.0000	.0000	.0000	.0000	.0000	.0000	.0000	.0000	.0000	.0001	.0001	.0001	.0001	.0001	.0002	.0003	.0003
8	.0000	.0000	.0000	.0000	.0000	.0000	.0000	.0000	.0000	.0000	.0000	.0000	.0000	.0000	.0000	.0000	.0000	.0000	.0001

m

APPENDIX E. (cont'd)

x	1.4	1.5	1.6	1.7	1.8	1.9	2.0	2.1	2.2	2.3	2.4	2.5	2.6	2.7	2.8	2.9	3.0	3.1	3.2
0	.2466	.2231	.2019	.1827	.1653	.1496	.1353	.1225	.1108	.1003	.0907	.0821	.0743	.0672	.0608	.0550	.0498	.0450	.0408
1	.3452	.3347	.3230	.3106	.2975	.2842	.2707	.2572	.2438	.2306	.2177	.2052	.1931	.1815	.1703	.1596	.1494	.1397	.1304
2	.2417	.2510	.2584	.2640	.2678	.2700	.2707	.2700	.2681	.2652	.2613	.2565	.2510	.2450	.2384	.2314	.2240	.2165	.2087
3	.1128	.1255	.1378	.1496	.1607	.1710	.1804	.1890	.1966	.2033	.2090	.2138	.2176	.2205	.2225	.2237	.2240	.2237	.2226
4	.0395	.0471	.0551	.0636	.0723	.0812	.0902	.0992	.1082	.1169	.1254	.1336	.1414	.1488	.1557	.1622	.1680	.1733	.1781
5	.0111	.0141	.0176	.0216	.0260	.0309	.0361	.0417	.0476	.0538	.0602	.0668	.0735	.0804	.0872	.0940	.1008	.1075	.1140
6	.0026	.0035	.0047	.0061	.0078	.0098	.0120	.0146	.0174	.0206	.0241	.0278	.0319	.0362	.0407	.0455	.0504	.0555	.0608
7	.0005	.0008	.0011	.0015	.0020	.0027	.0034	.0044	.0055	.0068	.0083	.0099	.0118	.0139	.0163	.0188	.0216	.0246	.0278
8	.0001	.0001	.0002	.0003	.0005	.0006	.0009	.0011	.0015	.0019	.0025	.0031	.0038	.0047	.0057	.0068	.0081	.0095	.0111
9	.0000	.0000	.0000	.0001	.0001	.0001	.0002	.0003	.0004	.0005	.0007	.0009	.0011	.0014	.0018	.0022	.0027	.0033	.0040
10	.0000	.0000	.0000	.0000	.0000	.0000	.0000	.0001	.0001	.0001	.0002	.0002	.0003	.0004	.0005	.0006	.0008	.0010	.0013
11	.0000	.0000	.0000	.0000	.0000	.0000	.0000	.0000	.0000	.0000	.0000	.0000	.0001	.0001	.0001	.0002	.0002	.0003	.0004
12	.0000	.0000	.0000	.0000	.0000	.0000	.0000	.0000	.0000	.0000	.0000	.0000	.0000	.0000	.0000	.0000	.0001	.0001	.0001

x	3.3	3.4	3.5	3.6	3.7	3.8	3.9	4.0	4.1	4.2	4.3	4.4	4.5	4.6	4.7	4.8	4.9	5.0	5.1
0	.0369	.0334	.0302	.0273	.0247	.0224	.0202	.0183	.0166	.0150	.0136	.0123	.0111	.0101	.0091	.0082	.0074	.0067	.0061
1	.1217	.1135	.1057	.0984	.0915	.0850	.0789	.0733	.0679	.0630	.0583	.0540	.0500	.0462	.0427	.0395	.0365	.0337	.0311
2	.2008	.1929	.1850	.1771	.1692	.1615	.1539	.1465	.1393	.1323	.1254	.1188	.1125	.1063	.1005	.0948	.0894	.0842	.0793
3	.2209	.2186	.2158	.2125	.2087	.2046	.2001	.1954	.1904	.1852	.1798	.1743	.1687	.1631	.1574	.1517	.1460	.1404	.1348
4	.1823	.1858	.1888	.1912	.1931	.1944	.1951	.1954	.1951	.1944	.1933	.1917	.1898	.1875	.1849	.1820	.1789	.1755	.1719
5	.1203	.1264	.1322	.1377	.1429	.1477	.1522	.1563	.1600	.1633	.1662	.1687	.1708	.1725	.1738	.1747	.1753	.1755	.1753
6	.0662	.0716	.0771	.0826	.0881	.0936	.0989	.1042	.1093	.1143	.1191	.1237	.1281	.1323	.1362	.1398	.1432	.1462	.1490
7	.0312	.0348	.0385	.0425	.0466	.0508	.0551	.0595	.0640	.0686	.0732	.0778	.0824	.0869	.0914	.0959	.1002	.1044	.1086
8	.0129	.0148	.0169	.0191	.0215	.0241	.0269	.0298	.0328	.0360	.0393	.0428	.0463	.0500	.0537	.0575	.0614	.0653	.0692
9	.0047	.0056	.0066	.0076	.0089	.0102	.0116	.0132	.0150	.0168	.0188	.0209	.0232	.0255	.0280	.0307	.0334	.0363	.0392
10	.0016	.0019	.0023	.0028	.0033	.0039	.0045	.0053	.0061	.0071	.0081	.0092	.0104	.0118	.0132	.0147	.0164	.0181	.0200
11	.0005	.0006	.0007	.0009	.0011	.0013	.0016	.0019	.0023	.0027	.0032	.0037	.0043	.0049	.0056	.0064	.0073	.0082	.0093
12	.0001	.0002	.0002	.0003	.0004	.0004	.0005	.0006	.0008	.0009	.0011	.0013	.0016	.0019	.0022	.0026	.0030	.0034	.0039
13	.0000	.0000	.0001	.0001	.0001	.0001	.0002	.0002	.0002	.0003	.0004	.0005	.0006	.0007	.0008	.0009	.0011	.0013	.0015
14	.0000	.0000	.0000	.0000	.0000	.0000	.0000	.0001	.0001	.0001	.0001	.0001	.0002	.0002	.0003	.0003	.0004	.0005	.0006
15	.0000	.0000	.0000	.0000	.0000	.0000	.0000	.0000	.0000	.0000	.0000	.0000	.0001	.0001	.0001	.0001	.0001	.0002	.0002
16	.0000	.0000	.0000	.0000	.0000	.0000	.0000	.0000	.0000	.0000	.0000	.0000	.0000	.0000	.0000	.0000	.0000	.0000	.0001

APPENDIX E. (cont'd)

x	5.2	5.3	5.4	5.5	5.6	5.7	5.8	5.9	6.0	6.1	6.2	6.3	6.4	6.5	6.6	6.7	6.8	6.9	7.0
0	.0055	.0050	.0045	.0041	.0037	.0033	.0030	.0027	.0025	.0022	.0020	.0018	.0017	.0015	.0014	.0012	.0011	.0010	.0009
1	.0287	.0265	.0244	.0225	.0207	.0191	.0176	.0162	.0149	.0137	.0126	.0116	.0106	.0098	.0090	.0082	.0076	.0070	.0064
2	.0746	.0701	.0659	.0618	.0580	.0544	.0509	.0477	.0446	.0417	.0390	.0364	.0340	.0318	.0296	.0276	.0258	.0240	.0223
3	.1293	.1239	.1185	.1133	.1082	.1033	.0985	.0938	.0892	.0848	.0806	.0765	.0726	.0688	.0652	.0617	.0584	.0552	.0521
4	.1681	.1641	.1600	.1558	.1515	.1472	.1428	.1383	.1339	.1294	.1249	.1205	.1162	.1118	.1076	.1034	.0992	.0952	.0912
5	.1748	.1740	.1728	.1714	.1697	.1678	.1656	.1632	.1606	.1579	.1549	.1519	.1487	.1454	.1420	.1385	.1349	.1314	.1277
6	.1515	.1537	.1555	.1571	.1584	.1594	.1601	.1605	.1606	.1605	.1601	.1595	.1586	.1575	.1562	.1546	.1529	.1511	.1490
7	.1125	.1163	.1200	.1234	.1267	.1298	.1326	.1353	.1377	.1399	.1418	.1435	.1450	.1462	.1472	.1480	.1486	.1489	.1490
8	.0731	.0771	.0810	.0849	.0887	.0925	.0962	.0998	.1033	.1066	.1099	.1130	.1160	.1188	.1215	.1240	.1263	.1284	.1304
9	.0423	.0454	.0486	.0519	.0552	.0586	.0620	.0654	.0688	.0723	.0757	.0791	.0825	.0858	.0891	.0923	.0954	.0985	.1014
10	.0220	.0241	.0262	.0285	.0309	.0334	.0359	.0386	.0413	.0441	.0469	.0498	.0528	.0558	.0588	.0618	.0649	.0679	.0710
11	.0104	.0116	.0129	.0143	.0157	.0173	.0190	.0207	.0225	.0244	.0265	.0285	.0307	.0330	.0353	.0377	.0401	.0426	.0452
12	.0045	.0051	.0058	.0065	.0073	.0082	.0092	.0102	.0113	.0124	.0137	.0150	.0164	.0179	.0194	.0210	.0227	.0245	.0263
13	.0018	.0021	.0024	.0028	.0032	.0036	.0041	.0046	.0052	.0058	.0065	.0073	.0081	.0089	.0099	.0108	.0119	.0130	.0142
14	.0007	.0008	.0009	.0011	.0013	.0015	.0017	.0019	.0022	.0025	.0029	.0033	.0037	.0041	.0046	.0052	.0058	.0064	.0071
15	.0002	.0003	.0003	.0004	.0005	.0006	.0007	.0008	.0009	.0010	.0012	.0014	.0016	.0018	.0020	.0023	.0026	.0029	.0033
16	.0001	.0001	.0001	.0001	.0002	.0002	.0002	.0003	.0003	.0004	.0005	.0005	.0006	.0007	.0008	.0010	.0011	.0013	.0014
17	.0000	.0000	.0000	.0000	.0001	.0001	.0001	.0001	.0001	.0001	.0002	.0002	.0002	.0003	.0003	.0004	.0004	.0005	.0006
18	.0000	.0000	.0000	.0000	.0000	.0000	.0000	.0000	.0000	.0000	.0001	.0001	.0001	.0001	.0001	.0001	.0002	.0002	.0002
19	.0000	.0000	.0000	.0000	.0000	.0000	.0000	.0000	.0000	.0000	.0000	.0000	.0000	.0000	.0000	.0001	.0001	.0001	.0001

x	7.5	8.0	8.5	9.0	9.5	10.0	10.5	11.0	11.5	12.0	12.5	13.0	13.5	14.0	14.5	15.0	20.0	25.0	30.0
0	.0006	.0003	.0002	.0001	.0001	.0000	.0000	.0000	.0000	.0000	.0000	.0000	.0000	.0000	.0000	.0000	.0000	.0000	.0000
1	.0041	.0027	.0017	.0011	.0007	.0005	.0003	.0002	.0001	.0001	.0000	.0000	.0000	.0000	.0000	.0000	.0000	.0000	.0000
2	.0156	.0107	.0074	.0050	.0034	.0023	.0015	.0010	.0007	.0004	.0003	.0002	.0001	.0001	.0001	.0000	.0000	.0000	.0000
3	.0389	.0286	.0208	.0150	.0107	.0076	.0053	.0037	.0026	.0018	.0012	.0008	.0006	.0004	.0003	.0002	.0000	.0000	.0000
4	.0729	.0573	.0443	.0337	.0254	.0189	.0139	.0102	.0074	.0053	.0038	.0027	.0019	.0013	.0009	.0006	.0000	.0000	.0000
5	.1094	.0916	.0752	.0607	.0483	.0378	.0293	.0224	.0170	.0127	.0095	.0070	.0051	.0037	.0027	.0019	.0001	.0000	.0000
6	.1367	.1221	.1066	.0911	.0764	.0631	.0513	.0411	.0325	.0255	.0197	.0152	.0115	.0087	.0065	.0048	.0002	.0000	.0000
7	.1465	.1396	.1294	.1171	.1037	.0901	.0769	.0646	.0535	.0437	.0353	.0281	.0222	.0174	.0135	.0104	.0005	.0000	.0000
8	.1373	.1396	.1375	.1318	.1232	.1126	.1009	.0888	.0769	.0655	.0551	.0457	.0375	.0304	.0244	.0194	.0013	.0001	.0000
9	.1144	.1241	.1299	.1318	.1300	.1251	.1177	.1085	.0982	.0874	.0765	.0661	.0563	.0473	.0394	.0324	.0029	.0001	.0000
10	.0858	.0993	.1104	.1186	.1235	.1251	.1236	.1194	.1129	.1048	.0956	.0859	.0760	.0663	.0571	.0486	.0058	.0004	.0000
11	.0585	.0722	.0853	.0970	.1067	.1137	.1180	.1194	.1181	.1144	.1087	.1015	.0932	.0844	.0753	.0663	.0106	.0008	.0000
12	.0366	.0481	.0604	.0728	.0844	.0948	.1032	.1094	.1131	.1144	.1132	.1099	.1049	.0984	.0910	.0829	.0176	.0017	.0001
13	.0211	.0296	.0395	.0504	.0617	.0729	.0834	.0926	.1001	.1056	.1089	.1099	.1089	.1060	.1014	.0956	.0271	.0033	.0002
14	.0113	.0169	.0240	.0324	.0419	.0521	.0625	.0728	.0822	.0905	.0972	.1021	.1050	.1060	.1051	.1024	.0387	.0059	.0005

474

APPENDIX E. (cont'd)

15	.0057	.0090	.0136	.0194	.0265	.0347	.0438	.0534	.0630	.0724	.0810	.0885	.0945	.0989	.1016	.1024	.0516	.0099	.0010	
16	.0026	.0045	.0072	.0109	.0157	.0217	.0287	.0367	.0453	.0543	.0633	.0719	.0798	.0866	.0920	.0960	.0646	.0155	.0019	
17	.0012	.0021	.0036	.0058	.0088	.0128	.0177	.0237	.0306	.0383	.0465	.0550	.0633	.0713	.0785	.0847	.0760	.0227	.0034	
18	.0005	.0009	.0017	.0029	.0046	.0071	.0104	.0145	.0196	.0255	.0323	.0397	.0475	.0554	.0632	.0706	.0844	.0316	.0057	
19	.0002	.0004	.0008	.0014	.0023	.0037	.0057	.0084	.0119	.0161	.0213	.0272	.0337	.0409	.0483	.0557	.0888	.0415	.0089	
20	.0001	.0002	.0003	.0006	.0011	.0019	.0030	.0046	.0068	.0097	.0133	.0177	.0228	.0286	.0350	.0418	.0888	.0519	.0134	
21	.0000	.0001	.0001	.0003	.0005	.0009	.0015	.0024	.0037	.0055	.0079	.0109	.0146	.0191	.0242	.0299	.0846	.0618	.0192	
22	.0000	.0000	.0001	.0001	.0002	.0004	.0007	.0012	.0020	.0030	.0045	.0065	.0090	.0121	.0159	.0204	.0769	.0702	.0261	
23	.0000	.0000	.0000	.0000	.0001	.0002	.0003	.0006	.0010	.0016	.0024	.0037	.0053	.0074	.0100	.0133	.0669	.0763	.0341	
24	.0000	.0000	.0000	.0000	.0000	.0001	.0001	.0003	.0005	.0008	.0013	.0020	.0030	.0043	.0061	.0083	.0557	.0795	.0426	
25	.0000	.0000	.0000	.0000	.0000	.0000	.0001	.0001	.0002	.0004	.0006	.0010	.0016	.0024	.0035	.0050	.0446	.0795	.0511	
26	.0000	.0000	.0000	.0000	.0000	.0000	.0000	.0000	.0001	.0002	.0003	.0005	.0008	.0013	.0020	.0029	.0343	.0765	.0590	
27	.0000	.0000	.0000	.0000	.0000	.0000	.0000	.0000	.0000	.0001	.0001	.0002	.0004	.0007	.0011	.0016	.0254	.0708	.0655	
28	.0000	.0000	.0000	.0000	.0000	.0000	.0000	.0000	.0000	.0000	.0001	.0001	.0002	.0003	.0005	.0009	.0181	.0632	.0702	
29	.0000	.0000	.0000	.0000	.0000	.0000	.0000	.0000	.0000	.0000	.0000	.0001	.0001	.0002	.0003	.0004	.0125	.0545	.0726	
30	.0000	.0000	.0000	.0000	.0000	.0000	.0000	.0000	.0000	.0000	.0000	.0000	.0001	.0001	.0001	.0002	.0083	.0454	.0726	
31	.0000	.0000	.0000	.0000	.0000	.0000	.0000	.0000	.0000	.0000	.0000	.0000	.0000	.0000	.0001	.0001	.0054	.0366	.0703	
32	.0000	.0000	.0000	.0000	.0000	.0000	.0000	.0000	.0000	.0000	.0000	.0000	.0000	.0000	.0000	.0001	.0034	.0286	.0659	
33	.0000	.0000	.0000	.0000	.0000	.0000	.0000	.0000	.0000	.0000	.0000	.0000	.0000	.0000	.0000	.0000	.0020	.0217	.0599	
34	.0000	.0000	.0000	.0000	.0000	.0000	.0000	.0000	.0000	.0000	.0000	.0000	.0000	.0000	.0000	.0000	.0012	.0159	.0529	
35	.0000	.0000	.0000	.0000	.0000	.0000	.0000	.0000	.0000	.0000	.0000	.0000	.0000	.0000	.0000	.0000	.0007	.0114	.0453	
36	.0000	.0000	.0000	.0000	.0000	.0000	.0000	.0000	.0000	.0000	.0000	.0000	.0000	.0000	.0000	.0000	.0004	.0079	.0378	
37	.0000	.0000	.0000	.0000	.0000	.0000	.0000	.0000	.0000	.0000	.0000	.0000	.0000	.0000	.0000	.0000	.0002	.0053	.0306	
38	.0000	.0000	.0000	.0000	.0000	.0000	.0000	.0000	.0000	.0000	.0000	.0000	.0000	.0000	.0000	.0000	.0001	.0035	.0242	
39	.0000	.0000	.0000	.0000	.0000	.0000	.0000	.0000	.0000	.0000	.0000	.0000	.0000	.0000	.0000	.0000	.0001	.0023	.0186	
40	.0000	.0000	.0000	.0000	.0000	.0000	.0000	.0000	.0000	.0000	.0000	.0000	.0000	.0000	.0000	.0000	.0000	.0014	.0139	
41	.0000	.0000	.0000	.0000	.0000	.0000	.0000	.0000	.0000	.0000	.0000	.0000	.0000	.0000	.0000	.0000	.0000	.0009	.0102	
42	.0000	.0000	.0000	.0000	.0000	.0000	.0000	.0000	.0000	.0000	.0000	.0000	.0000	.0000	.0000	.0000	.0000	.0005	.0073	
43	.0000	.0000	.0000	.0000	.0000	.0000	.0000	.0000	.0000	.0000	.0000	.0000	.0000	.0000	.0000	.0000	.0000	.0003	.0051	
44	.0000	.0000	.0000	.0000	.0000	.0000	.0000	.0000	.0000	.0000	.0000	.0000	.0000	.0000	.0000	.0000	.0000	.0002	.0035	
45	.0000	.0000	.0000	.0000	.0000	.0000	.0000	.0000	.0000	.0000	.0000	.0000	.0000	.0000	.0000	.0000	.0000	.0001	.0023	
46	.0000	.0000	.0000	.0000	.0000	.0000	.0000	.0000	.0000	.0000	.0000	.0000	.0000	.0000	.0000	.0000	.0000	.0001	.0015	
47	.0000	.0000	.0000	.0000	.0000	.0000	.0000	.0000	.0000	.0000	.0000	.0000	.0000	.0000	.0000	.0000	.0000	.0000	.0010	
48	.0000	.0000	.0000	.0000	.0000	.0000	.0000	.0000	.0000	.0000	.0000	.0000	.0000	.0000	.0000	.0000	.0000	.0000	.0006	
49	.0000	.0000	.0000	.0000	.0000	.0000	.0000	.0000	.0000	.0000	.0000	.0000	.0000	.0000	.0000	.0000	.0000	.0000	.0004	
50	.0000	.0000	.0000	.0000	.0000	.0000	.0000	.0000	.0000	.0000	.0000	.0000	.0000	.0000	.0000	.0000	.0000	.0000	.0002	
51	.0000	.0000	.0000	.0000	.0000	.0000	.0000	.0000	.0000	.0000	.0000	.0000	.0000	.0000	.0000	.0000	.0000	.0000	.0001	
52	.0000	.0000	.0000	.0000	.0000	.0000	.0000	.0000	.0000	.0000	.0000	.0000	.0000	.0000	.0000	.0000	.0000	.0000	.0001	

APPENDIX F. Poisson Distribution — Cumulative Probabilities

m

x	0.001	0.002	0.003	0.004	0.005	0.006	0.007	0.008	0.009	0.010	0.011	0.012	0.013	0.014	0.015
0	1.0000	1.0000	1.0000	1.0000	1.0000	1.0000	1.0000	1.0000	1.0000	1.0000	1.0000	1.0000	1.0000	1.0000	1.0000
1	0.0010	0.0020	0.0030	0.0040	0.0050	0.0060	0.0070	0.0080	0.0090	0.0099	0.0109	0.0119	0.0129	0.0139	0.0149
2	0.0000	0.0000	0.0000	0.0000	0.0000	0.0000	0.0000	0.0000	0.0000	0.0000	0.0001	0.0001	0.0001	0.0001	0.0001

x	0.016	0.017	0.018	0.019	0.020	0.030	0.040	0.050	0.060	0.070	0.080	0.090	0.100	0.110	0.120
0	1.0000	1.0000	1.0000	1.0000	1.0000	1.0000	1.0000	1.0000	1.0000	1.0000	1.0000	1.0000	1.0000	1.0000	1.0000
1	0.0159	0.0169	0.0178	0.0188	0.0198	0.0296	0.0392	0.0488	0.0582	0.0676	0.0769	0.0861	0.0952	0.1042	0.1131
2	0.0001	0.0001	0.0002	0.0002	0.0002	0.0004	0.0008	0.0012	0.0017	0.0023	0.0030	0.0038	0.0047	0.0056	0.0066
3	0.0000	0.0000	0.0000	0.0000	0.0000	0.0000	0.0000	0.0000	0.0000	0.0001	0.0001	0.0001	0.0002	0.0002	0.0003

x	0.130	0.140	0.150	0.160	0.170	0.180	0.190	0.200	0.210	0.220	0.230	0.240	0.250	0.260	0.270
0	1.0000	1.0000	1.0000	1.0000	1.0000	1.0000	1.0000	1.0000	1.0000	1.0000	1.0000	1.0000	1.0000	1.0000	1.0000
1	0.1219	0.1306	0.1393	0.1479	0.1563	0.1647	0.1730	0.1813	0.1894	0.1975	0.2055	0.2134	0.2212	0.2289	0.2366
2	0.0078	0.0089	0.0102	0.0115	0.0129	0.0144	0.0159	0.0175	0.0192	0.0209	0.0227	0.0246	0.0265	0.0285	0.0305
3	0.0003	0.0004	0.0005	0.0006	0.0007	0.0008	0.0010	0.0011	0.0013	0.0015	0.0017	0.0019	0.0022	0.0024	0.0027
4	0.0000	0.0000	0.0000	0.0000	0.0000	0.0000	0.0000	0.0001	0.0001	0.0001	0.0001	0.0001	0.0001	0.0001	0.0002

x	0.280	0.290	0.300	0.310	0.320	0.330	0.340	0.350	0.360	0.370	0.380	0.390	0.400	0.450	0.500
0	1.0000	1.0000	1.0000	1.0000	1.0000	1.0000	1.0000	1.0000	1.0000	1.0000	1.0000	1.0000	1.0000	1.0000	1.0000
1	0.2442	0.2517	0.2592	0.2666	0.2738	0.2811	0.2882	0.2953	0.3023	0.3093	0.3161	0.3229	0.3297	0.3624	0.3935
2	0.0326	0.0347	0.0369	0.0392	0.0415	0.0438	0.0462	0.0487	0.0512	0.0537	0.0563	0.0589	0.0615	0.0754	0.0902
3	0.0030	0.0033	0.0036	0.0039	0.0043	0.0047	0.0051	0.0055	0.0059	0.0064	0.0069	0.0074	0.0079	0.0109	0.0144
4	0.0002	0.0002	0.0003	0.0003	0.0003	0.0004	0.0004	0.0005	0.0005	0.0006	0.0006	0.0007	0.0008	0.0012	0.0017
5	0.0000	0.0000	0.0000	0.0000	0.0000	0.0000	0.0000	0.0000	0.0000	0.0000	0.0000	0.0001	0.0001	0.0001	0.0002

x	0.550	0.600	0.650	0.700	0.750	0.800	0.850	0.900	0.950	1.000	1.050	1.100	1.150	1.200	1.250
0	1.0000	1.0000	1.0000	1.0000	1.0000	1.0000	1.0000	1.0000	1.0000	1.0000	1.0000	1.0000	1.0000	1.0000	1.0000
1	0.4231	0.4512	0.4780	0.5034	0.5276	0.5507	0.5726	0.5934	0.6133	0.6321	0.6500	0.6671	0.6833	0.6988	0.7134
2	0.1057	0.1219	0.1386	0.1558	0.1734	0.1912	0.2093	0.2275	0.2458	0.2642	0.2826	0.3009	0.3192	0.3373	0.3553
3	0.0185	0.0231	0.0283	0.0341	0.0405	0.0474	0.0549	0.0629	0.0713	0.0803	0.0897	0.0996	0.1098	0.1205	0.1315
4	0.0025	0.0034	0.0044	0.0058	0.0073	0.0091	0.0111	0.0135	0.0161	0.0190	0.0222	0.0257	0.0296	0.0337	0.0382
5	0.0003	0.0004	0.0006	0.0008	0.0011	0.0014	0.0018	0.0023	0.0029	0.0036	0.0045	0.0054	0.0065	0.0077	0.0091
6	0.0000	0.0000	0.0001	0.0001	0.0001	0.0002	0.0003	0.0003	0.0004	0.0006	0.0007	0.0009	0.0012	0.0015	0.0018
7	0.0000	0.0000	0.0000	0.0000	0.0000	0.0000	0.0000	0.0000	0.0001	0.0001	0.0001	0.0001	0.0002	0.0002	0.0003

APPENDIX F. (cont'd)

x	1.300	1.400	1.500	1.600	1.700	1.800	1.900	2.000	2.100	2.200	2.300	2.400	2.500	2.600	2.700
0	1.0000	1.0000	1.0000	1.0000	1.0000	1.0000	1.0000	1.0000	1.0000	1.0000	1.0000	1.0000	1.0000	1.0000	1.0000
1	0.7275	0.7534	0.7769	0.7981	0.8173	0.8347	0.8504	0.8647	0.8775	0.8892	0.8997	0.9093	0.9179	0.9257	0.9328
2	0.3732	0.4082	0.4422	0.4751	0.5068	0.5372	0.5662	0.5940	0.6204	0.6454	0.6691	0.6915	0.7127	0.7326	0.7513
3	0.1429	0.1665	0.1912	0.2166	0.2428	0.2694	0.2963	0.3233	0.3504	0.3773	0.4040	0.4303	0.4562	0.4815	0.5063
4	0.0431	0.0537	0.0656	0.0788	0.0932	0.1087	0.1253	0.1429	0.1614	0.1806	0.2006	0.2213	0.2424	0.2640	0.2859
5	0.0107	0.0143	0.0186	0.0237	0.0296	0.0364	0.0441	0.0527	0.0621	0.0725	0.0837	0.0959	0.1088	0.1226	0.1371
6	0.0022	0.0032	0.0045	0.0060	0.0080	0.0104	0.0132	0.0166	0.0204	0.0249	0.0300	0.0357	0.0420	0.0490	0.0567
7	0.0004	0.0006	0.0009	0.0013	0.0019	0.0026	0.0034	0.0045	0.0059	0.0075	0.0094	0.0116	0.0142	0.0172	0.0205
8	0.0001	0.0001	0.0002	0.0003	0.0004	0.0006	0.0008	0.0011	0.0015	0.0020	0.0026	0.0033	0.0042	0.0053	0.0066
9	0.0000	0.0000	0.0000	0.0000	0.0001	0.0001	0.0002	0.0002	0.0003	0.0005	0.0006	0.0009	0.0011	0.0015	0.0019
10	0.0000	0.0000	0.0000	0.0000	0.0000	0.0000	0.0000	0.0000	0.0001	0.0001	0.0001	0.0002	0.0003	0.0004	0.0005
11	0.0000	0.0000	0.0000	0.0000	0.0000	0.0000	0.0000	0.0000	0.0000	0.0000	0.0000	0.0000	0.0000	0.0001	0.0001

x	2.800	2.900	3.000	3.100	3.200	3.300	3.400	3.500	3.600	3.700	3.800	3.900	4.000	4.100	4.200
0	1.0000	1.0000	1.0000	1.0000	1.0000	1.0000	1.0000	1.0000	1.0000	1.0000	1.0000	1.0000	1.0000	1.0000	1.0000
1	0.9392	0.9450	0.9502	0.9549	0.9592	0.9631	0.9666	0.9698	0.9727	0.9753	0.9776	0.9797	0.9817	0.9834	0.9850
2	0.7689	0.7854	0.8009	0.8153	0.8288	0.8414	0.8532	0.8641	0.8743	0.8838	0.8926	0.9008	0.9084	0.9155	0.9220
3	0.5305	0.5540	0.5768	0.5988	0.6201	0.6406	0.6603	0.6791	0.6972	0.7146	0.7311	0.7469	0.7619	0.7762	0.7897
4	0.3081	0.3304	0.3528	0.3752	0.3975	0.4197	0.4416	0.4634	0.4848	0.5058	0.5265	0.5467	0.5665	0.5858	0.6046
5	0.1523	0.1682	0.1847	0.2018	0.2194	0.2374	0.2558	0.2745	0.2936	0.3128	0.3321	0.3516	0.3711	0.3907	0.4101
6	0.0651	0.0742	0.0839	0.0943	0.1054	0.1171	0.1295	0.1424	0.1559	0.1699	0.1844	0.1994	0.2148	0.2307	0.2468
7	0.0244	0.0287	0.0335	0.0388	0.0446	0.0510	0.0578	0.0653	0.0733	0.0818	0.0909	0.1005	0.1107	0.1213	0.1325
8	0.0081	0.0099	0.0119	0.0142	0.0168	0.0198	0.0231	0.0267	0.0308	0.0352	0.0401	0.0454	0.0511	0.0573	0.0639
9	0.0024	0.0031	0.0038	0.0047	0.0057	0.0069	0.0083	0.0099	0.0117	0.0137	0.0160	0.0185	0.0213	0.0245	0.0279
10	0.0007	0.0009	0.0011	0.0014	0.0018	0.0022	0.0027	0.0033	0.0040	0.0048	0.0058	0.0069	0.0081	0.0095	0.0111
11	0.0002	0.0002	0.0003	0.0004	0.0005	0.0006	0.0008	0.0010	0.0013	0.0016	0.0019	0.0023	0.0028	0.0034	0.0040
12	0.0000	0.0001	0.0001	0.0001	0.0001	0.0002	0.0002	0.0003	0.0004	0.0005	0.0006	0.0007	0.0008	0.0011	0.0013
13	0.0000	0.0000	0.0000	0.0000	0.0000	0.0000	0.0001	0.0001	0.0001	0.0001	0.0002	0.0002	0.0003	0.0003	0.0004
14	0.0000	0.0000	0.0000	0.0000	0.0000	0.0000	0.0000	0.0000	0.0000	0.0000	0.0000	0.0000	0.0001	0.0001	0.0001

x	4.300	4.400	4.500	4.600	4.700	4.800	4.900	5.000	5.100	5.200	5.300	5.400	5.500	5.600	5.700
0	1.0000	1.0000	1.0000	1.0000	1.0000	1.0000	1.0000	1.0000	1.0000	1.0000	1.0000	1.0000	1.0000	1.0000	1.0000
1	0.9864	0.9877	0.9889	0.9899	0.9909	0.9918	0.9925	0.9933	0.9939	0.9945	0.9950	0.9955	0.9959	0.9963	0.9966
2	0.9281	0.9337	0.9389	0.9437	0.9482	0.9523	0.9561	0.9596	0.9628	0.9658	0.9685	0.9711	0.9734	0.9756	0.9776
3	0.8026	0.8149	0.8264	0.8374	0.8477	0.8575	0.8667	0.8753	0.8835	0.8912	0.8984	0.9052	0.9116	0.9176	0.9232
4	0.6228	0.6406	0.6577	0.6743	0.6903	0.7058	0.7206	0.7350	0.7487	0.7619	0.7746	0.7867	0.7983	0.8094	0.8199

477

APPENDIX F. (cont'd)

x	5.800	5.900	6.000	6.100	6.200	6.300	6.400	6.500	6.600	6.700	6.800	6.900	7.000	7.500	8.000
5	0.4296	0.4488	0.4679	0.4868	0.5054	0.5237	0.5418	0.5595	0.5769	0.5939	0.6105	0.6267	0.6425	0.6578	0.6728
6	0.2633	0.2801	0.2971	0.3142	0.3316	0.3490	0.3665	0.3840	0.4016	0.4191	0.4365	0.4539	0.4711	0.4881	0.5050
7	0.1442	0.1564	0.1689	0.1820	0.1954	0.2092	0.2233	0.2378	0.2526	0.2676	0.2829	0.2983	0.3139	0.3297	0.3456
8	0.0710	0.0786	0.0866	0.0950	0.1040	0.1133	0.1231	0.1334	0.1440	0.1551	0.1665	0.1783	0.1905	0.2030	0.2158
9	0.0317	0.0358	0.0403	0.0451	0.0503	0.0558	0.0618	0.0681	0.0748	0.0819	0.0894	0.0973	0.1056	0.1143	0.1234
10	0.0129	0.0149	0.0171	0.0195	0.0222	0.0251	0.0283	0.0318	0.0356	0.0397	0.0440	0.0487	0.0538	0.0591	0.0648
11	0.0048	0.0057	0.0067	0.0078	0.0090	0.0104	0.0120	0.0137	0.0156	0.0177	0.0200	0.0225	0.0252	0.0282	0.0314
12	0.0017	0.0020	0.0024	0.0029	0.0034	0.0040	0.0047	0.0054	0.0063	0.0073	0.0084	0.0096	0.0110	0.0125	0.0141
13	0.0005	0.0007	0.0008	0.0010	0.0012	0.0014	0.0017	0.0020	0.0024	0.0028	0.0033	0.0038	0.0044	0.0051	0.0059
14	0.0002	0.0002	0.0003	0.0003	0.0004	0.0005	0.0006	0.0007	0.0008	0.0010	0.0012	0.0014	0.0017	0.0020	0.0023
15	0.0000	0.0001	0.0001	0.0001	0.0001	0.0001	0.0002	0.0002	0.0003	0.0003	0.0004	0.0005	0.0006	0.0007	0.0008
16	0.0000	0.0000	0.0000	0.0000	0.0000	0.0000	0.0001	0.0001	0.0001	0.0001	0.0001	0.0001	0.0002	0.0002	0.0003
17	0.0000	0.0000	0.0000	0.0000	0.0000	0.0000	0.0000	0.0000	0.0000	0.0000	0.0000	0.0000	0.0000	0.0001	0.0001

x	5.800	5.900	6.000	6.100	6.200	6.300	6.400	6.500	6.600	6.700	6.800	6.900	7.000	7.500	8.000
0	1.0000	1.0000	1.0000	1.0000	1.0000	1.0000	1.0000	1.0000	1.0000	1.0000	1.0000	1.0000	1.0000	1.0000	1.0000
1	0.9970	0.9973	0.9975	0.9978	0.9980	0.9982	0.9983	0.9985	0.9986	0.9988	0.9989	0.9990	0.9991	0.9994	0.9996
2	0.9794	0.9811	0.9826	0.9841	0.9854	0.9866	0.9877	0.9887	0.9897	0.9905	0.9913	0.9920	0.9927	0.9953	0.9969
3	0.9285	0.9334	0.9380	0.9423	0.9464	0.9502	0.9537	0.9570	0.9600	0.9629	0.9656	0.9680	0.9704	0.9797	0.9862
4	0.8300	0.8396	0.8488	0.8575	0.8658	0.8736	0.8811	0.8881	0.8948	0.9012	0.9072	0.9129	0.9182	0.9408	0.9576
5	0.6873	0.7013	0.7149	0.7281	0.7408	0.7531	0.7649	0.7763	0.7873	0.7978	0.8080	0.8177	0.8270	0.8679	0.9003
6	0.5217	0.5381	0.5543	0.5702	0.5859	0.6012	0.6163	0.6310	0.6453	0.6593	0.6730	0.6863	0.6993	0.7585	0.8087
7	0.3616	0.3776	0.3937	0.4098	0.4258	0.4418	0.4577	0.4735	0.4892	0.5047	0.5201	0.5353	0.5503	0.6218	0.6866
8	0.2290	0.2424	0.2560	0.2699	0.2840	0.2982	0.3127	0.3272	0.3419	0.3567	0.3715	0.3864	0.4013	0.4753	0.5470
9	0.1328	0.1425	0.1528	0.1633	0.1741	0.1852	0.1967	0.2084	0.2204	0.2327	0.2452	0.2580	0.2709	0.3380	0.4074
10	0.0708	0.0772	0.0839	0.0910	0.0984	0.1061	0.1142	0.1226	0.1314	0.1404	0.1498	0.1595	0.1695	0.2236	0.2833
11	0.0349	0.0386	0.0426	0.0469	0.0514	0.0563	0.0614	0.0668	0.0726	0.0786	0.0849	0.0916	0.0985	0.1377	0.1841
12	0.0159	0.0179	0.0201	0.0224	0.0250	0.0277	0.0307	0.0339	0.0373	0.0409	0.0448	0.0490	0.0533	0.0792	0.1119
13	0.0068	0.0078	0.0088	0.0100	0.0113	0.0127	0.0143	0.0160	0.0179	0.0199	0.0221	0.0245	0.0270	0.0427	0.0638
14	0.0027	0.0031	0.0036	0.0042	0.0048	0.0055	0.0062	0.0071	0.0080	0.0091	0.0102	0.0114	0.0128	0.0216	0.0341
15	0.0010	0.0012	0.0014	0.0016	0.0019	0.0022	0.0026	0.0030	0.0034	0.0039	0.0044	0.0050	0.0057	0.0102	0.0172
16	0.0004	0.0004	0.0005	0.0006	0.0007	0.0008	0.0010	0.0012	0.0013	0.0016	0.0018	0.0021	0.0024	0.0046	0.0082
17	0.0001	0.0001	0.0001	0.0002	0.0002	0.0003	0.0004	0.0004	0.0005	0.0006	0.0007	0.0008	0.0010	0.0019	0.0037
18	0.0000	0.0000	0.0001	0.0001	0.0001	0.0001	0.0001	0.0001	0.0002	0.0002	0.0003	0.0003	0.0004	0.0008	0.0016
19	0.0000	0.0000	0.0000	0.0000	0.0000	0.0000	0.0000	0.0000	0.0001	0.0001	0.0001	0.0001	0.0001	0.0003	0.0006
20	0.0000	0.0000	0.0000	0.0000	0.0000	0.0000	0.0000	0.0000	0.0000	0.0000	0.0000	0.0000	0.0000	0.0001	0.0002
21	0.0000	0.0000	0.0000	0.0000	0.0000	0.0000	0.0000	0.0000	0.0000	0.0000	0.0000	0.0000	0.0000	0.0000	0.0001

APPENDIX F. (cont'd)

x	8.500	9.000	9.500	10.000	10.500	11.000	11.500	12.000	12.500	13.000	13.500	14.000	14.500	15.000	20.000
0	1.0000	1.0000	1.0000	1.0000	1.0000	1.0000	1.0000	1.0000	1.0000	1.0000	1.0000	1.0000	1.0000	1.0000	1.0000
1	0.9998	0.9999	0.9999	1.0000	1.0000	1.0000	1.0000	1.0000	1.0000	1.0000	1.0000	1.0000	1.0000	1.0000	1.0000
2	0.9981	0.9988	0.9992	0.9995	0.9997	0.9998	0.9999	0.9999	0.9999	1.0000	1.0000	1.0000	1.0000	1.0000	1.0000
3	0.9907	0.9938	0.9958	0.9972	0.9982	0.9988	0.9992	0.9995	0.9997	0.9998	0.9999	0.9999	0.9999	1.0000	1.0000
4	0.9699	0.9788	0.9851	0.9897	0.9928	0.9951	0.9966	0.9977	0.9984	0.9989	0.9993	0.9995	0.9997	0.9998	1.0000
5	0.9256	0.9450	0.9597	0.9707	0.9789	0.9849	0.9893	0.9924	0.9947	0.9963	0.9974	0.9982	0.9987	0.9991	1.0000
6	0.8504	0.8843	0.9115	0.9329	0.9496	0.9625	0.9723	0.9797	0.9852	0.9893	0.9923	0.9945	0.9961	0.9972	1.0000
7	0.7438	0.7932	0.8350	0.8699	0.8984	0.9214	0.9397	0.9542	0.9654	0.9741	0.9807	0.9858	0.9895	0.9924	0.9998
8	0.6144	0.6761	0.7313	0.7798	0.8215	0.8568	0.8863	0.9105	0.9302	0.9460	0.9585	0.9684	0.9761	0.9820	0.9993
9	0.4769	0.5443	0.6082	0.6672	0.7206	0.7680	0.8094	0.8450	0.8751	0.9002	0.9210	0.9379	0.9516	0.9625	0.9980
10	0.3470	0.4126	0.4782	0.5421	0.6029	0.6595	0.7112	0.7576	0.7986	0.8342	0.8647	0.8906	0.9122	0.9301	0.9950
11	0.2366	0.2940	0.3547	0.4170	0.4793	0.5401	0.5983	0.6528	0.7029	0.7483	0.7888	0.8243	0.8551	0.8815	0.9892
12	0.1513	0.1970	0.2480	0.3032	0.3613	0.4207	0.4802	0.5384	0.5942	0.6468	0.6955	0.7400	0.7799	0.8152	0.9787
13	0.0909	0.1242	0.1636	0.2084	0.2580	0.3113	0.3671	0.4240	0.4810	0.5369	0.5907	0.6415	0.6889	0.7324	0.9610
14	0.0514	0.0739	0.1019	0.1355	0.1746	0.2187	0.2670	0.3185	0.3722	0.4270	0.4818	0.5355	0.5875	0.6368	0.9339
15	0.0274	0.0415	0.0600	0.0835	0.1121	0.1460	0.1847	0.2280	0.2750	0.3249	0.3767	0.4296	0.4824	0.5343	0.8952
16	0.0138	0.0220	0.0335	0.0487	0.0683	0.0926	0.1217	0.1556	0.1940	0.2364	0.2822	0.3306	0.3808	0.4319	0.8435
17	0.0066	0.0111	0.0177	0.0270	0.0396	0.0559	0.0764	0.1013	0.1307	0.1645	0.2025	0.2441	0.2888	0.3359	0.7790
18	0.0030	0.0053	0.0089	0.0143	0.0219	0.0322	0.0458	0.0630	0.0842	0.1095	0.1391	0.1728	0.2103	0.2511	0.7030
19	0.0013	0.0024	0.0043	0.0072	0.0115	0.0177	0.0262	0.0374	0.0519	0.0698	0.0916	0.1174	0.1470	0.1805	0.6186
20	0.0005	0.0011	0.0020	0.0035	0.0058	0.0093	0.0143	0.0213	0.0306	0.0427	0.0579	0.0765	0.0988	0.1248	0.5298
21	0.0002	0.0004	0.0009	0.0016	0.0028	0.0047	0.0075	0.0116	0.0173	0.0250	0.0351	0.0479	0.0638	0.0830	0.4410
22	0.0001	0.0002	0.0004	0.0007	0.0013	0.0023	0.0038	0.0061	0.0094	0.0141	0.0204	0.0288	0.0396	0.0531	0.3563
23	0.0000	0.0001	0.0001	0.0003	0.0006	0.0010	0.0018	0.0030	0.0049	0.0076	0.0115	0.0167	0.0237	0.0327	0.2794
24	0.0000	0.0000	0.0001	0.0001	0.0002	0.0005	0.0008	0.0015	0.0025	0.0040	0.0062	0.0093	0.0137	0.0195	0.2126
25	0.0000	0.0000	0.0000	0.0000	0.0001	0.0002	0.0004	0.0007	0.0012	0.0020	0.0032	0.0050	0.0076	0.0112	0.1568
26	0.0000	0.0000	0.0000	0.0000	0.0000	0.0001	0.0002	0.0003	0.0006	0.0010	0.0016	0.0026	0.0041	0.0062	0.1122
27	0.0000	0.0000	0.0000	0.0000	0.0000	0.0000	0.0001	0.0001	0.0003	0.0005	0.0008	0.0013	0.0021	0.0033	0.0779
28	0.0000	0.0000	0.0000	0.0000	0.0000	0.0000	0.0000	0.0001	0.0001	0.0002	0.0004	0.0006	0.0011	0.0017	0.0525
29	0.0000	0.0000	0.0000	0.0000	0.0000	0.0000	0.0000	0.0000	0.0000	0.0001	0.0002	0.0003	0.0005	0.0009	0.0344
30	0.0000	0.0000	0.0000	0.0000	0.0000	0.0000	0.0000	0.0000	0.0000	0.0000	0.0001	0.0001	0.0002	0.0004	0.0219
31	0.0000	0.0000	0.0000	0.0000	0.0000	0.0000	0.0000	0.0000	0.0000	0.0000	0.0000	0.0001	0.0001	0.0002	0.0135
32	0.0000	0.0000	0.0000	0.0000	0.0000	0.0000	0.0000	0.0000	0.0000	0.0000	0.0000	0.0000	0.0000	0.0001	0.0081
33	0.0000	0.0000	0.0000	0.0000	0.0000	0.0000	0.0000	0.0000	0.0000	0.0000	0.0000	0.0000	0.0000	0.0000	0.0048
34	0.0000	0.0000	0.0000	0.0000	0.0000	0.0000	0.0000	0.0000	0.0000	0.0000	0.0000	0.0000	0.0000	0.0000	0.0027

APPENDIX F. (cont'd)

X	25.000	30.000												
0	1.0000	1.0000	0.0000	0.0000	0.0000	0.0000	0.0000	0.0000	0.0000	0.0000	0.0000	0.0000	0.0000	0.0000
1	1.0000	1.0000	0.0000	0.0000	0.0000	0.0000	0.0000	0.0000	0.0000	0.0000	0.0000	0.0000	0.0000	0.0000
2	1.0000	0.9999	0.0000	0.0000	0.0000	0.0000	0.0000	0.0000	0.0000	0.0000	0.0000	0.0000	0.0000	0.0000
3	1.0000	0.9998	0.0000	0.0000	0.0000	0.0000	0.0000	0.0000	0.0000	0.0000	0.0000	0.0000	0.0000	0.0000
4	1.0000	0.9996	0.0000	0.0000	0.0000	0.0000	0.0000	0.0000	0.0000	0.0000	0.0000	0.0000	0.0000	0.0000
5	1.0000	0.9991	0.0000	0.0000	0.0000	0.0000	0.0000	0.0000	0.0000	0.0000	0.0000	0.0000	0.0000	0.0000
6	1.0000	0.9980	0.0000	0.0000	0.0000	0.0000	0.0000	0.0000	0.0000	0.0000	0.0000	0.0000	0.0000	0.0000
7	1.0000	0.9961	0.0000	0.0000	0.0000	0.0000	0.0000	0.0000	0.0000	0.0000	0.0000	0.0000	0.0000	0.0000
8	1.0000	0.9927	0.0000	0.0000	0.0000	0.0000	0.0000	0.0000	0.0000	0.0000	0.0000	0.0000	0.0000	0.0000
9	0.9999	0.9871	0.0000	0.0000	0.0000	0.0000	0.0000	0.0000	0.0000	0.0000	0.0000	0.0000	0.0000	0.0000
10	0.9998	0.9781	0.0000	0.0000	0.0000	0.0000	0.0000	0.0000	0.0000	0.0000	0.0000	0.0000	0.0000	0.0000
11	0.9994	0.9647	0.0000	0.0000	0.0000	0.0000	0.0000	0.0000	0.0000	0.0000	0.0000	0.0000	0.0000	0.0000
12	0.9986	0.9456	0.0000	0.0000	0.0000	0.0000	0.0000	0.0000	0.0000	0.0000	0.0000	0.0000	0.0000	0.0000
13	0.9968	0.9194	0.0000	0.0000	0.0000	0.0000	0.0000	0.0000	0.0000	0.0000	0.0000	0.0000	0.0000	0.0000
14	0.9935	0.8854	0.0000	0.0000	0.0000	0.0000	0.0000	0.0000	0.0000	0.0000	0.0000	0.0000	0.0000	0.0000
15	0.9876	0.9781												
16	0.9777	0.9647												
17	0.9622	0.9456												
18	0.9395	0.9194												
19	0.9079	0.8854												
20	0.8664	0.9781												
21	0.8145	0.9647												
22	0.7527	0.9456												
23	0.6825	0.9194												
24	0.6061	0.8854												
35	0.0000	0.0000	0.0000	0.0000	0.0000	0.0000	0.0000	0.0000	0.0000	0.0000	0.0000	0.0000	0.0000	0.0015
36	0.0000	0.0000	0.0000	0.0000	0.0000	0.0000	0.0000	0.0000	0.0000	0.0000	0.0000	0.0000	0.0000	0.0009
37	0.0000	0.0000	0.0000	0.0000	0.0000	0.0000	0.0000	0.0000	0.0000	0.0000	0.0000	0.0000	0.0000	0.0005
38	0.0000	0.0000	0.0000	0.0000	0.0000	0.0000	0.0000	0.0000	0.0000	0.0000	0.0000	0.0000	0.0000	0.0003
39	0.0000	0.0000	0.0000	0.0000	0.0000	0.0000	0.0000	0.0000	0.0000	0.0000	0.0000	0.0000	0.0000	0.0002
40	0.0000	0.0000	0.0000	0.0000	0.0000	0.0000	0.0000	0.0000	0.0000	0.0000	0.0000	0.0000	0.0000	0.0001

APPENDIX F. (cont'd)

25	0.5266	0.8428	
26	0.4471	0.7916	
27	0.3706	0.7327	
28	0.2998	0.6671	
29	0.2366	0.5969	
30	0.1821	0.5243	
31	0.1367	0.4516	
32	0.1001	0.3814	
33	0.0715	0.3155	
34	0.0498	0.2556	
35	0.0338	0.2027	
36	0.0225	0.1574	
37	0.0146	0.1196	
38	0.0092	0.0890	
39	0.0057	0.0649	
40	0.0034	0.0463	
41	0.0020	0.0323	
42	0.0012	0.0221	
43	0.0007	0.0148	
44	0.0004	0.0097	
45	0.0002	0.0063	
46	0.0001	0.0040	
47	0.0001	0.0025	
48	0.0000	0.0015	
49	0.0000	0.0009	
50	0.0000	0.0005	
51	0.0000	0.0003	
52	0.0000	0.0002	
53	0.0000	0.0001	

APPENDIX G. Chi-square Distribution

Degrees of Freedom	p					
	.70	.80	.90	.95	.975	.99
1	1.07	1.64	2.71	3.84	5.02	6.63
2	2.41	3.22	4.61	5.99	7.38	9.21
3	3.66	4.64	6.25	7.81	9.35	11.3
4	4.88	5.99	7.78	9.49	11.1	13.3
5	6.06	7.29	9.24	11.1	12.8	15.1
6	7.23	8.56	10.6	12.6	14.4	16.8
7	8.38	9.80	12.0	14.1	16.0	18.5
8	9.52	11.0	13.4	15.5	17.5	20.1
9	10.7	12.2	14.7	16.9	19.0	21.7
10	11.8	13.4	16.0	18.3	20.5	23.2
11	12.9	14.6	17.3	19.7	21.9	24.7
12	14.0	15.8	18.5	21.0	23.3	26.2
13	15.1	17.0	19.8	22.4	24.7	27.7
14	16.2	18.2	21.1	23.7	26.1	29.1
15	17.3	19.3	22.3	25.0	27.5	30.6
16	18.4	20.5	23.5	26.3	28.8	32.0
17	19.5	21.6	24.8	27.6	30.2	33.4
18	20.6	22.8	26.0	28.9	31.5	34.8
19	21.7	23.9	27.2	30.1	32.9	36.2
20	22.8	25.0	28.4	31.4	34.2	37.6
21	23.9	26.2	29.6	32.7	35.5	38.9
22	24.9	27.3	30.8	33.9	36.8	40.3
23	26.0	28.4	32.0	35.2	38.1	41.6
24	27.1	29.6	33.2	36.4	39.4	43.0
25	28.2	30.7	34.4	37.7	40.6	44.3
26	29.2	31.8	35.6	38.9	41.9	45.6
27	30.3	32.9	36.7	40.1	43.2	47.0
28	31.4	34.0	37.9	41.3	44.5	48.3
29	32.5	35.1	39.1	42.6	45.7	49.6
30	33.5	36.2	40.3	43.8	47.0	50.9
40	44.2	47.3	51.8	55.8	59.3	63.7
50	54.7	58.2	63.2	67.5	71.4	76.2
60	65.2	69.0	74.4	79.1	83.3	88.4

APPENDIX H. F Distribution

		Numerator Degrees of Freedom											
		1	2	3	4	5	6	8	10	12	15	20	24	30
Denominator Degrees of Freedom	1	161	200	216	225	230	234	239	242	244	246	248	249	250
	2	18.5	19.0	19.2	19.2	19.3	19.3	19.4	19.4	19.4	19.4	19.4	19.5	19.5
	3	10.1	9.55	9.28	9.12	9.01	8.94	8.85	8.79	8.74	8.70	8.66	8.64	8.62
	4	7.71	6.94	6.59	6.39	6.26	6.16	6.04	5.96	5.91	5.86	5.80	5.77	5.75
	5	6.61	5.79	5.41	5.19	5.05	4.95	4.82	4.74	4.68	4.62	4.56	4.53	4.50
	6	5.99	5.14	4.76	4.53	4.39	4.28	4.15	4.06	4.00	3.94	3.87	3.84	3.81
	7	5.59	4.74	4.35	4.12	3.97	3.87	3.73	3.64	3.57	3.51	3.44	3.41	3.38
	8	5.32	4.46	4.07	3.84	3.69	3.58	3.44	3.35	3.28	3.22	3.15	3.12	3.08
	9	5.12	4.26	3.86	3.63	3.48	3.37	3.23	3.14	3.07	3.01	2.94	2.90	2.86
	10	4.96	4.10	3.71	3.48	3.33	3.22	3.07	2.98	2.91	2.85	2.77	2.74	2.70
	11	4.84	3.98	3.59	3.36	3.20	3.09	2.95	2.85	2.79	2.72	2.65	2.61	2.57
	12	4.75	3.89	3.49	3.26	3.11	3.00	2.85	2.75	2.69	2.62	2.54	2.51	2.47
	13	4.67	3.81	3.41	3.18	3.03	2.92	2.77	2.67	2.60	2.53	2.46	2.42	2.38
	14	4.60	3.74	3.34	3.11	2.96	2.85	2.70	2.60	2.53	2.46	2.39	2.35	2.31
	15	4.54	3.68	3.29	3.06	2.90	2.79	2.64	2.54	2.48	2.40	2.33	2.29	2.25
	16	4.49	3.63	3.24	3.01	2.85	2.74	2.59	2.49	2.42	2.35	2.28	2.24	2.19
	17	4.45	3.59	3.20	2.96	2.81	2.70	2.55	2.45	2.38	2.31	2.23	2.19	2.15
	18	4.41	3.55	3.16	2.93	2.77	2.66	2.51	2.41	2.34	2.27	2.19	2.15	2.11
	19	4.38	3.52	3.13	2.90	2.74	2.63	2.48	2.38	2.31	2.23	2.16	2.11	2.07
	20	4.35	3.49	3.10	2.87	2.71	2.60	2.45	2.35	2.28	2.20	2.12	2.08	2.04
	21	4.32	3.47	3.07	2.84	2.68	2.57	2.42	2.32	2.25	2.18	2.10	2.05	2.01
	22	4.30	3.44	3.05	2.82	2.66	2.55	2.40	2.30	2.23	2.15	2.07	2.03	1.98
	23	4.28	3.42	3.03	2.80	2.64	2.53	2.37	2.27	2.20	2.13	2.05	2.01	1.96
	24	4.26	3.40	3.01	2.78	2.62	2.51	2.36	2.25	2.18	2.11	2.03	1.98	1.94
	25	4.24	3.39	2.99	2.76	2.60	2.49	2.34	2.24	2.16	2.09	2.01	1.96	1.92
	30	4.17	3.32	2.92	2.69	2.53	2.42	2.27	2.16	2.09	2.01	1.93	1.89	1.84
	40	4.08	3.23	2.84	2.61	2.45	2.34	2.18	2.08	2.00	1.92	1.84	1.79	1.74
	60	4.00	3.15	2.76	2.53	2.37	2.25	2.10	1.99	1.92	1.84	1.75	1.70	1.65

APPENDIX H. (cont'd)

		Numerator Degrees of Freedom												
		1	2	3	4	5	6	8	10	12	15	20	24	30
Denominator Degrees of Freedom	1	4050	5000	5400	5620	5760	5860	5980	6060	6110	6160	6210	6235	6260
	2	98.5	99.0	99.2	99.2	99.3	99.3	99.4	99.4	99.4	99.4	99.4	99.5	99.5
	3	34.1	30.8	29.5	28.7	28.2	27.9	27.5	27.3	27.1	26.9	26.7	26.6	26.5
	4	21.2	18.0	16.7	16.0	15.5	15.2	14.8	14.5	14.4	14.2	14.0	13.9	13.8
	5	16.3	13.3	12.1.	11.4	11.0	10.7	10.3	10.1	9.89	9.72	9.55	9.47	9.38
	6	13.7	10.9	9.78	9.15	8.75	8.47	8.10	7.87	7.72	7.56	7.40	7.31	7.23
	7	12.2	9.55	8.45	7.85	7.46	7.19	6.84	6.62	6.47	6.31	6.16	6.07	5.99
	8	11.3	8.65	7.59	7.01	6.63	6.37	6.03	5.81	5.67	5.52	5.36	5.28	5.20
	9	10.6	8.02	6.99	6.42	6.06	5.80	5.47	5.26	5.11	4.96	4.81	4.73	4.65
	10	10.0	7.56	6.55	5.99	5.64	5.39	5.06	4.85	4.71	4.56	4.41	4.33	4.25
	11	9.65	7.21	6.22	5.67	5.32	5.07	4.74	4.54	4.40	4.25	4.10	4.02	3.94
	12	9.33	6.93	5.95	5.41	5.06	4.82	4.50	4.30	4.16	4.01	3.86	3.78	3.70
	13	9.07	6.70	5.74	5.21	4.86	4.62	4.30	4.10	3.96	3.82	3.66	3.59	3.51
	14	8.86	6.51	5.56	5.04	4.69	4.46	4.14	3.94	3.80	3.66	3.51	3.43	3.35
	15	8.68	6.36	5.42	4.89	4.56	4.32	4.00	3.80	3.67	3.52	3.37	3.29	3.21
	16	8.53	6.23	5.29	4.77	4.44	4.20	3.89	3.69	3.55	3.41	3.26	3.18	3.10
	17	8.40	6.11	5.18	4.67	4.34	4.10	3.79	3.59	3.46	3.31	3.16	3.08	3.00
	18	8.29	6.01	5.09	4.58	4.25	4.01	3.71	3.51	3.37	3.23	3.08	3.00	2.92
	19	8.18	5.93	5.01	4.50	4.17	3.94	3.63	3.43	3.30	3.15	3.00	2.92	2.84
	20	8.10	5.85	4.94	4.43	4.10	3.87	3.56	3.37	3.23	3.09	2.94	2.86	2.78
	21	8.02	5.78	4.87	4.37	4.04	3.81	3.51	3.31	3.17	3.03	2.88	2.80	2.72
	22	7.95	5.72	4.82	4.31	3.99	3.76	3.45	3.26	3.12	2.98	2.83	2.75	2.67
	23	7.88	5.66	4.76	4.26	3.94	3.71	3.41	3.21	3.07	2.93	2.78	2.70	2.62
	24	7.82	5.61	4.72	4.22	3.90	3.67	3.36	3.17	3.03	2.89	2.74	2.66	2.58
	25	7.77	5.57	4.68	4.18	3.86	3.63	3.32	3.13	2.99	2.85	2.70	2.62	2.54
	30	7.56	5.39	4.51	4.02	3.70	3.47	3.17	2.98	2.84	2.70	2.55	2.47	2.39
	40	7.31	5.18	4.31	3.83	3.51	3.29	2.99	2.80	2.66	2.52	2.37	2.29	2.20
	60	7.08	4.98	4.13	3.65	3.34	3.12	2.82	2.63	2.50	2.35	2.20	2.12	2.03

APPENDIX I. Squares and Square Roots

n	n²	√n	√10n	n	n²	√n	√10n
1.0	1.00	1.000	3.162	5.5	30.25	2.345	7.416
1.1	1.21	1.049	3.317	5.6	31.36	2.366	7.483
1.2	1.44	1.095	3.464	5.7	32.49	2.387	7.550
1.3	1.69	1.140	3.606	5.8	33.64	2.408	7.616
1.4	1.96	1.183	3.742	5.9	34.81	2.429	7.681
1.5	2.25	1.225	3.873	6.0	36.00	2.449	7.746
1.6	2.56	1.265	4.000	6.1	37.21	2.470	7.810
1.7	2.89	1.304	4.123	6.2	38.44	2.490	7.874
1.8	3.24	1.342	4.243	6.3	39.69	2.510	7.937
1.9	3.61	1.378	4.359	6.4	40.96	2.530	8.000
2.0	4.00	1.414	4.472	6.5	42.25	2.550	8.062
2.1	4.41	1.449	4.583	6.6	43.56	2.569	8.124
2.2	4.84	1.483	4.690	6.7	44.89	2.588	8.185
2.3	5.29	1.517	4.796	6.8	46.24	2.608	8.246
2.4	5.76	1.549	4.899	6.9	47.61	2.627	8.307
2.5	6.25	1.581	5.000	7.0	49.00	2.646	8.367
2.6	6.76	1.612	5.099	7.1	50.41	2.665	8.426
2.7	7.29	1.643	5.196	7.2	51.84	2.683	8.485
2.8	7.84	1.673	5.292	7.3	53.29	2.702	8.544
2.9	8.41	1.703	5.385	7.4	54.76	2.720	8.602
3.0	9.00	1.732	5.477	7.5	56.25	2.739	8.660
3.1	9.61	1.761	5.568	7.6	57.76	2.757	8.718
3.2	10.24	1.789	5.657	7.7	59.29	2.775	8.775
3.3	10.89	1.817	5.745	7.8	60.84	2.793	8.832
3.4	11.56	1.844	5.831	7.9	62.41	2.811	8.888
3.5	12.25	1.871	5.916	8.0	64.00	2.828	8.944
3.6	12.96	1.897	6.000	8.1	65.61	2.846	9.000
3.7	13.69	1.924	6.083	8.2	67.24	2.864	9.055
3.8	14.44	1.949	6.164	8.3	68.89	2.881	9.110
3.9	15.21	1.975	6.245	8.4	70.56	2.898	9.165
4.0	16.00	2.000	6.325	8.5	72.25	2.915	9.220
4.1	16.81	2.025	6.403	8.6	73.96	2.933	9.274
4.2	17.64	2.049	6.481	8.7	75.69	2.950	9.327
4.3	18.49	2.074	6.557	8.8	77.44	2.966	9.381
4.4	19.36	2.098	6.633	8.9	79.21	2.983	9.434
4.5	20.25	2.121	6.708	9.0	81.00	3.000	9.487
4.6	21.16	2.145	6.782	9.1	82.81	3.017	9.539
4.7	22.09	2.168	6.856	9.2	84.64	3.033	9.592
4.8	23.04	2.191	6.928	9.3	86.49	3.050	9.644
4.9	24.01	2.214	7.000	9.4	88.36	3.066	9.695
5.0	25.00	2.236	7.071	9.5	90.25	3.082	9.747
5.1	26.01	2.258	7.141	9.6	92.16	3.098	9.798
5.2	27.04	2.280	7.211	9.7	94.09	3.114	9.849
5.3	28.09	2.302	7.280	9.8	96.04	3.130	9.899
5.4	29.16	2.324	7.348	9.9	98.01	3.146	9.950

APPENDIX J. Random Digits

00000	10097	32533	74520	13586	34673	54876	80959	09117	39292	74945
00001	37542	04805	64894	74296	24805	24037	20636	10402	00822	91665
00002	08422	68953	19645	09303	23209	02560	15953	34764	35080	33606
00003	99019	02529	09376	70715	38311	31165	88676	74397	04436	27659
00004	12807	99970	80157	36147	64032	36653	98951	16877	12171	76833
00005	66065	74717	34072	76850	36697	36170	65813	39885	11199	29170
00006	31060	10805	45571	82406	35303	42614	86799	07439	23403	09732
00007	85269	77602	02051	65692	68665	74818	73053	85247	18623	88579
00008	63573	32135	05325	47048	90553	57548	28468	28709	83491	25624
00009	73796	45753	03529	64778	35808	34282	60935	20344	25273	88435
00010	98520	17767	14905	68607	22109	40558	60970	93433	50500	73998
00011	11805	05431	39808	27732	50725	68248	29405	24201	52775	67851
00012	83452	99634	06288	98083	13746	70078	18475	40610	28711	77817
00013	88685	40200	86507	58401	36766	67951	90364	76493	29609	11062
00014	99594	67348	87517	64969	91826	08928	93785	61368	23478	34113
00015	65481	17674	17468	50950	58047	76974	73039	57186	40218	16544
00016	80124	35635	17727	08015	45318	22374	21115	78253	14385	53763
00017	74350	99817	77402	77214	43236	00210	45521	64237	96286	02655
00018	69916	26803	66252	29148	36936	87203	76621	13990	94400	56418
00019	09893	20505	14225	68514	46427	56788	96297	78822	54382	14598
00020	91499	14523	68479	27686	46162	83554	94750	89923	37089	20048
00021	80336	94598	26940	36858	70297	34135	53140	33340	42050	82341
00022	44104	81949	85157	47954	32979	26575	57600	40881	22222	06413
00023	12550	73742	11100	02040	12860	74697	96644	89439	28707	25815
00024	63606	49329	16505	34484	40219	52563	43651	77082	07207	31790
00025	61196	90446	26457	47774	51924	33729	65394	59593	42582	60527
00026	15474	45266	95270	79953	59367	83848	82396	10118	33211	59466
00027	94557	28573	67897	54387	54622	44431	91190	42592	92927	45973
00028	42481	16213	97344	08721	16868	48767	03071	12059	25701	46670
00029	23523	78317	73208	89837	68935	91416	26252	29663	05522	82562
00030	04493	52494	75246	33824	45862	51025	61962	79335	65337	12472
00031	00549	97654	64051	88159	96119	63896	54692	82391	23287	29529
00032	35963	15307	26898	09354	33351	35462	77974	50024	90103	39333
00033	59808	08391	45427	26842	83609	49700	13021	24892	78565	20106
00034	46058	85236	01390	92286	77281	44077	93910	83647	70617	42941
00035	32179	00597	87379	25241	05567	07007	86743	17157	85394	11838
00036	69234	61406	20117	45204	15956	60000	18743	92423	97118	96338
00037	19565	41430	01758	75379	40419	21585	66674	36806	84962	85207
00038	45155	14938	19476	07246	43667	94543	59047	90033	20826	69541
00039	94864	31994	36168	10851	34888	81553	01540	35456	05014	51176
00040	98086	24826	45340	28404	44999	08896	39094	73407	35441	31880
00041	33185	16232	41941	50949	89435	48581	88695	41994	37548	73043
00042	80951	00406	96382	70774	20151	23387	25016	25298	94624	61171
00043	79752	49140	71961	28296	69861	02591	74852	20539	00387	59579
00044	18633	32537	98145	06571	31010	24674	05455	61427	77938	91936

* Source: Rand (1955), 1–4. Used by permission of the publisher.

APPENDIX J. (cont'd)

00045	74029	43902	77557	32270	97790	17119	52527	58021	80814	51748
00046	54178	45611	80993	37143	05335	12969	56127	19255	36040	90324
00047	11664	49883	52079	84827	59381	71539	09973	33440	88461	23356
00048	48324	77928	31249	64710	02295	36870	32307	57546	15020	09994
00049	69074	94138	87637	91976	35584	04401	10518	21615	01848	76938
00050	09188	20097	32825	39527	04220	86304	83389	87374	64278	58044
00051	90045	85497	51981	50654	94938	81997	91870	76150	68476	64659
00052	73189	50207	47677	26269	62290	64464	27124	67018	41361	82760
00053	75768	76490	20971	87749	90429	12272	95375	05871	93823	43178
00054	54016	44056	66281	31003	00682	27398	20714	53295	07706	17813
00055	08358	69910	78542	42785	13661	58873	04618	97553	31223	08420
00056	28306	03264	81333	10591	40510	07893	32604	60475	94119	01840
00057	53840	86233	81594	13628	51215	90290	28466	68795	77762	20791
00058	91757	53741	61613	62269	50263	90212	55781	76514	83483	47055
00059	89415	92694	00397	58391	12607	17646	48949	72306	94541	37408
00060	77513	03820	86864	29901	68414	82774	51908	13980	72893	55507
00061	19502	37174	69979	20288	55210	29773	74287	75251	65344	67415
00062	21818	59313	93278	81757	05686	73156	07082	85046	31853	38452
00063	51474	66499	68107	23621	94049	91345	42836	09191	08007	45449
00064	99559	68331	62535	24170	69777	12830	74819	78142	43860	72834
00065	33713	48007	93584	72869	51926	64721	58303	29822	93174	93972
00066	85274	86893	11303	22970	28834	34137	73515	90400	71148	43643
00067	84133	89640	44035	52166	73852	70091	61222	60561	62327	18423
00068	56732	16234	17395	96131	10123	91622	85496	57560	81604	18880
00069	65138	56806	87648	85261	34313	65861	45875	21069	85644	47277
00070	38001	02176	81719	11711	71602	92937	74219	64049	65584	49698
00071	37402	96397	01304	77586	56271	10086	47324	62605	40030	37438
00072	97125	40348	87083	31417	21815	39250	75237	62047	15501	29578
00073	21826	41134	47143	34072	64638	85902	49139	06441	03856	54552
00074	73135	42742	95719	09035	85794	74296	08789	88156	64691	19202
00075	07638	77929	03061	18072	96207	44156	23821	99538	04713	66994
00076	60528	83441	07954	19814	59175	20695	05533	52139	61212	06455
00077	83596	35655	06958	92983	05128	09719	77433	53783	92301	50498
00078	10850	62746	99599	10507	13499	06319	53075	71839	06410	19362
00079	39820	98952	43622	63147	64421	80814	43800	09351	31024	73167
00080	59580	06478	75569	78800	88835	54486	23768	06156	04111	08408
00081	38508	07341	23793	48763	90822	97022	17719	04207	95954	49953
00082	30692	70668	94688	16127	56196	80091	82067	63400	05462	69200
00083	65443	95659	18288	27437	49632	24041	08337	65676	96299	90836
00084	27267	50264	13192	72294	07477	44606	17985	48911	97341	30358
00085	91307	06991	19072	24210	36699	53728	28825	35793	28976	66252
00086	68434	94688	84473	13622	62126	98408	12843	82590	09815	93146
00087	48908	15877	54745	24591	35700	04754	83824	52692	54130	55160
00088	06913	45197	42672	78601	11883	09528	63011	98901	14974	40344
00089	10455	16019	14210	33712	91342	37821	88325	80851	43667	70883

APPENDIX K. Common Logarithms

N	0	1	2	3	4	5	6	7	8	9
10	0000	0043	0086	0128	0170	0212	0253	0294	0334	0374
11	0414	0453	0492	0531	0569	0607	0645	0682	0719	0755
12	0792	0828	0864	0899	0934	0969	1004	1038	1072	1106
13	1139	1173	1206	1239	1271	1303	1335	1367	1399	1430
14	1461	1492	1523	1553	1584	1614	1644	1673	1703	1732
15	1761	1790	1818	1847	1875	1903	1931	1959	1987	2014
16	2041	2068	2095	2122	2148	2175	2201	2227	2253	2279
17	2304	2330	2355	2380	2405	2430	2455	2480	2504	2529
18	2553	2577	2601	2625	2648	2672	2695	2718	2742	2765
19	2788	2810	2833	2856	2878	2900	2923	2945	2967	2989
20	3010	3032	3054	3075	3096	3118	3139	3160	3181	3201
21	3222	3243	3263	3284	3304	3324	3345	3365	3385	3404
22	3424	3444	3464	3483	3502	3522	3541	3560	3579	3598
23	3617	3636	3655	3674	3692	3711	3729	3747	3766	3784
24	3802	3820	3838	3856	3874	3892	3909	3927	3945	3962
25	3979	3997	4014	4031	4048	4065	4082	4099	4116	4133
26	4150	4166	4183	4200	4216	4232	4249	4265	4281	4298
27	4314	4330	4346	4362	4378	4393	4409	4425	4440	4456
28	4472	4487	4502	4518	4533	4548	4564	4579	4594	4609
29	4624	4639	4654	4669	4683	4698	4713	4728	4742	4757
30	4771	4786	4800	4814	4829	4843	4857	4871	4886	4900
31	4914	4928	4942	4955	4969	4983	4997	5011	5024	5038
32	5051	5065	5079	5092	5105	5119	5132	5145	5159	5172
33	5185	5198	5211	5224	5237	5250	5263	5276	5289	5302
34	5315	5328	5340	5353	5366	5378	5391	5403	5416	5428
35	5441	5453	5465	5478	5490	5502	5514	5527	5539	5551
36	5563	5575	5587	5599	5611	5623	5635	5647	5658	5670
37	5682	5694	5705	5717	5729	5740	5752	5763	5775	5786
38	5798	5809	5821	5832	5843	5855	5866	5877	5888	5899
39	5911	5922	5933	5944	5955	5966	5977	5988	5999	6010
40	6021	6031	6042	6053	6064	6075	6085	6096	6107	6117
41	6128	6138	6149	6160	6170	6180	6191	6201	6212	6222
42	6232	6243	6253	6263	6274	6284	6294	6304	6314	6325
43	6335	6345	6355	6365	6375	6385	6395	6405	6415	6425
44	6435	6444	6454	6464	6474	6484	6493	6503	6513	6522
45	6532	6542	6551	6561	6571	6580	6590	6599	6609	6618
46	6628	6637	6646	6656	6665	6675	6684	6693	6702	6712
47	6721	6730	6739	6749	6758	6767	6776	6785	6794	6803
48	6812	6821	6830	6839	6848	6857	6866	6875	6884	6893
49	6902	6911	6920	6928	6937	6946	6955	6964	6972	6981
50	6990	6998	7007	7016	7024	7033	7042	7050	7059	7067
51	7076	7084	7093	7101	7110	7118	7126	7135	7143	7152
52	7160	7168	7177	7185	7193	7202	7210	7218	7226	7235
53	7243	7251	7259	7267	7275	7284	7292	7300	7308	7316
54	7324	7332	7340	7348	7356	7364	7372	7380	7388	7396
N	0	1	2	3	4	5	6	7	8	9

APPENDIX K. (cont'd)

N	0	1	2	3	4	5	6	7	8	9
55	7404	7412	7419	7427	7435	7443	7451	7459	7466	7474
56	7482	7490	7497	7505	7513	7520	7528	7536	7543	7551
57	7559	7566	7574	7582	7589	7597	7604	7612	7619	7627
58	7634	7642	7649	7657	7664	7672	7679	7686	7694	7701
59	7709	7716	7723	7731	7738	7745	7752	7760	7767	7774
60	7782	7789	7796	7803	7810	7818	7825	7832	7839	7846
61	7853	7860	7868	7875	7882	7889	7896	7903	7910	7917
62	7924	7931	7938	7945	7952	7959	7966	7973	7980	7987
63	7993	8000	8007	8014	8021	8028	8035	8041	8048	8055
64	8062	8069	8075	8082	8089	8096	8102	8109	8116	8122
65	8129	8136	8142	8149	8156	8162	8169	8176	8182	8189
66	8195	8202	8209	8215	8222	8228	8235	8241	8248	8254
67	8261	8267	8274	8280	8287	8293	8299	8306	8312	8319
68	8325	8331	8338	8344	8351	8357	8363	8370	8376	8382
69	8388	8395	8401	8407	8414	8420	8426	8432	8439	8445
70	8451	8457	8463	8470	8476	8482	8488	8494	8500	8506
71	8513	8519	8525	8531	8537	8543	8549	8555	8561	8567
72	8573	8579	8585	8591	8597	8603	8609	8615	8621	8627
73	8633	8639	8645	8651	8657	8663	8669	8675	8681	8686
74	8692	8698	8704	8710	8716	8722	8727	8733	8739	8745
75	8751	8756	8762	8768	8774	8779	8785	8791	8797	8802
76	8808	8814	8820	8825	8831	8837	8842	8848	8854	8859
77	8865	8871	8876	8882	8887	8893	8899	8904	8910	8915
78	8921	8927	8932	8938	8943	8949	8954	8960	8965	8971
79	8976	8982	8987	8993	8998	9004	9009	9015	9020	9025
80	9031	9036	9042	9047	9053	9058	9063	9069	9074	9079
81	9085	9090	9096	9101	9106	9112	9117	9122	9128	9133
82	9138	9143	9149	9154	9159	9165	9170	9175	9180	9186
83	9191	9196	9201	9206	9212	9217	9222	9227	9232	9238
84	9243	9248	9253	9258	9263	9269	9274	9279	9284	9289
85	9294	9299	9304	9309	9315	9320	9325	9330	9335	9340
86	9345	9350	9355	9360	9365	9370	9375	9380	9385	9390
87	9395	9400	9405	9410	9415	9420	9425	9430	9435	9440
88	9445	9450	9455	9460	9465	9469	9474	9479	9484	9489
89	9494	9499	9504	9509	9513	9518	9523	9528	9533	9538
90	9542	9547	9552	9557	9562	9566	9571	9676	9581	9586
91	9590	9595	9600	9605	9609	9614	9619	9624	9628	9633
92	9638	9643	9647	9652	9657	9661	9666	9671	9675	9680
93	9685	9689	9694	9699	9703	9708	9713	9717	9722	9727
94	9731	9736	9741	9745	9750	9754	9859	9763	9768	9773
95	9777	9782	9786	9791	9795	9800	9805	9809	9814	9818
96	9823	9827	9832	9836	9841	9845	9850	9854	9859	9863
97	9868	9872	9877	9881	9886	9890	9894	9899	9903	9908
98	9912	9917	9921	9926	9930	9934	9939	9943	9948	9952
99	9956	9961	9965	9969	9974	9978	9983	9987	9991	9996
N	0	1	2	3	4	5	6	7	8	9

APPENDIX L. Greek Letters

The Greek alphabet

A	α	alpha
B	β	beta
Γ	γ	gamma
Δ	δ	delta
E	ϵ	epsilon
Z	ζ	zeta
H	η	eta
Θ	θ	theta
I	ι	iota
K	κ	kappa
Λ	λ	lambda
M	μ	mu
N	ν	nu
Ξ	ξ	xi
O	o	omicron
Π	π	pi
P	ρ	rho
Σ	σ	sigma
T	τ	tau
Υ	υ	upsilon
Φ	ϕ	phi
X	χ	chi
Ψ	ψ	psi
Ω	ω	omega

Index

Aggregate price indexes, 425, 426
Alpha, 221, 249–250
Alternative hypothesis, 220
Analysis of variance, 280–289, 330–332
   steps in conducting, 282–283, 284, 286
Arithmetic scale, 556
Association. *See* Correlation analysis
Average. *See* Mean
   arithmetic, 6
Average deviation, 71
Average of price relatives, 427–429

Bar chart, 36
Base period for index numbers, 421, 424
Bayes, Thomas, 118
Bayesian analysis, 117–122, 385, 404–417
   benefits of, 415–416
   limitations of, 416–417
Beta, 249–250
Beta weights, 332
Bias of nonresponse, 20
Bilateral causation, 293
Bimodal distribution, 60
Binomial distribution, 139, 140–155
   characteristics of, 148
   cumulative, 147
   expansion, 145
   mean of, 150
   normal approximation to, 151–154
   use of tables, 148–149
Bivariate regression, 294–319
Business indicators, 421

Causal relationships, 292–293
Census, 4, 91
Central limit theorem, 186
Central tendency, measures of, 54–65
   limitations of, 61–65
Certainty, 386
Chi-square analysis
   cautions in use, 278
   degrees of freedom for, 274, 375
   in goodness-of-fit, 371–375
   for independence, 271–279
   statistics and test, 271
   steps for conducting, 275
Class, 35
   limit, 39
   open-end, 40
   size, 36
   width, 39
Classification of data, 35, 39
Cluster sampling, 173
Coefficient of determination, 309, 330–332
   calculation of, 309, 331
   compared to correlation coefficient, 309
   interpretation, 309, 331–332
   multiple regression, 331
   simple regression, 309
Coefficient of variation, 78
Collectively exhaustive events, 90
Composite index, 422
Computers, 314
Conditional probability distributions, 100
Confidence coefficient, 202
Confidence interval, 202, 305–307
Confidence interval estimate
   of regression estimate, 306–307
   of slope of regression line, 305
Constant cause system, 791
Contingency table, 97, 99, 118, 120, 271
   degrees of freedom for, 274–275
Continuity correction, 289
Continuous data, 27
Convenience sample, 174
Correlation analysis, 307–309
   multiple, 331
   partial, 308
   perfect, 308
Critical value, 221, 250, 253
Cross-sectional data, 294
Cross tabulations, 264
Cumulative frequency distribution, 46–48
Cyclical time series variations, 440

Data
   categorical, 338
   continuous, 27
   dichotomous, 23
   discrete, 27

Data (cont.)
  grouped, 23
  integer, 29
  interval scaled, 30, 338
  multiple response, 29
  nominal-scaled, 30, 338
  ordinal-scaled, 30
  ratio-scaled, 30
Data classifications, 35
Decision theory, 385, 389
Decision tree, 389–391, 414
Deflation, 432
Degrees of freedom, number of
  for contingency table, 274
  for F test, 283–284, 286
  in goodness-of-fit test, 375
  in regression analysis, 299, 328, 336
Dependence, 94
Dependent event, 99
Dependent variable, 292
Descriptive statistics, 9
Deviation
  average, 71
  standard, 72–78
Dichotomous data, 23
Discrete data, 27
Dispersion, 69–81
Distribution, sampling, 177–187
  binomial, 140–155
  normal, 130–137
  Poisson, 156–160
Distributions. *See* Frequency distribution; Probability distributions
Double sampling, 175

Error. *See* Measurement error; Sampling error
Estimation, 9–10, 197–215
  finite population, 208
  interval, 200–215
  interval estimate, 9–10
  large sample, 207–210
  normal population, 206
  point, 199–200
  point estimate, 9–10
  population not normal, 203–206
  population σ known, 210
  population σ unknown, 207
  proportion, 208–210
  of sample size, 206

  selection of equation, 214
  small sample, 210–212
Estimator, 212–213
  criteria for choosing
    consistency, 213
    efficiency, 213
    unbiased, 212
Event, 89
  collectively exhaustive, 90
  complex, 90
  dependent, 99
  independent, 99
  mutually exclusive, 90
  simple, 90
Event sets, 89–90
Expected frequencies, in chi square, 271, 373
Expected opportunity loss, 394–399
Expected value, 129
Expected value of perfect information (EVPI), 399–400
Experimentation, 18
Explained variation, in regression, 309, 330–332
Extrapolation error, 313
Extreme values, 63

F distribution
  degrees of freedoms for, 284, 330–332
F statistic, 281
F test. *See* Analysis of variance
Finite population correction factor, 190
Frequency distribution, 35–81
  cumulative, 46
  relative, 48
Frequency polygon, 44

Goodness-of-fit test, 371–375
Graphs, 36
Grouped data, 23

Histogram, 43–44
Homogeneous group, 172
Hypergeometric distribution, 150
Hypotheses, 220
  accept or reject, 223
  alternative, 220–234
  formulation of, 220
  null, 220, 234
Hypothesis testing, 217–257, 271–289

  in analysis of variance, 280–289
  application of, 217–218
  between two samples, 234
  in chi square analysis, 271–279
  critical value, 221, 250, 253–255
  establishing, 220
  large sample, 224, 229, 234
  limitations of, 256
  need for, 218
  one sample, 224–232
  one-tailed, 220, 235
  population distribution unknown, 224, 230
  population normal, 224, 229
  proportions, 230–232
  in regression, 304–305, 326
  risk, 221, 243–248
  significance level, 221, 249
  small sample, 224, 230, 237
  standard deviation known, 224
  standard deviation unknown, 229, 230
  steps in, 219–220
  trade off of α and β, 249–250
  two proportions, 238
  two samples, 234, 237
  two-tailed tests, 227
Hypothesis tests, for equality of means. *See* Analysis of variance
  for goodness-of-fit, 371–375
  for independence, 271–279
  for slope of regression line, 304

Independence, 93–94
Independent events, 99
Independent variables, 292
Indexes, 421–435
  base, 424
  construction of an index, 423, 435
  price relatives, 427–428
  quantity, 429
  shifting base, 431
  simple aggregative, 425
  splicing, 431
  types, 425
  value, 430
  weighted aggregative, 426
Indifference probability, 408
Inference, 9
  See also Statistical inference
Integer data, 29

Index

Interfractile range, 70
Interquartile range, 71
Intersection, 92
Interval estimate, 9–10, 200–215
    advantages of, 201
Interval scale, 30

Joint probability, 98
Judgment sample, 175

Least squares method, 301, 318–319, 355
    in multiple regression, 355
Length-of-runs test, 369–371
Linear regression. *See* Bivariate regression; Least squares method
Logarithms, table of (Appendix K)

Mann-Whitney U test, 363–366
Marginal probability, 97
Mean, 55–58, 74–78
    calculation of, 55
    for grouped data, 56–58, 74–78
Mean square, 282–286
Measurement error, 8
Median, 58–59
Modal class, 60
Mode, 59–60
Model, 19
Multiple coefficient of determination, 330–332
Multiple regression, 320–355
    dummy variable techniques in, 338–340
    validation, 340–343
Multiple response data, 29
Multiple sampling, 175
Multiplication law, 109–111
Mutually exclusive events, 90

Node, 414
Nominal scaled data, 30
Nonparametric statistics, 356–383
    advantages, 358
    disadvantages, 358
Nonprobability sampling, 167, 174–175
Nonrandom sampling, 167, 174–175
Nonreplacement, 168
Nonresponse, 20
Nonresponse error, 20
Normal curve, 130–131
Normal deviate, 132

Normal distribution, 130–134, 179–182
    characteristics of, 130
Normal equations, 309–355

Objective probability, 95, 386
Ogive, 48
One-tailed test, 220
Opportunity loss, 394–399
Ordinal scaled data, 30
Ordinary least squares, 301, 318–319
Ordinary length of runs test, 369–371

Parametric statistics, 357–358
Payoff table, 391–400
Payoffs, 386, 391–400
Percentile, 70, 71
Perfect information, 399–400
Point estimate, 9, 199–200
Poisson distribution, 156–160
    applications of, 156–157
    as approximation of the binomial, 156, 158
    mean of, 157
    variance of, 157
Polygon, frequency, 440
Population, 5, 91, 166
    description of, 5
    finite, 190
Population mean, 5
Population parameter, 5
Posterior analysis, 405, 408
Posterior probability, 121, 405
Predictions in regression analysis, 305–307
Preposterior analysis, 121, 405, 411
Price index, 421, 422, 425–429
Primary data, 4, 16
Prior analysis, 405
Prior probability, 121
Probability, 87–122
    conditional, 100
    definitions of, 87
    experimental, 89
    importance of, 87
    indifference, 408
    joint, 98
    laws
        of addition, 106–109
        of multiplication, 109–111

    marginal, 97
    objective, 95
    posterior, 121
    prior, 121
    subjective, 95–96
Probability distribution, 127–128
    objective, 386
    subjective, 386
Probability distributions, 127–161
    See also under individual listings: Binomial distribution; Chi-square analysis; F distribution; Hypergeometric distribution; Normal distribution; Poisson distribution; Student t statistic
Probability sampling, 167–174
Probability tree, 114
Proportionate sampling, 172

Quantity index, 421
Quartiles, 70
Quota sampling, 174

Random numbers, 169–171
Random samples, 5, 91, 167–174
    restricted, 5, 171–174
    systematic, 171
    unrestricted, 5, 6
Random variable, 128, 141
    continuous, 128
    discrete, 128
Range, 70
Rank-order correlation, 375–378
    significance tests, 377–378
    Spearman's rho, 375–376
Ratio scaled data, 30
Raw data, 34
Regression analysis, 291–355
    assumptions of, 310–313
    cautions in use, 313–314, 323–326, 333
    collinearity, 329
    interpretation, 297–307, 334–338, 343
    spurious results, 314
    validation of, 340–342
Regression coefficients, 296, 298
    estimation of, 296, 302–303, 321, 355
    interpretation of, 298, 327–328
    net, 327–328
    relative importance of, 332
    true, 296, 322

Regression equation, 296–297
   estimated, 296
   fitting, 294–297
   for multiple regression, 332, 339, 342
Relative frequency distribution, 48–49
Replacement, 92, 140, 168
Restricted random sampling, 171–174
Risk, 243–248, 387
   in estimation, 197
   in hypothesis testing, 243–248

Sample, 4, 5, 91, 163
   See also Random samples
   convenience sample, 174
   judgment sample, 175
   quota, 174
   random, 91, 167
   restricted random, 171–174
   simple random, 167–168
   stratified, 172
   systematic random, 171
Sample space, 90
Sample standard deviation, 73–76
Sampling, 163–175
   advantages of, 164
   cluster, 173
   convenience, 174
   double, 175
   judgment, 175
   nonprobability, 167, 174–175
   nonrandom, 167
   probability, 167–174
   quota, 174
   random, 167–174
   reasons for, 164
   sequential, 175
   single, 175
   stratified, 172
   systematic, 171
Sampling distribution of mean, 177–187
   for normal population, 179–182
   when population is not normal, 183–186
   for small populations, 189–191
   standard deviation of, 177
      See also Standard error of mean

Sampling distribution of proportion, 191–193
   mean of, 192
   standard deviation of, 192
Sampling error, 8
Sampling unit, 166
Scatter diagram, 294–295, 325
Seasonal time series variations, 439
Secondary data, 4, 16
Secular trend, 437
Sequential analysis, 406
Sequential sampling, 175
Set, 89
Shifting base of index numbers, 431
Sign test, 360–363
Significance level, 221
Simple aggregative index, 425–426
Simple index, 422
Simple random sample, 167
Single sampling, 175
Skewness
   negative (leftward), 64
   positive (rightward), 64
Slope of regression line, estimation of, 296–298
Splicing index numbers, 431
Standard deviation, 72–78
   calculated from grouped data, 74–78
   estimation of, 182
   of sample, 73–76
   of sample mean, 181–182
Standard error of estimate, 299–300, 303–304
   in multiple regression, 330–333
Standard error of mean, 181–182
   distinguished from s, 182
Standard error of proportion, 192
   for small populations, 192
Standard error of regression coefficient, 304, 328
Statistic, 6, 12
Statistical inference, 197
Statistics, defined, 6
Strata, 172
Stratified sampling, 172
Stratum, 172
Student t statistic, compared to sign test statistic, 305, 328
   used in regression analysis, 305, 328
Subjective probability, 95–96, 386

Sum of squares, 281–287
   among groups, 282
   total, 282
   within groups, 282
Survey, 19–21
Systematic random sample, 171

Tabulation, 259
Time series, 294, 436–443
   classical model for, 441
   deflating using index numbers, 432
Time-series analysis, 441–443
Time-series components, 439
   cyclical movement, 440
   irregular variation, 440
   seasonal fluctuation, 439
   secular trend, 437
Total sum of squares, 282
Total variation, in regression, 309, 330–333
Trends, 437
   secular, 437

Uncertainty, 385–419
Ungrouped data, 23
Unrestricted random sample, 167
Unweighted index, 422

Value index, 421
Variability, 69
   See also Dispersion
Variable
   dependent, 292
   discrete, 36
   dummy, 338
   independent, 292
Variance, 72–78
   calculated from grouped data, 74–75
   See also Standard deviation
Variation, between-groups, 78, 282
   coefficient of, 78
   within-groups, 282
Venn diagrams, 92–93

Weighted index, 422, 424, 426
Within-groups mean square, 282
Within-groups sum of squares, 282
Within-groups variation, 282

Y intercept of line, 296–298

List of Symbols

a	Y intercept for sample regression equation
α (alpha)	Probability of a Type I error in hypothesis testing Area under tail of a distribution Population intercept in regression
b	Slope of the sample regression line in bivariate (simple) regression Net regression coefficient in multiple regression
β (beta)	Probability of a Type II error in hypothesis testing Slope of population regression line (bivariate regression) Population net regression coefficient (multiple regression)
$_nC_r$	Possible combinations from a total of n items taken r at a time
χ^2	Chi-square statistic
e	Sample error term in regression analysis
$E(X)$	Expected value of X
E (epsilon)	Population error term in regression analysis
EVPI	Expected value of perfect information
ϵ (epsilon)	Population error term for regression analysis model
f	Frequency (number of observations)
f_o	Frequency observed
f_e	Frequency expected
$f(X)$	Function of X
F	F statistic used in analysis of variance
H_0	Null hypothesis
H_a	Alternative hypothesis
L	Lower limit of median class
m	Mean of the Poisson distribution
Md	Median
μ (mu)	Population mean
n	Number of items or observations in a sample
N	Number of items or observations in a population
P	Probability of a particular outcome in a Bernoulli trial (overall)
p	Sample proportion Probability of success for one trial in a Bernoulli process